SECOND EDITION

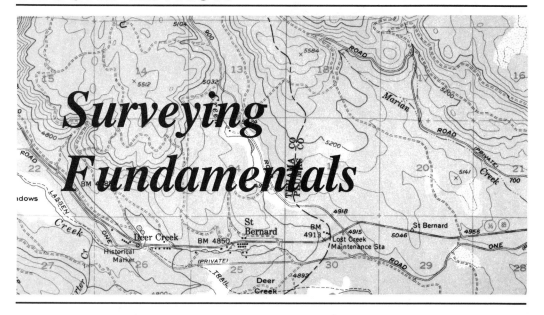

Surveying Fundamentals

Jack C. McCormac

Department of Civil Engineering
Clemson University

PRENTICE HALL, *Englewood Cliffs, New Jersey 07632*

Library of Congress Cataloging-in-Publication Data

McCormac, Jack C.
    Surveying fundamentals / by Jack McCormac. — 2nd ed.
      p. cm.
    Includes index.
    ISBN 0-13-878026-9
    1. Surveying. I. Title.
TA545.M33 1991
526.9—dc20   90-43312
                                    CIP

Editorial/production supervision: *Merrill Peterson*
Interior design: *Joan Stone*
Cover design: *Butler/Udell Design*
Prepress buyer: *Linda Behrens*
Manufacturing buyer: *Dave Dickey*

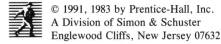
© 1991, 1983 by Prentice-Hall, Inc.
A Division of Simon & Schuster
Englewood Cliffs, New Jersey 07632

Printed in the United States of America

10 9 8 7 6 5 4 3 2 1

ISBN 0-13-878026-9

PRENTICE-HALL INTERNATIONAL (UK) LIMITED, *London*
PRENTICE-HALL OF AUSTRALIA PTY. LIMITED, *Sydney*
PRENTICE-HALL CANADA INC., *Toronto*
PRENTICE-HALL HISPANOAMERICANA, S.A., *Mexico*
PRENTICE-HALL OF INDIA PRIVATE LIMITED, *New Delhi*
PRENTICE-HALL OF JAPAN, INC., *Tokyo*
SIMON & SCHUSTER ASIA PTE. LTD., *Singapore*
EDITORA PRENTICE-HALL DO BRASIL, LTDA., *Rio de Janeiro*

Contents

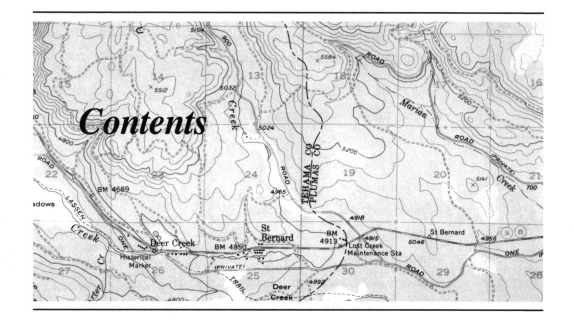

CHAPTER ONE

Introduction   1

CHAPTER TWO

Introduction to Measurements and Random Error Theory   11

CHAPTER THREE

Introduction to Distance Measurement   36

CHAPTER FOUR

Distance Corrections and Miscellaneous Taping Operations   *57*

CHAPTER FIVE

Electronic Distance Measuring Instruments   *85*

CHAPTER SIX

Introduction to Leveling 106

CHAPTER SEVEN

Differential Leveling 125

Contents

Contents

vii

Contents   vii

7-4   Earth's Curvature and Atmospheric Refraction   *129*

Contents vii

7-4 Earth's Curvature and Atmospheric Refraction *129*

7-5 Verniers *130*

7-6 Level Rod Targets *132*

7-7 Common Leveling Mistakes *134*

7-8 Leveling Errors *134*

7-9 Suggestions for Good Leveling *137*

7-10 Comments on Telescope Readings *137*

7-11 Precision of Differential Leveling *138*

7-12 Hand Signals *140*

7-13 Adjustments of Dumpy Levels *141*

7-14 Principle of Reversion *144*

7-15 Adjustments of Automatic Levels *145*

Problems *146*

CHAPTER EIGHT

Leveling Continued 152

8-1 Reciprocal Leveling *152*

8-2 Adjustments of Level Circuits *153*

8-3 Precise Leveling *157*

8-4 Profile Leveling *160*

8-5 Profiles *162*

8-6 Cross Sections *163*

8-7 Mistakes in Nonclosed Leveling Routes *166*

Problems *166*

CHAPTER NINE

Angles and Directions 170

9-1 Meridians *170*

9-2 Units for Measuring Angles *171*

CHAPTER TEN

Angles and Directions with Transits and Theodolites *191*

CHAPTER ELEVEN

Miscellaneous Angle Discussion   216

CHAPTER TWELVE

Traverse Adjustments and Area Computation   233

Computer Calculations and Miscellaneous Traverse Computations   *262*

CHAPTER FOURTEEN

Topographic Surveying 284

CHAPTER FIFTEEN

Total Stations and the Global Positioning System 311

CHAPTER SIXTEEN

Surveying Astronomy   322

Contents xiii

CHAPTER SEVENTEEN

Construction Surveying   354

CHAPTER EIGHTEEN

CHAPTER NINETEEN

Contents

Contents   XV

Contents

XV

Contents XV

Contents XV

Contents   **XV**

19-15 Detailed Subdivision 411

Contents XV

19-15 Detailed Subdivision *411*
19-16 Corner Markers *418*
19-17 Witness Corners *419*
19-18 Restoring Lost or Obliterated Corners *419*
19-19 Deed Descriptions of Land *419*

CHAPTER TWENTY

Control Surveys *421*

20-1 Introduction *421*
20-2 Horizontal Control *422*
20-3 Triangulation *424*
20-4 Accuracy Standards and Specifications for Control Surveys *425*
20-5 Triangulation Stations *428*
20-6 Strength of Figures *430*
20-7 Triangulation Systems *430*
20-8 Measurement of Angles and Baselines *432*
20-9 Work Involved in Triangulation *433*
20-10 Adjustment of a Chain of Single Triangles *433*
20-11 Adjustment of a Quadrilateral *435*
20-12 Trilateration *439*
Problems *441*

CHAPTER TWENTY-ONE

State Plane Coordinates *444*

21-1 Introduction *444*
21-2 Comments on Costs of Establishing and Using a State Plane Coordinate System *445*
21-3 Types of Coordinate Systems *445*
21-4 Disadvantages of State Plane Coordinates *446*

CHAPTER TWENTY-TWO

Horizontal Curves *467*

CHAPTER TWENTY-THREE

Vertical Curves *483*

CHAPTER TWENTY-FOUR

Photogrammetry   *500*

CHAPTER TWENTY-FIVE

Surveying—The Profession   *525*

APPENDIX A

Surveying Tables and Formulas   **532**

APPENDIX B

The Method of Least Squares   **555**

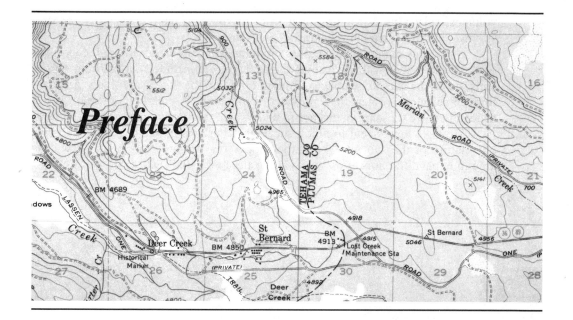

Preface

The purpose of this book is unchanged: to serve as an introduction to surveying and to present the elementary principles in such a manner as to encourage the reader toward further study of the subject. It is written for a one- or two-term course for students in civil engineering, forestry, building and construction science, architecture, agriculture, geography, and related areas. Hopefully, it will be useful for practicing surveyors as well.

With this edition the author has added sections concerning total station instruments, the Global Positioning System, and the hour-angle method for making astronomic observations. Expanded coverage is provided for several topics, including traversing and the use of electronic distance-measuring instruments. Almost all of the homework problems have been revised and many new ones have been added.

A computer disk, enclosed with the book and written for IBM and IBM-compatible machines, will enable the user to handle quickly some of the more tedious mathematical problems involved in surveying. Among the problems that may be solved are the computation of latitudes, departures and precisions for closed traverses, the balancing of errors and the computation of areas and coordinates for closed traverses, the calculation of lengths and directions for missing lines, stadia calculations for topography, and the azimuth calculations for hour angle observations of the sun and other stars.

The measurement of distances, elevations, and directions are considered in Chapters 1 to 11, while Chapters 12 to 25 are devoted to the applications of those measurements to the determination of land areas, preparations of topographic maps, practice of land surveying, public land surveys, computation of earthwork volumes, surveying astronomy, construction surveys, photogrammetry, control surveys, and professional registration and ethics.

Numerous numerical examples and homework problems are included in the book. If the reader will solve several of each of the included sets of home problems, he or she should be able to firmly fix in mind the theory involved. Answers to even selected problems are provided throughout. A high level of math is not needed to solve these problems nor to understand the surveying theory presented.

Most students seem to enjoy surveying. Perhaps this is due to the "hands on" nature of the subject. The author hopes that through this book he will be able to expand the appeal of surveying to the interested reader.

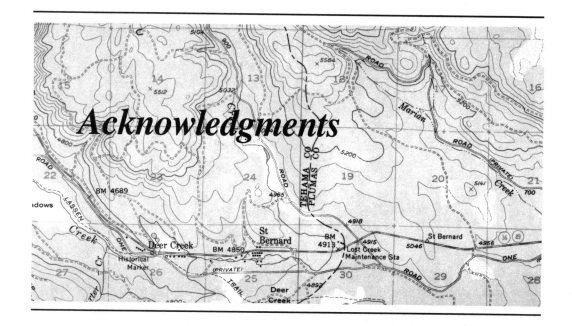

Acknowledgments

The author gratefully acknowledges the aid received from several sources. I am indebted to John W. Adcox, Jr., Ben Benson, R. Burtch, W. P. Byrd, C. J. Crandall, Robert C. Darling, Frank J. Hatfield, L. G. Holderly, Michael Orlando, F. T. Quiett, and Jerome L. Spurr, who have by their suggestions and criticisms directly contributed to the preparation of this book; to my own surveying professor, the late Colonel John Anderson, who patiently instructed me in his surveying class; to my associates who have helped in my study of surveying, including George R. Glenn, J. P. Rostron, Donald B. Stafford (who wrote Chapter 24 on Photogrammetry), I. A. Trively, and the late J. M. Ford, Jr. Thanks is also due to Nadim Aziz for his work on the computer programs.

JACK C. McCORMAC

Surveyor using Zeiss Level 1 Ni 2. (Courtesy of Carl Zeiss, Inc., Thornwood, NY 10594)

CHAPTER ONE

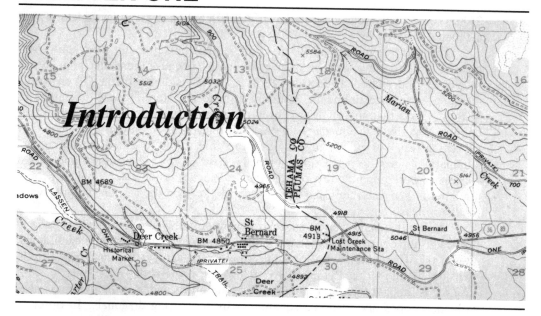

Introduction

1-1 FAMOUS SURVEYORS

Many famous persons in our history have engaged in surveying at some period in their lives. Particularly notable among these are several presidents—Washington, Jefferson, and Lincoln. Although the practice of surveying will not provide a sure road to the White House, many members of the profession like to think that the characteristics of the surveyor (honesty, perseverance, self-reliance, etc.) contributed to the development of these leaders. Today, surveying is an honored and widely respected profession. A knowledge of its principles and ethics is useful to a person whatever his or her future endeavors will be.[1]

1-2 EARLY HISTORY OF SURVEYING

It is impossible to determine when surveying was first used, but in its simplest form it is surely as old as recorded civilization. As long as there has been property ownership there have been means of measuring the property or distinguishing one person's land from another. Even in the Old Testament there are frequent references to property ownership, property corners, and property transfer. For instance, Proverbs 22:28: "Remove not the ancient landmark, which thy fathers have set." The Babylonians surely practiced some type of surveying as early as

---

[1] For purposes of convenience, the terms *rodman*, *instrumentman*, and *chainman* are used in the text, as the terms *rodperson*, *instrumentperson*, and *chainperson* are not widely used. The author does not mean this to minimize the role of women in the surveying profession.

2500 B.C. because archaeologists have found Babylonian maps on tablets of that estimated age.

The early development of surveying cannot be separated from the development of astronomy, astrology, or mathematics because these disciplines were so closely interrelated. In fact, the term *geometry* is derived from Greek words meaning earth measurements. The Greek historian Herodotus ("the father of history") says that surveying was used in Egypt as early as 1400 B.C. when that country was divided into plots for taxation purposes. Apparently, geometry or surveying was particularly necessary in the Nile valley to establish and control landmarks. When the yearly floods of the Nile swept away many of the landmarks, surveyors were appointed to replace them. These surveyors were called "rope-stretchers" because they used ropes (with markers on them at certain intervals) for their measurements.

During this same period surveyors were certainly needed for assistance in the design and construction of irrigation systems, huge pyramids, public buildings, and so on. Their work was apparently quite satisfactory. For instance, the dimensions of the Great Pyramid of Gizeh are in error about 8 in. over a 750-ft base.[2] It is thought that the rope-stretchers laid off the sides of the pyramid bases with their ropes and checked squareness by measuring diagonals. In order to obtain the almost level foundations of these great structures, the Egyptians probably either poured water into long, narrow clay troughs (an excellent method) or used triangular frames with plumb bobs or other weights suspended from their apexes as shown in Fig. 1-1.[3]

Each frame apparently had a mark on its lower bar which showed where the plumb line should be when that bar was horizontal. These frames, which were probably used for leveling for many centuries, could easily be checked for proper adjustment by reversing them end for end. If the plumb lines returned to the same points, the instruments were in proper adjustment and the tops of the supporting stakes (see Fig. 1-1) would be at the same elevation.

The practical-minded Romans introduced many advances in surveying by

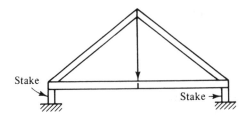

Stake   Stake →   **Figure 1-1** Ancient leveling frame.

[2] C. M. Brown, W. G. Robillard, and D. A. Wilson, *Evidence and Procedures for Boundary Location*, 2nd ed. (New York: John Wiley & Sons, Inc., 1981), p. 145.

[3] A. R. Legault, H. M. McMaster, and R. R. Marlette, *Surveying* (Englewood Cliffs, N.J.: Prentice-Hall, Inc., 1956), p. 5.

an amazing series of engineering projects constructed throughout their empire. They laid out projects such as cities, military camps, and roads by using a system of rectangular coordinates. They surveyed the principal routes used for military operations on the European continent, in the British Isles, in northern Africa, and even in parts of Asia.

Two instruments used by the Romans were the *odometer*, or measuring wheel, and the *groma*. The groma, from which Roman surveyors received their name of *gromatici*, was used for laying off right angles. It consisted of two cross-arms fastened together at right angles in the shape of a horizontal cross with plumb lines hanging from each of the four ends. The groma, which was pivoted eccentrically on a vertical staff, could be leveled and sights taken along its cross-arms.

From Roman times until the modern era there were few advances in the art of surveying, but the last few centuries have seen the introduction of the telescope, the vernier, the theodolite, electronic distance-measuring equipment, and many other excellent devices. These developments will be mentioned in subsequent chapters. For a detailed historical list of early instrument development, the reader is referred to "Historical Notes upon Ancient and Modern Surveying and Surveying Instruments."[4]

1-3 SURVEYING DEFINED

Surveying is the science of determining the dimensions and contour (or three-dimensional characteristics) of the earth's surface by the measurements of distances, directions, and elevations. It also involves staking out the lines and grades needed for the construction of buildings, roads, dams, and other engineering structures. In addition to these field measurements, surveying includes the computation of areas, volumes, and other quantities, as well as the preparation of necessary maps and diagrams. Surveying has many industrial applications: for example, setting equipment, assembling aircraft, laying out assembly lines, and so on.

1-4 PLANE SURVEYS

In large-scale mapping, adjustments are made for the curvature of the earth and for the fact that north–south lines converge to meet at the poles. Plane surveys, however, are made on such small areas that the effect of these factors may be neglected. In plane surveying the earth is considered to be a flat surface and north–south lines are assumed to be parallel. Calculations for a plane surface are relatively simple, since the surveyor is able to use plane geometry and plane trigonometry.

[4] *Transactions of the ASCE,* 1893, vol. 30, pp. 135–154.

The great majority of plane surveys are sufficiently accurate for all but the largest areas. Surveys for farms, subdivisions, buildings, highways, railroads, and in fact most constructed works are plane surveys. Plane surveys are not generally considered to be sufficiently accurate for establishing state and national boundaries.

It can be shown that an arc along the earth's curved surface of 11.5 miles in length is only approximately 0.05 ft longer than the plane or chord distance between its ends. As a result, it probably seems to the reader that such discrepancies are insignificant. Direction discrepancies due to the convergence of north–south lines are, however, much more significant than are distance discrepancies.

1-5 GEODETIC SURVEYS

Geodetic surveys are those that are adjusted for the curved shape of the earth's surface. (The earth is an oblate spheroid whose radius at the equator is about 13.5 miles greater than its polar radius.) Since they allow for earth's curvature, geodetic surveys can be applied to both small and large areas. The equipment used and the methods of measurement applied are about the same as they are for plane surveys. Elevations are handled in the same manner for both plane and geodetic surveys. They are expressed in terms of vertical distances above or below a reference curved surface, usually mean sea level.

Most geodetic surveys are made by government agencies, such as the former U.S. Coast and Geodetic Survey (now the National Geodetic Survey, National Ocean Survey, National Oceanic and Atmospheric Administration of the U.S. Department of Commerce). Although only a relatively small number of surveyors are employed by the National Geodetic Survey (NGS), their work is extremely important to all other surveyors. They have established a network of reference points around the United States that provides very precise information on horizontal and vertical locations. On this network all sorts of other surveys (plane and geodetic) of lesser precision are based.

1-6 TYPES OF SURVEYS

This section is devoted to a brief description of the various types of surveys. Most of these types of surveys employ plane rather than geodetic techniques.

Land surveys are the oldest type of survey and have been performed since earliest recorded history. They are normally plane surveys made for locating property lines, subdividing land into smaller parts, determining land areas, and any other information involving the transfer of land from one owner to another. These surveys are also called *property surveys*, *boundary surveys*, or *cadastral surveys*. Today, the term *cadastral* is usually used with regard to surveys of public lands.

Figure 1-2 SET electronic total station instrument. (Courtesy of the Leitz Company.)

Topographic surveys are made for locating objects and measuring the relief, roughness, or three-dimensional variations of the earth's surface. Detailed information is obtained pertaining to elevations as well as to the locations of constructed and natural features (buildings, roads, streams, etc.) and the entire information is plotted on maps (called topographic maps).

Route surveys involve the determination of the relief and the location of natural and artificial objects along a proposed route for a highway, railroad, canal, pipeline, power line, or other utility. They may further involve the location or staking out of the facility and the calculation of earthwork quantities.

City or *municipal surveys* are made within a given municipality for the purpose of laying out streets, planning sewer systems, preparing maps, and so on. When the term is used, it usually brings to mind topographic surveys in or near a city for the purpose of planning urban expansions or improvements.

Construction surveys are made for purposes of locating structures and providing required elevation points during their construction. They are needed to control every type of construction project. It has been estimated that 60% of the surveying done in the United States is construction surveying.[5]

---

[5] R. C. Brinker and R. Minnick, editors, *The Surveying Handbook* (New York: Van Nostrand Reinhold Company, Inc., 1987), p. 791.

Hydrographic surveys pertain to lakes, streams, and other bodies of water. Shorelines are charted, shapes of areas beneath water surfaces are determined, water flow of streams is estimated, and other information needed relative to navigation, flood control and development of water resources, is obtained. These surveys are usually made by a governmental agency, for example, the National Geodetic Survey, the U.S. Geological Survey, or the U.S. Army Corps of Engineers.

Marine surveys are related to hydrographic surveys, but they are thought to cover a broader area. They include the surveying necessary for offshore platforms, the science of navigation, the theory of tides, and the preparation of hydrographic maps and charts.

Mine surveys are made to obtain the relative positions and elevations of underground shafts, geological formations, and so on, and to determine quantities and establish lines and grades for work to be done.

Forestry and geological surveys are probably much more common than the average layperson realizes. Foresters use surveying for boundary locations, timber cruising, topography, and so on. Similarly, surveying has much application in the preparation of geological maps.

Photogrammetric surveys are those in which photographs (generally aerial) are used in conjunction with limited ground surveys (used to establish or locate certain control points visible from the air). Photogrammetry is extremely valuable because of the speed with which it can be applied, the economy, the applications to areas difficult of access, the great detail provided, and so on. Its uses are becoming more extensive each year.

As-built surveys are made after a construction project is complete to provide the positions and dimensions of the features of the project as they were actually constructed. Such surveys not only provide a record of what was constructed but also provide a check to see if the work proceeded according to the design plan.

The usual construction project is subject to numerous changes from the original plans due to design changes as well as to problems encountered in the field, such as underground pipes and conduits, unexpected foundation conditions, and other situations. As a result, the as-built survey becomes a very important document that must be preserved for future repairs, expansions, and modifications. For example just imagine how important it is to know the precise location of water and sewer lines.

Control surveys are reference surveys. For a particular control survey a number of points are established and their horizontal and vertical positions very accurately determined. The points are established so that other work can conveniently be referenced or oriented to them.

The horizontal and vertical controls form a network over the area to be surveyed. For a particular project the horizontal control is probably tied to property lines, road center lines, and other prominent features. Vertical control consists of a set of relatively permanent points whose elevations above or below sea level have been carefully determined. (These points are called bench marks.)

In the decades to come, undoubtedly other special types of surveying will develop. Surveyors might very well have to establish boundaries under the ocean, in the Arctic and Antarctic, and even on the moon and other planets. Great skill and judgment by the surveying profession will undoubtedly be required to handle these tasks.

1-7 FIRST-, SECOND-, AND THIRD-ORDER SURVEYS

Surveys are frequently specified as being first-, second-, or third-order. Because these terms are occasionally used in various places in the text, a brief description of them is presented in this section so that the reader will not be puzzled when they are used.

The Federal Geodetic Control Committee has established a set of accuracy standards for horizontal and vertical control surveys. There are three major classifications and they are given in descending order as to accuracy requirements. A very brief description of each of these classifications is given here, and more detailed information is provided in Chapters 7 and 20.

1. First-order surveys are made for the primary national control network, metropolitan area surveys, and scientific studies. (In detail they are very accurate surveys used for military defense, sophisticated engineering projects, dams, tunnels, and studies of regional earth crustal movements.)
2. Second-order surveys are done somewhat less accurately than are first-order surveys. They are used to densify the national network as well as for subsidiary metropolitan control. (In detail they are used for control along tidal boundaries, for large construction projects, for interstate highways, for monitoring local crustal movements, for urban renewal and small reservoirs.)
3. Third-order surveys are done somewhat less accurately than are second-order ones. They are general control surveys referred to the national network. (In detail they are used for local control surveys, small engineering projects, small-scale topographic maps, and boundary surveys.)

1-8 IMPORTANCE OF SURVEYING

As described in Section 1-2, it has been necessary since the earliest civilizations to determine property boundaries and divide sections of land into smaller pieces. Through the centuries the uses of surveying have expanded until today it is difficult to imagine any type of construction project that does not involve some type of surveying.

All types of engineers, as well as architects, foresters, and geologists, are concerned with surveying as a means of planning and laying out their projects.

Surveying is needed for subdivisions, buildings, bridges, highways, railroads, canals, piers, wharves, dams, irrigation and drainage networks, and many other projects. In addition, surveying is required to lay out industrial equipment, set machinery, hold tolerances in ships and airplanes, prepare forestry and geological maps, and numerous other applications.

The study of surveying is an important part of the training of a technical student even though that student may never actually practice surveying. It will appreciably help him or her to learn to think logically, to plan, to take pride in working carefully and accurately, and to record his or her work in a neat and orderly fashion. He or she will learn a great deal about the relative importance of measurements, develop some sense of proportion as to what is important and what is not, and acquire essential habits of checking numerical calculations and measurements (a necessity for anyone working in an engineering or scientific field).

1-9 LIABILITY INSURANCE

We are all familiar with stories of almost unbelievable court settlements which seem to occur often in cases of personal injury. Similar settlements are frequently made in cases where surveying mistakes cause financial damages to clients. These cases are more common for construction projects than for rural land surveys.

For this discussion just imagine a surveyor who in preparing a topographic map is 2 ft off on elevations and that this mistake causes the contractor to have to fill in an acre or two of land with 2 ft of soil. What do you think the contractor is going to try to do with the bill for this extra expense? In a similar vein, imagine that a surveyor makes a mistake in locating a property line and as a result a building is erected which runs a few feet onto someone else's land. Who do you think is a prime candidate for paying damages?

From the preceding discussion the reader can see why it is today so common for industrial clients to require surveying companies to provide evidence of suitable liability insurance coverage. Unfortunately, this insurance is very expensive, deductibles on claims are quite large, and as in other insurance situations, frequent claims lead to much higher premiums.

In some states if a surveyor causes financial injury to a client, that surveyor may lose his or her license unless satisfactory financial restitution is made to the client. A surveyor who carries a suitable liability policy will certainly be in a position to get more industrial jobs than a surveyor who does not have such insurance.

This discussion can be continued to cover personal injuries. The reader can understand that if a surveying employee (worker's compensation may be involved here) or an outsider is injured as a result of surveying activities, the surveyor or the insurance company will be right in line for paying damages.

1-10 SAFETY

A consideration of safety for surveying employees, clients, and outsiders is an extremely important topic not only from the point of view of physical suffering but also from that of economic loss. Personal injury and sickness cause a loss of efficiency for the company and greater expenses from the standpoint of less production and higher insurance costs. The reader understands that the more accidents that occur to a particular firm, the higher will be their insurance premiums. These premiums are today no small item in the cost of operating a business (even one with a splendid safety record).

In Chapter 2 the author has quite a bit to say about the importance of planning in obtaining good surveys. Planning is also of extraordinary importance when we think of achieving safety. The surveyor needs to conduct periodic safety meetings and needs to continuously discuss with his or her employees the dangers in the various types of surveys they conduct. There are many hazards that may be encountered when working near or on highways, on construction projects, near power lines, and on remote pieces of land with dangerous terrain features. It is absolutely essential for surveying parties to have first-aid kits and to have some personnel trained in first-aid procedures.

Surveyors should wear conspicuous clothing such as orange hunting vests. When in the area of construction projects, hard hats and safety shoes are a necessity. In snake-infested areas it is necessary to wear boots and/or leggings. Stings from wasps, bees, and yellow-jackets, however, are a much more likely possibility. As a result, first-aid kits should include antidotes for persons who are allergic to such bites.

Ticks seem to reside everywhere trees are located, and their bites can cause life-threatening diseases. As a result, surveyors in tick-infested areas should use insect repellant, wear long-sleeved shirts, and tuck their pants into their boots. Furthermore, it is desirable for personnel to wear light-colored clothing so that ticks can easily be seen. Employees should also check each other for the presence of these pests.

One of the most dangerous situations for surveyors is work along highways. For such jobs it is absolutely necessary to make use of warning signs and flagmen. Of course, the best prevention of all is to stay off the roads and work with offsets if possible.

The purpose of this section is not to list every possible safety precaution that should be taken but rather, to make surveyors conscious of safety and to try to make them think through the work to be done and the possible hazards that might be present. A few of many possible safety precautions follow:

1. Do not look at the sun through instrument telescopes unless special filters are used, because serious and permanent eye injury may result.
2. Be careful of the danger of sunstrokes.

3. Wear gloves when working in briars or poison oak.
4. When cutting bushes, be extremely careful of the location of other persons.
5. Do not use steel or fiberglass tapes near electric lines.
6. Do not throw range poles or chaining pins.

1-11 OPPORTUNITIES IN SURVEYING

There are few professions that need qualified people as much as does the surveying profession. In the United States, tremendous physical developments (subdivisions, factories, dams, power lines, cities, etc.) have created a need for surveyors at a faster rate than our schools are producing them. Construction, our largest industry, requires a constant supply of new surveyors.

For a person with a liking for a combination of outdoor and indoor work, surveying offers attractive opportunities. It now appears that beginning salaries for qualified surveyors, which have lagged behind civil engineering salaries in the past, are catching up and will perhaps pass them in the next few years. Despite this fact, most surveying programs across the country are faced with decreasing enrollments.

It is necessary for a person going into private practice to meet the licensing requirements of his or her state. These requirements, which have become stiffer in recent years, are discussed in detail in Chapter 25.

CHAPTER TWO

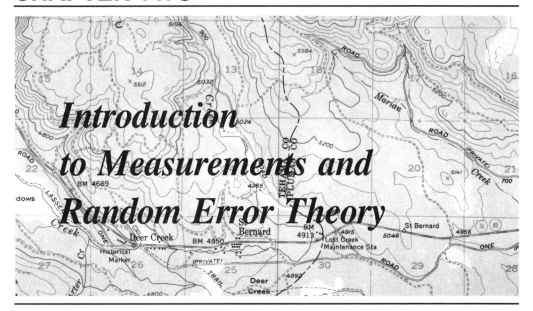

Introduction to Measurements and Random Error Theory

2-1 MEASUREMENT

In ordinary life, most of us are accustomed to *counting* but not as much to *measuring*. If the number of people in a room is counted, the result is an exact number without a decimal, say 9 people. It would be ridiculous to say that there are 9.23 people in a room. In a similar fashion a person might count the amount of money in his or her pocket. Although the result may contain a decimal as $5.65, the result is still an exact value.

Surveying is concerned with measurements of quantities whose exact or true values may not be determined, such as distances, elevations, volumes, directions, and weights. If a person were to measure the width of his desk with a ruler divided into tenths of an inch, he could estimate the width to hundredths of an inch. If he were to use a ruler graduated in hundredths of an inch, he could estimate the width to thousandths of an inch; and so on. Obviously, with better equipment he can estimate an answer that is closer to the exact value but will never be able to determine the value absolutely. Thus a fundamental principle of surveying is that no measurement is exact and the true value of the quantity being measured is never known. (Exact or true values do exist, but they cannot be determined.) *Measurement is the principal concern of a surveyor.*

2-2 NECESSITY FOR ACCURATE SURVEYS

The surveyor must have the skill and judgment necessary to make very accurate measurements. This fact is obvious when one is thinking in terms of the construction of long bridges, tunnels, tall buildings, and missile sites or with the setting of delicate machinery, but it can be just as important in land surveying.

As late as a few decades ago land prices were not extremely high except in and around the largest cities. If the surveyor gained or lost a few feet in a lot or a few acres in a farm, it was usually not considered to be a matter of great importance. The instruments commonly used for surveying before this century were not very good compared to today's equipment, and it was probably impossible for the surveyor to do the quality of work expected of today's surveyor. (What will surveyors in future centuries think of twentieth-century surveying, and what wonderful equipment will they have to work with?)

It is said that early surveyors in the United States had to complete their work very quickly when working in Indian country. The Indians were well aware that numerous settlers followed on the heels of surveyors, and the Indians apparently thought if they eliminated the surveyors, the settlers might not come to steal their land. Surveyors working under these conditions were said rather frequently to neglect to set some property corners, particularly in forbidding forests, and their rapid measurements were not always the best. In addition, many of their values were more than likely computed in the office and not measured on the ground.

Today, land prices are in most areas very high, and evidently the climb has only begun. In many areas of high population and in many popular resort areas, land is sold by so many dollars per square foot or so many hundreds or even thousands of dollars per front foot; therefore, the surveyor must be able to do splendid work. Even in rural areas, land is frequently "sky high."

2-3 ACCURACY AND PRECISION

The terms *accuracy* and *precision* are constantly used in surveying, yet their correct meanings are a little difficult to grasp. In an attempt to clarify the distinction, the following definitions are presented:

Accuracy refers to the degree of perfection obtained in measurements. It denotes how close a given measurement is to the true value of the quantity.

Precision or *apparent accuracy* is the degree of refinement with which a given quantity is measured. In other words, it is the closeness of one measurement to another. If a quantity is measured several times and the values obtained are very close to each other, the precision is said to be high.

It does not necessarily follow that better precision means better accuracy. Consider the case in which a surveyor carefully measured a distance three times with a 100-ft steel tape and obtained the values 984.72 ft, 984.69 ft, and 984.73 ft. He did a very precise job and apparently a very accurate one. Should, however, the tape be found to actually be 100.30 ft long instead of 100.00 ft, the values obtained are not accurate, although they are precise. (The measurements could

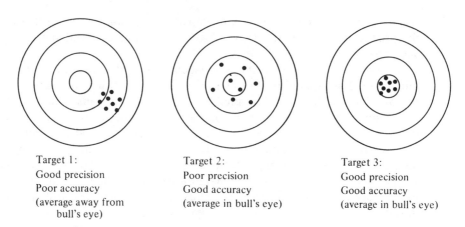

Target 1:
Good precision
Poor accuracy
(average away from
    bull's eye)

Target 2:
Poor precision
Good accuracy
(average in bull's eye)

Target 3:
Good precision
Good accuracy
(average in bull's eye)

Figure 2-1 Accuracy and precision.

be made accurate by making a numerical correction of 0.30 ft for each tape length.) It is possible for the surveyor to obtain both accuracy and precision by exercising care, patience, and using good instruments and procedures.

In measuring distance, precision is defined as the ratio of the error of the measurement to the distance measured and it is reduced to a fraction having a numerator of unity. If a distance of 4200 ft is measured and the error is later estimated to equal 0.7 ft, the precision of the measurement is 0.7/4200 = 1/6000. This means that for every 6000 ft measured, the error would be 1 ft if the work were done with this same degree of precision.

A method frequently used by surveying professors to define and distinguish between precision and accuracy is illustrated in Fig. 2-1. It is assumed that a person has been having a little target practice with his rifle. His results with the first target were very precise because the bullet holes were quite close to each other. They were not accurate, however, as they were located some little distance from the bull's eye.

The marksman fired accurately at target 2, as the holes were placed relatively close to the bull's eye. The shots were not precise, however, as they were scattered quite a bit with respect to each other.

Finally, in target 3, the shots were both precise and accurate, as they were placed in the bull's eye close to each other. *The objective of the surveyor is to make measurements that are both precise and accurate.*

2-4 ERRORS AND MISTAKES

There is no one whose senses are sufficiently perfect to measure any quantity exactly and there are no perfect instruments with which to do the measuring. The result is that all measurements are imperfect. A major concern in surveying is the precision of the work. This subject is repeatedly mentioned as we discuss each phase of surveying.

The ever-present differences between measured quantities and the true magnitudes of those quantities are classified herein as either mistakes or errors. A *mistake* (or blunder) is a difference from a true value caused by the inattention of the surveyor. For instance, he may read a number as 6 when it is actually 9, may record the wrong quantities in the field notes, or may add a column of numbers incorrectly. The important point here is that mistakes are caused by the carelessness of the surveyor, and carelessness *can be eliminated* by careful checking. Every surveyor will make occasional mistakes, but if he learns to apply carefully to his work the checks described in subsequent chapters, he will eliminate these mistakes. Any true professional surveyor will not be satisfied with his work until he has made sure that any blunders are detected and eliminated.

An *error* is a difference from a true value caused by the imperfection of a person's senses, by the imperfection of his equipment, or by weather effects. Errors *cannot be eliminated but they can be minimized* by careful work, combined with the application of certain numerical corrections.

2-5 SOURCES OF ERRORS

There are three sources of errors: people, instruments, and nature. Accordingly, errors in measurement are generally said to be personal, instrumental, and natural. Some errors, however, do not clearly fit into just one of these categories and may be due to a combination of factors.

Personal errors occur because no surveyor has perfect senses of sight and touch. For instance, in estimating the fractional part of a scale, the surveyor cannot read it perfectly and will always be either a little large or a little small.

Instrumental errors occur because instruments cannot be manufactured perfectly and the different parts of the instruments cannot be adjusted exactly with respect to each other. Moreover, with time the wear and tear of the instruments causes errors. Although the past few decades have seen the development of more precise equipment, the goal of perfection remains elusive. A reading from a scale will undoubtedly contain both a personal and an instrumental error. The observer cannot read the scale perfectly nor can the manufacturer make a perfect scale.

Natural errors are caused by temperature, wind, moisture, magnetic variations, and so on. On a summer day a 100-ft steel tape may increase in length by a few hundredths of a foot. Each time this tape is used to measure 100 ft there will be a temperature error of those few hundredths of a foot. The surveyor cannot normally remove the cause of errors such as these, but can minimize their effects by using good judgment and making proper corrections of the results.

2-6 SYSTEMATIC AND ACCIDENTAL ERRORS

Errors are said to be systematic or accidental. A *systematic* or *cumulative error* is one which, for constant conditions, remains the same as to sign and magnitude. For instance, if a steel tape is 0.10 ft too short, each time the tape is used the same error (because of that factor) is made. If the full tape length is used 10 times, the error accumulates and totals 10 times the error for one measurement.

   An *accidental*, *compensating*, or *random error* is one whose magnitude and direction are just an accident and beyond the control of the surveyor. For instance, when a person reads an angle with a surveying instrument, he or she cannot read it perfectly. One time he or she will read a value that is too large and the next time will read a value that is too small. Since these errors are just as likely to have one sign as the other, they tend to a certain degree to cancel each other or compensate for each other.

2-7 SIGNIFICANT FIGURES

This section might also be entitled "judgment" or "sense of proportion," and its comprehension is a very necessary part of the training of anyone who takes and/ or uses measured quantities of any kind. When measurements are made, the results can be precise only to the degree that the measuring instrument is precise. This means that numbers which represent measurements are all approximate values. For instance, a distance may be measured with a steel tape as being 465 ft or again more precisely as 465.3 ft or even with more care as 465.32 ft, but an exact answer can never be obtained. The value will always contain some error.

   The number of significant figures that a measured quantity has is not (as is frequently thought) the number of decimal places. Instead, it is the number of certain digits plus one digit that is estimated. For instance, in reading a steel tape a point may be between 34.2 and 34.3 ft (the scale being marked at the $\frac{1}{10}$-ft points) and the value is estimated as being 34.26 ft. The answer has four significant figures. Other examples of significant figures follow:

   36.00620 has seven significant figures.
   10.0 has three significant figures.
   0.003042 has four significant figures.

   The answer obtained by solving any problem can never be more accurate than the information used. If this principle is not completely understood, results will be slovenly. Notice that it is not reasonable to add 23.2 cu yd of concrete to 31 cu yd and get 54.2 cu yd. One cannot properly express the total to the nearest tenth of a yard as 54.2 because one of the quantities was not computed to the nearest tenth and the correct total should be 54 cu yd.

A few general rules regarding significant figures follow:

1. Zeros between other significant figures are significant, as, for example in the following numbers, each of which contains four significant figures: 23.07 and 3008.
2. For numbers less than unity, zeros immediately to the right of the decimal are not significant. They merely show the position of the decimal. The number 0.0034 has two significant figures.
3. Zeros placed at the end of decimal numbers, such as 24.3200, are significant.
4. When a number ends with one or more zeros to the left of the decimal it is necessary to indicate the exact number of significant figures. The number 352,000 could have three, four, five, or six significant figures. It could be written as 35$\overline{2}$,000, which has three significant figures, or as 352,00$\overline{0}$, which has six significant figures. It is also possible to handle the problem by using scientific notation. The number 2.500×10^3 has four significant figures, and the number 2.50×10^3 has three.
5. When numbers are multiplied or divided or both, the answer should not have more significant figures than those in the factor, which had the least number of significant figures. As an illustration, the following calculations should result in an answer having three significant figures, which is the number of significant figures in the term 3.25.

$$\frac{3.25 \times 4.6962}{8.1002 \times 6.152} = 0.306$$

It is desirable to carry calculations to one or more extra places during the various steps; as the final step, the answer is rounded off to the correct number of significant figures.

6. For addition or subtraction, the last significant figure in the final answer should correspond to the last column full of significant figures among the numbers. An addition example follows:

$$
\begin{array}{r}
33.842 \\
361.3 \\
81.24 \\
\hline
476.382 \ = \ 476.4
\end{array}
$$

last column of significant figures

2-8 PLANNING

If we are to make accurate surveys, we need to use precise equipment, good procedures, and good planning. Accurate surveys can be made with old and perhaps out-of-date equipment, but time and money can be saved with modern instruments. Good planning is the most important item necessary for achieving economy and is also very important for achieving accuracy.

Perhaps no topic in this book is of more importance than the few words presented in this chapter on the subject of planning. The underlying thought in the author's mind as he prepared this chapter was planning. Not only was he thinking of organizing the parties and equipment but also selecting the procedures to be used. In other words, if we want a survey to be done with a precision of 1/30,000, what procedures and equipment will we have to use?

A great deal of space in this book is devoted to the terms *accuracy* and *precision*. A thorough understanding of these subjects will enable the surveyor to improve his or her work and reduce the costs involved by careful planning.

The planning of a survey for a particular project includes the selection of equipment and methods to be used, reconnaissance of the area for existing surveying monuments, selection of possible locations for surveying stations, and so on.

The longer surveys take for an engineering project, the higher become costs such as interest expenses. The cost of accurate surveying is frequently 1 to 3% of the overall cost of an engineering project—but if it is done poorly, the percentages may be many times those values. Thus carefulness in a survey must not be slighted even though speed may be essential.

In general, surveyors would like to obtain precisions which are better than those required for the particular project at hand. However, economic limitations (i.e., what the employer is willing to pay) will frequently limit these desires.

2-9 DISCUSSION OF ACCIDENTAL OR RANDOM ERRORS

Often when the author talks to students about the information to follow on random errors and statistics, he is rewarded with a series of blank faces. These pages may initially affect you that way, too. It is hoped, however, that as random errors and their propagation and their effects on precision are discussed in later chapters, you will come back and study these pages again.

A surveyor is not able to make perfect readings. Each time a measurement is made, it will be either too large or too small. For every measurement there is some doubt. Even after corrections have been carefully made for systematic errors, random errors will remain that cannot be exactly determined or eliminated.

Although it is impossible to eliminate systematic errors completely, it is assumed for this discussion that they have been reduced to such small values as to make them negligible. The error theory discussed here has been found to apply very well to the accidental or random errors occurring in properly conducted surveys.

An introductory study of the theory of probability will enable a surveyor easily to grasp several useful principles relating to error theory, even though he or she does not have a detailed background in statistics and mathematics. These principles can be learned while avoiding the partial differentials and other advanced concepts needed to make absolute mathematical proofs.

In statistics, data are considered to be a sample of an infinitely large number of repetitions of a measurement for which the mean will be the true value. *Although probability theory is based on an infinite number of observations, it may in practice be applied with good results to situations where only a few observations are made.*

For many surveys the quality of a measurement can be expressed by stating a relative error. For example, a distance may be expressed as 835.82 ± 0.06 ft. Such a statement indicates that the true distance measured probably falls between 835.76 and 835.88 ft, and its most probable value is 835.82 ft. The sign or direction of the probable error is not known and thus no correction can be made. The probable error can either be plus or minus, between the limits within which the error is likely to fall. Note that this form does not specify the magnitude of the actual error, nor does it indicate the error most likely to occur.

If the distance referred to in the preceding paragraph had a probable error of ±0.06 ft, the probable precision is 0.06/835.82 = 1/13,390.

The relative error (such as the ±0.06 ft used above) cannot just be a subjective guess on the surveyor's part. It must be a logical estimate based on his or her previous experience, the methods and equipment used, and the field conditions encountered. If no statement is given as to the relative error, the person using the data will automatically assume that the uncertainty equals ± one-half of the last decimal place. (If a distance is recorded as 232.4 ft with no probable error given, the user will assume the probable error is one-half of ±0.1 or ±0.05 ft.) It is hoped that the surveyor who made the measurement understands this fact.

2-10 OCCURRENCE OF RANDOM ERRORS

If a coin is flipped 100 times, the probability is that there will be 50 heads and 50 tails. Each time the coin is flipped there is an equal chance of it being heads or tails, and it is apparent that the more times the coin is flipped, the more likely it will be that the total number of heads will equal the total number of tails.

When a quantity such as distance is being measured, random errors occur due to the imperfections of the observer and the equipment used. The observer is not able to make perfect readings. Each time a measurement is made it will be either too large or too small.

For this discussion it is assumed that a distance was measured 28 times, with the results shown in Table 2-1. In the first column of the table the values of the measurements are arranged for convenience in order of increasing value, while the number of times a particular measurement was obtained is given in the second

TABLE 2-1

Measurement	Number or frequency of each measurement	Residual or deviation
96.90	1	−0.04
96.91	2	−0.03
96.92	3	−0.02
96.93	5	−0.01
96.94	6	0.00
96.95	5	+0.01
96.96	3	+0.02
96.97	2	+0.03
96.98	1	+0.04
Average = 96.94		

column. The latter values are also referred to as the *frequencies* of the measurements.

The errors of the measurements are not known because the true value of the quantity is not known. The true value, however, is assumed to equal the arithmetic mean of the measurements. It is referred to as the *best value* or as the *most probable value*. The error of each measurement is then assumed to equal the difference between the measurement and the mean value. These are not really errors and they are referred to as *residuals* or *deviations*. For the measurements being considered here, the residuals are shown in the third column of Table 2-1.

For a particular set of measurements of the same item, it is possible to take the residuals and plot them in the form of a bar graph, as shown in Fig. 2-2. This

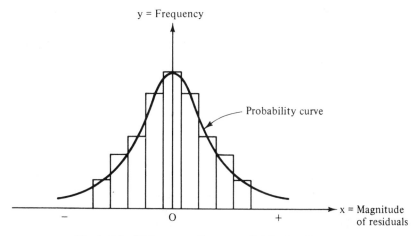

Figure 2-2 Histogram or frequency distribution diagram.

figure, called a *histogram* or a *frequency distribution diagram*, has the magnitudes of the residuals plotted along the horizontal axis. Their signs, plus or minus, are shown by plotting them to the right or left of the origin (O). In addition, the number or frequency of the residuals of a particular size are shown vertically.

2-11 PROBABILITY CURVE

If an infinite number of measurements of an item could be taken and if the residuals of those values were plotted as a histogram and a curve drawn through the values, the results would theoretically fall along a smooth bell-shaped curve called a *probability curve* (also called the *Gauss curve* or the *normal error distribution curve*). Almost all surveying measurements conform to such a curve. This curve shows the relationship between the size of an error and the probability of its occurrence.

The probability curve provides the most accurate method for studying the precision of surveys and also provides us with the best available means for estimating the precision of future planned surveys. The curve can be represented with a mathematical equation (given at the end of this section), but the equation is seldom used in surveying measurements.

Theoretically, the probability curve would be perfectly bell-shaped if an infinite number of measurements were taken and plotted. Even where a fairly large number of measurements of a particular quantity is made, the histogram can be plotted and a curve drawn as shown in Fig. 2-2. Such a curve will approximate the theoretical curve and can be used practically to estimate the most probable behavior of random errors. There is little approximation involved because a curve plotted for as few as five or six determinations of a quantity is practically the same as the theoretical curve based on an infinite number of measurements.

Several important items should be noted concerning the probability curve. These include:

1. Positive and negative errors occur with the same frequency.
2. Large errors do not occur often.
3. Small errors occur more often than large ones.

It is apparent that if the measurements are taken with a different degree of precision, the dimensions of the curve would be affected. The more precise the measurements, the more the errors or residuals will be concentrated near the center of the curve. For such a situation the curve will be higher and less spread out laterally. Should the measurements be taken with less precision, the reverse will be true. *The effect of changes in precision, however, will only be to change the scale of the curve vertically and horizontally. The other characteristics and the usefulness of the curve will not change.*

The probability curve of Fig. 2-2 is redrawn in Fig. 2-3 and a point P is shown which has an ordinate equal to y and an abscissa equal to x. The y value is a number that gives the probability of an error of magnitude x.

From a theoretical standpoint a probability curve approaches the x axis at plus infinity and minus infinity. It is therefore obvious that all of the readings (or rather their residuals) are represented underneath the curve. In other words, there are no values outside the curve and thus the sum of the values underneath the curve represents 100% of the values taken.

If the average of all the residuals for a particular set of measurements is computed and we assume that the curve is approximately symmetrical about the y axis, we see that approximately 50% of the values will be on the negative side and approximately 50% of the values will be on the positive side. (For the example presented in Fig. 2-2, the histogram is exactly symmetrical as to the number and sizes of plus and minus residuals. For many practical measurements, however, the diagram will not be perfectly symmetrical and is said to be *skewed*.)

Fifty percent of the area underneath the probability curve of Fig. 2-4 is shown shaded. There is a 50% chance that the error for a single measurement will fall within this area and a 50% chance that it will fall without. The value x_p shown in the figure is referred to as the *probable error* or 50% error. A particular measurement will have the same chance of having an error less than x_p as it does of being greater than x_p.

The average or 50% error can be determined by multiplying a constant 0.845 times the average numerical value of the residuals. The signs ($\pm$) of the residuals

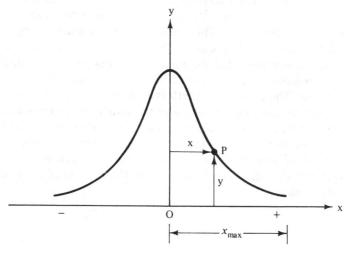

Figure 2-3

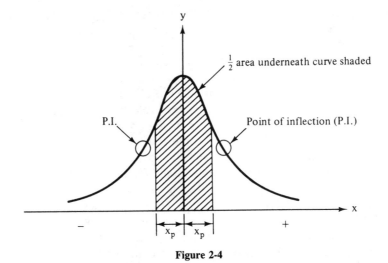

Figure 2-4

are not included in the average. Letting v equal a residual, the 50% error E_{50} is as follows:

$$E_{50} = \pm 0.845 v_{average}$$

The term *probable error* was commonly used in the past by surveyors and engineers but today is seldom used. Most of our references are now made to another value, called the standard error or standard deviation, a term that is defined in the next paragraph.

There are several ways in which errors may be denoted, but the most common one is to refer them to the *standard deviation* (σ), also called the *mean square error* or *standard error*. This value provides a practical means of indicating the reliability of a set of repeated measurements. If we examine the curve of Fig. 2-4, we will see that there are points of inflection (P.I.s) on each side of the curve, that is, points where the slope of the curve changes from concave to convex, or vice versa. The area underneath the probability curve between these points for a theoretical curve equals 68.3% of the total area. If a particular quantity is measured 10 times, it is anticipated that 68.3% or about 7 of the 10 measurements will fall between these values and 3 will not. The residuals at the P.I.s are called the *standard deviations* or *standard errors* and can be calculated from the following expression, which is derived by the method of least squares discussed in the Appendix:

$$\sigma = \pm \sqrt{\frac{\Sigma v^2}{n - 1}}$$

where Σv^2 is the sum of the squares of the residuals and n is the number of observations.

If the student will examine the literature published by manufacturers of surveying equipment, he or she will note that they usually provide the standard error associated with the use of their equipment. Actually, the 68.3% value is not often used as such by the surveyor. How many surveyors would be willing to make a statement that a particular measurement was 632.87 ± 0.05 ft when the ±0.05 ft was probably correct only 68.3% of the time? The answer is very few. He or she would be far happier if the value fell into the 90 or 95% or even higher range.

The probability of error at other positions on the curve can be determined from the following expression:

$$E_p = C_p \sigma$$

where E_p is the percentage error, C_p is a constant, and σ is the standard error. For instance, the 68.3% error occurs when $C_p = 1.00$ and the 95.4% error occurs when $C_p = 2.00$. It is impossible to establish an absolutely maximum error because this condition theoretically occurs at infinity, but many persons refer to the maximum error as being the 95.4% error (which occurs at 2.00σ), while others refer to the 99.9% error (which occurs at 3.29σ) as being the maximum. In the latter case, we see that 999 out of 1000 values would fall within this range. Several probabilities of error are summarized in Table 2-2 and represented graphically in Fig. 2-5. Sometimes the values 2σ and 3σ (which correspond to the 95.4% and 99.7% errors, respectively) are referred to as being two standard deviations or three standard deviations, respectively.

The general equation of the probability curve for an infinite number of ordinates may be expressed as follows:

$$y = ke^{-h^2x^2}$$

where y is the probability of occurrence of a random error of magnitude x, e is the base of the natural logarithms 2.718, and k and h are constants that determine the shape of the curve.

TABLE 2-2   PROBABILITIES FOR CERTAIN ERROR RANGE

Error (±)	Probability (%)	Probability that error will be larger
0.50σ	38.3	2 in 3
0.6745σ	50.0	1 in 2
1.00σ	68.3	1 in 3
1.6449σ	90.0	1 in 10
1.9599σ	95.0	1 in 20
2.00σ	95.4	1 in 23
3.00σ	99.7	1 in 333
3.29σ	99.9	1 in 1000

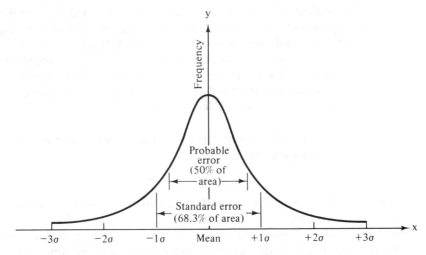

Figure 2-5   Probability curve.

2-12 PROPAGATION OF RANDOM ERRORS

This section is devoted to the presentation of a set of example problems that apply the preceding discussion of random errors to practical surveying problems. Included are calculations for single quantities as well as values for series of quantities. The accumulation of random errors for various calculations is referred to as the *propagation of random errors*. To be able to develop equations for random error propagation for any but the simplest cases requires a substantial background in calculus. It is assumed in this section that the readers of this book do not necessarily have such a background, so the simple algebraic equations that result from such derivations are given in the paragraphs to follow and applied to simple numerical cases.

Measurements of a Single Quantity

When a single quantity is measured several times, the probable error or 50% error of any one of the measurements can be determined from the following expression:

$$E_p = C_p \sqrt{\frac{\sum v^2}{n-1}}$$

where $C_p = C_{50} = 0.6745$. For other percent errors the values of C_p vary as shown in Table 2-2. Example 2-1 illustrates the application of the equation for different percent errors.

Example 2-1

Determine the 50%, 90%, and 95% errors for the list of distance measurements shown in the first column of the following table.

Solution

Measured value	Residual v	v^2
152.93	+0.02	0.0004
153.01	+0.10	0.0100
152.87	−0.04	0.0016
152.98	+0.07	0.0049
152.78	−0.13	0.0169
152.89	−0.02	0.0004
Avg = 152.91		$\sum v^2 = 0.0342$

$$E_{50} = \pm 0.6745 \sqrt{\frac{0.0342}{6-1}} = \pm 0.056 \text{ ft}$$

$$E_{90} = \pm 1.6449 \sqrt{\frac{0.0342}{6-1}} = \pm 0.136 \text{ ft}$$

$$E_{95} = \pm 1.9599 \sqrt{\frac{0.0342}{6-1}} = \pm 0.162 \text{ ft}$$

At this point we present a few remarks concerning measurements which seem to be completely out of line with the other values. When residuals are determined for a set of measurements they should be compared with the average value for those residuals. Should any of them be very large (say, four or five or more times the average value) they probably are mistakes and thus should be discarded and the calculations continued with the remaining ones. A value should not be eliminated just to make things neat or symmetrical. For a value to be deleted it should be set apart clearly from the other values so that a definite mistake is indicated.

A Series of Similar Measurements

Up to this point we have considered the measurement of single quantities only. Usually, however, the surveyor is involved with the measurement not just of one quantity but with a series of them. For instance, he or she usually measures at least several angles or distances or elevations while working on a particular project.

For this discussion it is assumed that a set of quantities are to be measured. It is desired that each measurement be made to the same probable error so that all of them will be equally reliable. Although random errors for these types of measurements tend to cancel each other to some extent, they will not do so completely for a practical number of measurements.

When a series of quantities are being measured, errors tend to accumulate in proportion to the square root of the number of measurements. This, the *law of compensation*, can be written as follows:

$$E_{series} = E \sqrt{n}$$

For instance, if a series of 12 angles is measured each with a 95% error of $\pm 20''$ of arc, the anticipated total 95% error for the 12 angles would be

$$E_{series} = \pm 20'' \sqrt{12} = \pm 69.3'' = \pm 1'09''$$

As another example, suppose that the 95% error in measuring a distance of one tape length (say 100 ft) is ± 0.01 ft. If a distance of 1600 ft is to be measured in 100-ft increments, the anticipated 95% error for the full distance will equal

$$E_{series} = \pm 0.01 \sqrt{16} = \pm 0.04 \text{ ft}$$

Notice that for this taping measurement the 50% probable error of measuring 100 ft can be determined by using the constants given in Table 2-2:

$$E_{50} = \frac{0.6745}{1.9599} (\pm 0.01) = \pm 0.00344 \text{ ft}$$

The 50% probable error for measuring the 1600-ft length will equal

$$E_{series} = \pm 0.00344 \sqrt{16} = \pm 0.0137 \text{ ft}$$

and the 95% error for measuring the 1600-ft length is

$$E_{series} = \pm 0.0137 \frac{1.9599}{0.6745} = \pm 0.04 \text{ ft}$$

As one moves further into his or her study of surveying, he or she may face the problem of needing to measure a distance, set of angles, and so on, with a total error not exceeding a certain limit. For instance, it might be necessary to measure a distance of 2000 ft with a total 99.7% error of no more than ± 0.10 ft. The question would then arise: How accurately should each 100 ft be measured so that the desired limit is not exceeded? The value can be determined as follows:

$$E_{series} = \pm E \sqrt{n}$$

$$\pm 0.10 = \pm E \sqrt{20}$$

$$E = \pm 0.022 \text{ ft} \quad \text{say,} \quad \pm 0.02 \text{ ft}$$

where E_{series} is ± 0.10 and $n = 20$.

A Set of Like Measurements

It has been shown that when a quantity is measured several times, the probable total error will equal the probable error in a single observation times the square root of the number of observations:

$$E_{\text{total}} = E_{\text{series}} \sqrt{n}$$

Since the mean equals the sum divided by the number of observations the standard error of the mean will equal

$$E_{\text{mean}} = \frac{E_{\text{total}}}{n}$$

If a distance is measured nine times with a probable accidental error of ± 0.10 ft in each measurement, the probable error in the distance will equal the total probable error in the nine observations, $\pm 0.10 \sqrt{9}$, divided by the number of observations:

$$\frac{\pm 0.10 \sqrt{9}}{9} = \frac{\pm 0.10}{\sqrt{9}} = \pm 0.03 \text{ ft}$$

In other words, the probable error in the mean of the nine readings equals the probable error of one observation divided by the square root of the number of observations. That is,

$$\pm \frac{0.10}{\sqrt{9}}$$

The preceding discussion clearly shows that the error of the mean varies inversely as the square root of the number of measurements. Thus, in order to double the precision of a particular measured quantity, four times as many measurements should be taken. To triple the precision, nine times as many measurements should be taken.

A Series of Unrepeated Measurements

When a series of independent measurements are made with probable errors of $E_1, E_2, E_3, \ldots$, respectively, the total probable error can be computed from the following expression:

$$E_{\text{sum}} = \sqrt{E_1^2 + E_2^2 + \cdots + E_n^2}$$

Example 2-2 illustrates the application of this expression.

Example 2-2

The four approximately equal sides of a tract of land were measured. These measurements included the following probable errors: ± 0.09 ft, ± 0.013 ft, ± 0.18 ft, and ± 0.40 ft, respectively. Determine the probable error for the total length or perimeter of the tract.

Solution

$$E_{\text{sum}} = \sqrt{(0.09)^2 + (0.013)^2 + (0.18)^2 + (0.40)^2}$$

$$= \pm 0.45 \text{ ft}$$

The reader should carefully note the results of the preceding calculations, where the uncertainty in the total distance measured (± 0.47 ft) is not very much different from the uncertainty given for the measurement of the fourth side alone (± 0.40 ft). *It should thus be obvious that there is little advantage in making very careful measurements for some of a group of quantities and not for the others.*

Probable Error for the Product of Two Quantities

Suppose that the sides of a rectangular piece of land have been measured each with certain estimated probable errors and it is desired to compute the area of the figure and the probable error in the resulting value. For such a situation the probable error in the area can be determined with the following expression:

$$E_{\text{product}} = \sqrt{A^2 E_b^2 + B^2 E_a^2}$$

where A and B are the measured lengths of the sides and E_a and E_b are the probable errors in those quantities.

Example 2-3 illustrates the calculation of the probable error for an area computation. This problem is one of the few cases involving random errors where a simple sketch clearly shows the theory involved in the solution. This is the idea of the Alternate Solution for this example.

Example 2-3

The measurements of the sides of a rectangle are, respectively, 232.60 ± 0.04 ft and 426.20 ± 0.06 ft (Fig. 2-6). What is the area of the figure and the probable error?

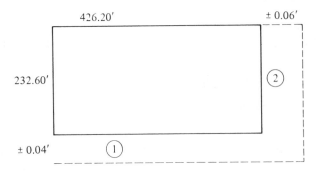

Figure 2-6

Solution

$$\text{Area} = (232.60)(426.20) = 99134.12 \text{ sq ft}$$

$$E_{\text{product}} = \pm \sqrt{(232.60)^2(0.06)^2 + (426.20)^2(0.04)^2}$$

$$= \pm 22.03 \text{ sq ft} \qquad \text{say,} \ \pm 22 \text{ sq ft}$$

Alternate Solution   From the sketch below the area errors numbered 1 and 2 can be calculated as follows:

$$E_1 = (\pm 0.04)(426.20) = \pm 17.048 \text{ sq ft}$$

$$E_2 = (\pm 0.06)(232.60) = \pm 13.956 \text{ sq ft}$$

Combining the area errors 1 and 2 by the law of compensation yields

$$E_{\text{series}} = \sqrt{(17.048)^2 + (13.956)^2} = \pm 22.03 \text{ sq ft} \qquad \text{say,} \ \pm 22 \text{ sq ft}$$

Weighted Observations

In the preceding examples the assumption has been made that each of the observations was equally reliable. A common problem which the surveyor faces is that of combining the results of different measurements that were made under different conditions and circumstances, and therefore have different degrees of reliability. For such a situation it is necessary to estimate the weight (or degree of reliability) for each of the measurements before they are combined.

Relative weights may be assigned to various observations based on (1) the judgment of the observer, (2) the number of like observations made, or (3) by assuming that the weights are inversely proportional to the square of the probable errors. If W_1 is the weight of quantity 1 and E_1 its probable error, the following expression may be written:

$$W_1 E_1^2 = W_2 E_2^2 = W_3 E_3^2 = \cdots = W_n E_n^2$$

A numerical example for this type of situation will be presented in Chapter 8, dealing with most probable elevation.

2-13 FIELD NOTES

Perhaps no other phase of surveying is as important as the proper recording of field data. No matter how much care is used in making field measurements, the effort is wasted unless a clear and legible record is kept of the work. For this reason we shall go to some length to emphasize this aspect of surveying.

The cost of keeping a surveying crew or party in the field is appreciable, perhaps a few hundred dollars per day. If the notes are wrong, incomplete, or confusing, much time and money have been wasted and the entire party may have to return to the job to repeat some or all of the work.

It is absolutely essential for good field notes to be prepared to document surveying work. The notes may be made manually and/or electronically. Both methods are discussed herein. *Notekeeping is such an important part of surveying that the most competent person in the party (i.e., the party chief) usually handles the task.*

Field books may be bound or loose-leaf, but bound books are usually used. Although it seems that the loose-leaf types have all the advantages (such as the capability of having their pages rearranged, filed, moved in with other sets of notes, shifted back and forth between the field and the office, etc.), the bound ones are more commonly used because of the possibility of losing some of the loose-leaf sheets. Long-life bound books capable of withstanding rough use and bad weather situations are the usual choice.

There are several kinds of field notebooks available, but the usual ones are $4\frac{5}{8}$ in. $\times$ $7\frac{1}{4}$ in., a size that can easily be carried in a pocket. This characteristic is quite important because the surveyor needs his hands for other work. Examples of field-book notes are shown throughout the text, the first one being Fig. 3-1. A general rule is that measured quantities are shown on the left-hand pages, and sketches and miscellaneous notes are shown on the right-hand pages.

In keeping notes the surveyor should bear in mind that on many occasions (particularly in large organizations), persons not familiar with the locality will make use of the notes. Some details may appear so obvious to the surveyor that he or she does not include them, but they may not be obvious at all to someone back in the office. Therefore, considerable effort should be made to record all the information necessary for others to understand the survey clearly. With practice the information needed in the notes will be learned.

An additional consideration is that surveying notes are on many occasions used for purposes other than the one for which they were originally developed, and therefore they should be carefully preserved. To maintain good field notes is not an easy task. Students are often embarrassed in their first attempts at making accurate and neat notes, but with practice the ability can be developed. The following items are absolutely necessary for the successful recording of surveying information:

1. The name, address, and phone number of the surveyor should be printed in india ink on the inside and outside of the field-book cover.
2. The title of the job, date, weather, and location should be recorded. When surveying notes are being used in the office, it may be helpful to know something of the weather conditions at the time the measurements were taken. This information will often be useful in judging the accuracy of a particular survey. Was it 110°F or $-10°F$? Was it raining? Were strong winds blowing? Was it foggy or dusty or snowing?

3. The names of the party members, together with their assignments as instrumentman ($\hbar$) rodman ($\emptyset$), and notekeeper (N), head tapeman or chainman (HT), rear tapeman or chainman (RT), should be recorded. Sometimes court cases require that these persons be interviewed many years after a survey is done.

4. Field notes should be organized in a form appropriate to the type of survey. Because other people may very well use these notes, generally standard forms are used for each of the different types of surveys. If each surveyor used his or her own individual forms for all surveys, there would be much confusion back in the office. Of course, there are situations where the notekeeper will have to improvise with some style of nonstandard notes.

5. Measurements must be recorded in the field when taken and not trusted to memory or written on scraps of paper to be recorded at a later date. Sometimes it is necessary to copy information from other field notes. In such cases the word COPY should be clearly marked on each page.

6. Frequent sketches are used where needed for clarity, preferably drawing lines with a straightedge. Because field books are relatively inexpensive compared to the other costs of surveying, crowding of sketches or other data does not really save money. (The sketches need not be drawn to scale, as distorted sketches may be better for clearing up questions.)

7. Field measurements must not be erased when incorrect entries are made. A line should be drawn through the incorrect number without destroying its legibility and the corrected value written above or below the old value. Erasures cause suspicion that there has been some dishonest alteration of values, but a crossed-out number is looked upon as an open admission of a blunder. (Imagine a property case coming up for court litigation and a surveying notebook containing frequent erasures being presented for evidence.) It is a good idea to use red ink for making additions to the notes back in the office to distinguish them clearly from values obtained in the field.

8. Notes are printed with a sharp medium-hard (3H or 4H) pencil so that the records will be relatively permanent and will not smear. Field books are generally used in damp and dirty situations and the use of hard pencils will preserve the notes. The lettering used on sketches is to be arranged so that it can be read from the bottom of the page or the right-hand side.

9. The instrument number should be recorded with each day's work. It may later be discovered that the measurements taken with that instrument contained errors that could not be accounted for in any other way. With the instrument identified the surveyor may be able to go back to the instrument and make satisfactory corrections.

10. A few other requirements include: numbering of pages, inclusion of a table of contents, drawing arrows on sketches indicating the general direction of north, and clear separation of each day's work by starting on a clean page

each day. Should a particular survey extend over several days, cross-references may be necessary between the various pages of that project. The numbering system generally used for surveying notes is to record the page number in the upper right corner of each right-hand page. A single page number is used for both the right- and left-hand sides.

Finally, it is essential that notes be checked before leaving the site of the survey to make sure that all required information has been obtained and recorded. Some surveyors keep checklists in their field books for different types of surveys. Before they leave a particular job, they refer to the appropriate list for a quick check. Imagine the expense in time and money of having a survey crew make an extra trip to the site of a job some distance away in order to obtain one or two minor bits of information that had been overlooked.

2-14 ELECTRONICALLY RECORDED NOTES

We discuss at various places in this book the availability and uses of modern surveying instruments such as electronic distance measuring devices, total station instruments, and similar equipment. Electronic data collectors are available which

Figure 2-7 Data recorder DR-2. (Courtesy of Nikon, Inc.)

when used with these other modern instruments will automatically display and record various distance and angle measurements at the touch of a button (Fig. 2-7). Furthermore, it is often possible to transfer the data stored to field or office computers. You can see that this capability enables us to reduce the blunders that might otherwise occur in recording information and in transferring it to the office.

It thus appears that data collectors make the notekeeper's work easier because he or she will be able to devote more time to preparing sketches and to writing descriptions and other nonnumerical information. There may not be as much extra time as you may think, however, because with modern equipment, measurements are made so much more rapidly.

There are some potential problems with automatically recorded notes in relation to legal situations. For instance, it is not yet clear whether the courts, which readily accept hand-prepared field books, will accept magnetic tapes. Furthermore, there are problems with the potentials of deliberate altering of the tapes and their possible inadvertent destruction by human mistakes or by other electronic equipment.

2-15 OFFICE WORK AND DIGITAL COMPUTERS

Field surveying measurements provide the basis for large amounts of office work. Some of this paperwork includes computations of precision, drawing of plats of land boundaries, calculations for and drawing of topographic maps, and computation of earthwork volumes. A large percentage of these items are commonly handled today with digital computers. Once a surveyor has used a successful computer program or a programmed desk or pocket calculator for some part of office work (such as the calculation of land areas), he or she will be reluctant ever again to make those calculations with an ordinary hand-held calculator.

It is doubtful if 5% of the adult population of the United States understands how cars actually operate, but just about 100% of those same people know what cars can do and how to make them do it. Similarly, it is doubtful if 5% of surveyors understand how electronic computers work, but all of them should know what computers can do and how they can be of service to the surveyor.

These devices can be used to perform the normal type of surveying calculations in a matter of minutes or even seconds while at the same time reducing the opportunities for mistakes to occur. The time required for looking up trigonometric functions, making interpolations, transferring numbers, and similar tasks is eliminated because this information is programmed into the machine.

Appropriate comments are made in various places throughout the text in an attempt to advise the student of areas in which computers or programmable calculators can be particularly helpful.

Enclosed with this book is a computer disk that may be used to handle several common types of surveying problems which involve some rather lengthy and tedious calculations. The use of this disk, which was prepared for IBM and IBM-compatible machines, is described in Chapters 13, 14, and 16.

PROBLEMS

2-1. A quantity was measured 10 times with the following results: 3.751, 3.781, 3.755, 3.749, 3.750, 3.747, 3.748, 3.754, 3.746, and 3.745 ft. Determine the following:
(a) Most probable value of the measured quantity.
(b) Probable error of a single measurement.
(c) 90% error.
(d) 95% error.
(Ans.: 3.753, ±0.007, ±0.017, and ±0.021)

2-2. The same questions apply as for Problem 2-1, but the following 12 quantities were measured: 157.20, 157.12, 157.16, 157.22, 157.28, 157.24, 157.22, 157.21, 157.15, 157.18, 157.31, and 157.25 ft.

2-3. The same questions apply as for Problem 2-1, but the following 11 readings were taken: 57.0, 57.5, 57.3, 57.1, 57.2, 56.8, 56.9, 57.0, 56.7, 57.4, and 57.5 minutes.
(Ans.: 57.1, ±0.2, ±0.5, and ±0.5)

2-4. A distance is measured to be 842.32 ft with an estimated standard deviation of ±0.16 ft. Determine the probable errors and the estimated probable precision at 2σ and at 3σ.

2-5. The following six independent length measurements were made (in feet) for a line: 642.349, 642.396, 642.381, 642.376, 642.368, and 642.343. Determine:
(a) The most probable value.
(b) The standard deviation of the measurements.
(c) The error at 3.29σ.
(Ans.: 642.369 ft, ±0.020 ft, ±0.066 ft)

In Problems 2-6 to 2-9, answer the same questions as for Problem 2-5.

Problem 2-6	Problem 2-7	Problem 2-8	Problem 2-9
842.872	196.349	202.62	754.36
842.903	196.506	202.28	754.35
842.901	196.339	202.78	754.39
842.888	196.362	202.61	754.37
842.876	196.347	202.63	754.40
842.867	196.341	202.72	754.29
842.858	196.352	202.56	754.33
842.863	196.501	202.65	754.27
842.869			754.32
842.870			754.42

(2.7 *Ans.*: 196.387, ±0.072, ±0.237)   (2.9 *Ans.*: 754.35, ±0.05, ±0.16)

2-10. If the estimated accidental error is ±0.009 ft for each of 31 separate measurements, what is the total probable error?

2-11. A random error of ±0.02 ft is estimated for each of 50 length measurements (which are to be added together to get the total length).
(a) What is the probable total error?
(b) What is the probable total error if the random error is ±0.05 ft per length?

(Ans.: ±0.14 ft and ±0.35 ft)

2-12. It is assumed that the probable error in measuring 100 ft with a steel tape is ±0.03 ft. The sides of a closed figure are measured with the following results: 262.5, 348.9, 102.8, and 136.6 ft. Determine the probable error in the perimeter of the figure. Assume that the probable error in each side is proportional to the square root of the number of 100-ft spans including fractional parts.

2-13. The same question as for Problem 2-12 applies, except that the probable error per 100 ft is ±0.02 ft and the measurements are 642.8, 549.6, 302.6, 849.8, and 503.7 ft.

(Ans.: ±0.11 ft)

2-14. Two sides of a rectangle were measured as being 162.32 ft ± 0.03 ft and 207.46 ft ± 0.04 ft. Determine the area of the figure and the probable error of the area.

2-15. A particular surveying party is capable of making 100-ft measurements with a standard deviation or standard error of ±0.01 ft.
(a) What total standard error is to be expected if a distance of 4000 ft is measured?
(b) What total 95% error is to be expected in the 4000 ft distance?

(Ans.: ±0.06 ft and ±0.13 ft)

2-16. The same questions apply as for Problem 2-15 except that the standard error is ±0.03 ft for a 100-ft measurement.

2-17. It is desired to tape a distance of 1100 ft with a total standard error of not more than ±0.10 ft.
(a) How accurately should each 100-ft distance be measured so that the permissible value is not exceeded?
(b) How accurately would each 100-ft distance have to be measured so that the 95% error would not exceed ±0.12 ft in a total distance of 2000 ft?

(Ans.: ±0.03 ft and ±0.01 ft)

2-18. For a nine-sided closed figure the sum of the interior angles is exactly 1620°. It is specified for a survey of this figure that if these angles are measured in the field, their sum should not miss 1620° by more than ±1'. How accurately should each angle be measured?

CHAPTER THREE

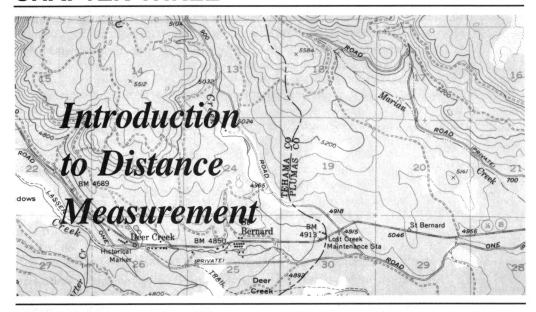

Introduction to Distance Measurement

3-1 INTRODUCTION

One of the most basic operations of surveying is the measurement of distance. In surveying, the distance between two points is understood to be the horizontal distance. The reason for this is that most of the surveyor's work is plotted on a drawing as some type of map. A map, of course, is plotted on a flat plane and the distances shown thereon are horizontal projections. Land areas are computed on the basis of the same horizontal measurements. This means that if a person wants to obtain the largest amount of actual land surface area for each acre of land he buys, he should buy it on the side of a very steep mountain.

Early measurements were made in terms of the dimensions of parts of the human body such as cubits, fathoms, and feet. The cubit (the unit Noah used in building his boat) was defined as the distance from the tip of a man's middle finger to the point of his elbow (about 18 in.); a fathom was the distance between the tips of a man's middle fingers when his arms were outstretched (approximately 6 ft). Other measurements were the foot (the distance from the tip of a man's big toe to the back of his heel) and the pole or rod or perch (the length of the pole used for driving oxen, later set at 16.5 ft). In England the "rood" (rod or perch) was once defined as being equal to the sum of the lengths of the left feet of 16 men, whether they were short or tall, as they came out of church one Sunday morning.[1] (Another historical distance unit of interest is the furlong. It is defined

---

[1] C. M. Brown, W. G. Robillard, and D. A. Wilson, *Evidence and Procedures for Boundary Location* 2nd ed (New York: John Wiley & Sons, Inc., 1981), p. 268.

as the length of the side of a square 10-acre field which is $\frac{1}{8}$ of a mile or 660 ft.) Today in the United States the foot is the basic unit of measurement, but most of the world uses the meter–decimal system and the American surveying profession is gradually becoming more familiar with metric units.

The meter is a unit of French origin and its application in the United States in the past has been limited almost entirely to geodetic surveys. In 1866 the U.S. Congress legalized the use of the metric system by which the meter was defined as being equal to 3.280833 ft (or 39.37 in.) and 1 inch was equal to 2.540005 centimeters. These values are based on the length at 0°C of an International Prototype Meter bar consisting of 90% platinum and 10% iridium which is kept in Sèvres, France, near Paris. In accordance with the treaty of May 20, 1875, the National Prototype Meter 27 identical with the International Prototype Meter was distributed to various countries. Two copies are kept by the U.S. Bureau of Standards at Gaithersburg, Maryland.

Reference to the prototype bar was ended in 1960 and the meter was redefined in terms of the wavelength of the orange-red light produced by burning the element krypton and set equal to $1,650,763.73\gamma_k$ or 3.280840 ft. In 1983 the meter definition was changed again to its present value: that is, the distance traveled by light in 1/229,792,458 sec. Supposedly, it is now possible to define the meter much more accurately. Furthermore, this enables us to use time (our most accurate basic measurement) to define length.[2]

The International Bureau of Weights and Measures has as its goal the establishment of a rational and coherent worldwide system of units. In 1960 this organization promulgated the "International System of Units," with the abbreviation SI in all languages. SI units are currently being adopted by quite a few countries, including most English-speaking nations (Britain, Australia, Canada, South Africa, and New Zealand). The purpose of the system is to standardize the units of measurements used throughout the world. Although the U.S. Congress has not officially adopted the system, it seems inevitable that they will eventually do so.

To the surveyor the most commonly used SI units are the meter (m) for linear measure, the square meter (m^2) for areas, the cubic meter (m^3) for volumes, and the radian (rad) for plane angles. In many countries the comma is used to indicate a decimal; thus to avoid confusion in the SI system, spaces rather than commas are used. For a number having four or more digits, the digits are separated into groups of threes, counting both right and left from the decimal. For instance, 4,642,261 is written as 4 642 261, and 2340.32165 is written as 2340.321 65.

To change surveying to SI units may seem at first glance to be quite simple. In a sense this is true in that if the surveyor can measure distance with a 100-ft tape, he or she can do just as well with a 30-m tape using the same procedures. Furthermore almost all electronic distance-measuring devices will provide dis-

[2] R. C. Brinker and R. Minnick, editors, *The Surveying Handbook* (New York: Van Nostrand Reinhold Company, Inc., 1987), p. 74.

tances in feet or meters as desired. However, we have several hundred years of land descriptions recorded in the English systems of units and stored in our various courthouses and other archives. Future generations of surveyors will therefore never get away completely from the English system of units. As an illustration, today's surveyor still frequently encounters old land descriptions made in terms of so many chains (where the 66-ft chain is referred to). In some parts of the United States a unit of Spanish origin called the *vara* is encountered. The vara is commonly used in Spanish-American countries. The Castilian vara is 32.8748 in., but in California it equals 32.372 in. and it is 33⅓ in. in Texas. Finally, it will take decades to replace the expensive surveying instruments graduated in English units that our surveyors presently own.

There are several methods used for actually measuring distance: pacing, odometer readings, stadia, taping, and the use of various electronic devices. These methods are discussed briefly in subsequent sections. The method to be used for a particular survey is dependent on the purpose of the work. An experienced surveyor will be able to select the best method to be used considering the objectives of the survey and the cost involved.

3-2 PACING

The ability to pace distances with reasonable precision is very useful to almost anyone. The surveyor in particular can use pacing to make approximate measurements quickly or to check measurements made by more precise means. By so doing he or she will often be able to detect large mistakes.

A person can determine the value of his or her average pace by counting the number of paces necessary to walk a distance that has previously been measured more accurately (e.g., with a steel tape). Here the author defines the pace as one step. Sometimes the term *stride* is used. A stride is considered to equal two steps or two paces. There is available on the market a device called the *passometer* which automatically counts the number of steps or paces a person takes. For most persons pacing is done most satisfactorily when taking natural steps. Others like to try to take paces of certain lengths (e.g., 3 ft), but this method is tiring for long distances and usually gives results of lower precision for short or long distances. As horizontal distances are needed, some adjustments should be made when pacing on sloping ground. Paces tend to be shorter on uphill slopes and longer on downhill ones. Thus the surveyor would do well to measure his or her pace on sloping ground as well as on level ground.

With a little practice a person can pace distances with a precision of roughly 1/50 to 1/200, depending on the ground conditions (slope, underbrush). For distances of more than a few hundred feet, a mechanical counter or *pedometer* can conveniently be used. Pedometers can be adjusted to the average pace of the user and automatically record the distance paced.

The notes shown in Fig. 3-1 are presented as a pacing example for students. A five-sided figure was laid out and each of the corners marked with a stake driven into the ground. The average pace of the surveyor was determined by pacing a known distance of 400 ft, as shown at the bottom of the left page of the field notes. Then the sides of the figure were paced and their lengths calculated. This same figure or traverse is used in later chapters as an example for measuring distances with a steel tape, measuring angles with a transit or theodolite, computing the precision of the latter measurements, and calculating the enclosed area of the figure.

A note should be added here regarding the placement of the stakes. It is good practice for the student to record in his or her field book information on the location of the stakes. The position of each stake should be determined in relation to at least two prominent objects, such as trees, walls, sidewalks, and so on, so that there will be no difficulty in relocating them a week or a month later. This information can be shown on the sketch as illustrated in Fig. 3-1.

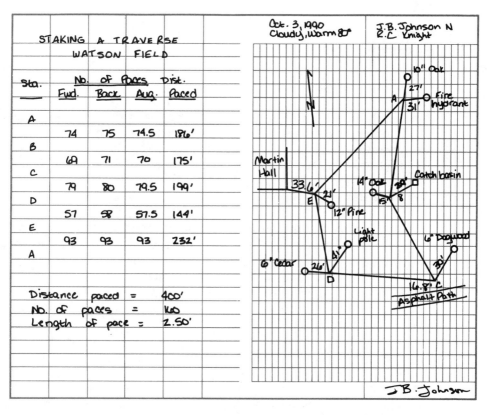

Figure 3-1

3-3 ODOMETERS AND MEASURING WHEELS

Distances can be roughly measured by rolling a wheel along the line in question and counting the number of revolutions. An *odometer* is a device attached to the wheel (similar to the distance recorder used in a car) which does the counting and from the circumference of the wheel converts the number of revolutions to a distance. Such a device provides a precision of approximately 1/200 when the ground is smooth, as along a highway, but results are much poorer when the surface is irregular.

The odometer may be useful for preliminary surveys, perhaps when pacing would take too long. It is occasionally used for initial route-location surveys and for quick checks on other measurements. A similar device is the *measuring wheel*, which is a wheel mounted on a rod. Its user can push the wheel along the line to be measured. It is frequently used for curved lines. Some odometers are available that can be attached to the rear end of a motor vehicle and used while the vehicle is moving at a speed of several miles per hour.

3-4 TACHYMETRY

The term *tachymetry* or *tacheometry*, which means "swift measurements," is derived from the Greek words *taklus*, meaning "swift," and *metron,* meaning "measurement." Actually, any measurement made swiftly could be said to be tacheometric, but the generally accepted practice is to list under this category only measurements made with subtense bars or by stadia. Thus the extraordinarily fast electronic distance-measuring devices are not listed in this section.

Subtense Bar

A tachymetric method that yields measurements of satisfactory precision for some rural property surveys makes use of the subtense bar (Fig. 3-2). This device has largely been replaced by electronic distance-measuring devices. In Europe, where the method is most commonly used, a horizontal bar with sighting marks on it, usually located 2 m apart, is mounted on a tripod. The tripod is centered over one end of the line to be measured and the bar is leveled and turned so that it is made roughly perpendicular to the line.

A theodolite (described in Section 10-12) is set up at the other end of the line and sighted on the subtense bar. It is used to align the bar precisely in a position perpendicular to the line of sight by observing the sighting mark on the subtense bar. This mark can be seen clearly only when the bar is perpendicular to the line of sight. The angle subtended between the marks on the bar is carefully measured (preferably with a theodolite measuring to the nearest second of arc) and the distance between the ends of the line is computed. The distance D from

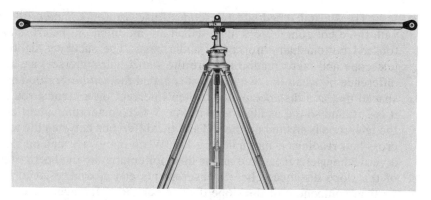

Figure 3-2   Subtense bar. (Courtesy of Kern Instruments, Inc.)

the theodolite to the subtense bar can be computed from the following expression:

$$D = \frac{1}{2} S \cot \frac{\alpha}{2}$$

in which S is the distance between the sighting marks on the bar and α is the subtended angle. These values are shown in Fig. 3-3.

For reasonably short distances, say less than 500 ft, errors are small and a precision of 1/1000 to 1/5000 is ordinarily obtained. The subtense bar is particularly useful for measuring distances across rivers, canyons, busy streets, and other difficult areas. It has an additional advantage in that the subtended angle is independent of the slope of the line of sight; thus the horizontal distance is obtained directly and no slope correction has to be made.

Stadia

Although the subtense bar has been used in the United States, the stadia method is far more common. Its development is generally credited to the Scotsman James Watt in 1771.[3] The word *stadia* is the plural of the Greek work *stadium*, which was the name given to a foot-race track approximately 600 ft in length.

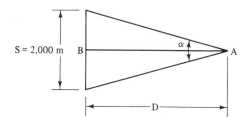

Figure 3-3

[3] A. R. Legault, H. M. McMaster, and R. R. Marlette, *Surveying* (Englewood Cliffs, N.J.: Prentice-Hall, Inc., 1956), p. 39.

Transit and theodolite telescopes (discussed in Chapter 10) are equipped with three horizontal cross hairs which are mounted on the cross-hair ring. The top and bottom hairs are called stadia hairs. The surveyor sights through the telescope and takes readings where the stadia hairs intersect a scaled rod. The difference between the two readings is called the *rod intercept*. The hairs are so spaced that at a distance of 100 ft their intercept on a vertical rod is 1 ft; at 200 it is 2 ft; and so on, as illustrated in Fig. 3-4. To determine a particular distance, the telescope is sighted on the rod and the difference between the top and bottom cross-hair readings is multiplied by 100. When one is working on sloping ground, a vertical angle is measured and is used for computing the horizontal component of the slope distance. These measurements can also be used to determine the vertical component of the slope distance or the difference in elevation between the two points.

The stadia method has its greatest value in locating details for maps, but it is also useful for making rough surveys and for checking more precise ones. Precisions of the order of approximately 1/250 to 1/1000 can be obtained with stadia. Such precision is not usually satisfactory for property surveys. The method is discussed in considerable detail in Chapter 14, where its usefulness in mapping is described.

The same principle can be used to estimate the heights of buildings, trees, or other objects, as illustrated in Fig. 3-5. A ruler is held upright at a given distance such as arm's length (about 2 ft for many people) in front of the observer's eye so that its top falls in line with the top of the tree. Then the thumb is moved down so that it coincides with the point where the line of sight strikes the ruler when the observer looks at the base of the tree. Finally, the distance to the base of the tree is measured by pacing or taping. By assuming the dimensions shown in Fig. 3-5, the height of the tree shown can be estimated as follows:

$$\text{height of tree} = \left(\frac{100}{2}\right)(6 \text{ in.}) = 300 \text{ in.} = 25 \text{ ft}$$

Artillery students were formerly trained in a similar fashion to estimate distances by estimating the size of objects when they were viewed between the knuckles of their hands held at arm's length.

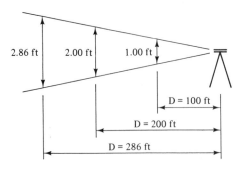

Figure 3-4 Stadia readings.

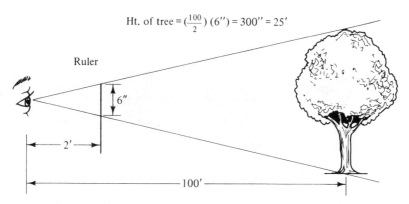

Figure 3-5 Estimating the height of a tree.

3-5 TAPING OR CHAINING

For many centuries, surveyors measured distances with ropes, lines, or cords that were treated with wax and calibrated in cubits or other ancient units. These devices are obsolete today, although precisely calibrated wires are sometimes used. For the first two-thirds of this century the 100-ft steel ribbon tape was the common device used for measuring distances. Such measuring is often called *chaining*, a carryover name from the time when Gunter's chain was introduced. The English mathematician Edmund Gunter (1581–1626) invented the surveyor's chain (Fig. 3-6) early in the seventeenth century. His chain, which was a great improvement over the ropes and rods used up until that time, was available in several lengths, including 33 ft, 66 ft, and 100 ft. The 66-ft length was the most common. (Gunter is also credited with the introduction of the words *cosine* and *cotangent* to trigonometry, the discovery of magnetic variations (discussed in Chapter 9), and other outstanding scientific accomplishments.[4])

The 66-ft chain, sometimes called the four-pole chain, consisted of 100 heavy wire links each 7.92 in. in length. In studying old deeds and plats the surveyor will often find distances measured with the 66-ft chain. He or she might very well see a distance given as 11 ch. 20.2 lks. or 11.202 ch. The usual area measurement in the United States is the acre, which equals 10 square chains. This is equivalent to 66 ft by 660 ft, or $\frac{1}{80}$ mi by $\frac{1}{8}$ mi, or 43,560 sq ft.

Many of the early two-lane roads in the United States and Canada were laid out to a width of one chain, resulting in a 66-ft right-of-way.

Steel tapes came into general use around the beginning of the twentieth century. They are available in lengths from a few feet to 1000 ft. For ordinary conditions, precisions of from 1/1000 to 1/5000 can be obtained, although much better work can be done by using procedures to be described later.

---

[4] C. M. Brown, W. G. Robillard, and D. A. Wilson, *Evidence and Procedures for Boundary Location* 2nd ed (New York: John Wiley & Sons, Inc., 1981), p. 151.

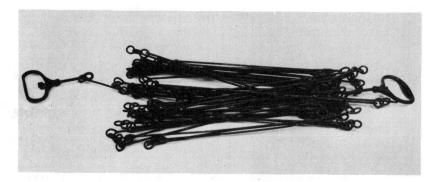

Figure 3-6   Surveyor's chain.

The general public, riding in their cars and seeing surveyors measuring distances with a steel tape, might think: "Anybody could do that. What could be simpler?" The truth is, however, that the measurement of distance with a steel tape, though simple in theory, is probably the most difficult part of good surveying. The efficient use of today's superbly manufactured surveying equipment for other surveying functions such as the measurement of angles is quickly learned. But correspondingly precise distance measurement with a steel tape requires thought, care, and experience. In theory, it is simple, but in practice it is not so easy. Sections 3-8 to 3-10 are devoted to a detailed description of taping equipment and field measurements made with steel tapes.

3-6 ELECTRONIC DISTANCE MEASUREMENTS

Sound waves have long been used for estimating distances. Nearly all of us have counted the number of seconds elapsing between the flash of a bolt of lightning and the arrival of the sound of the thunder and then multiplied the number of seconds by the speed of sound (about one-fifth of a mile per second). The speed of sound is 1129 ft/sec at 70°F and increases by a little more than 1 ft/sec for each degree Fahrenheit increase in temperature. For this reason alone, sound waves do not serve as a practical means of precise distance measurement because temperatures along a line being measured would have to be known to almost the nearest 0.01°F.

In the same way some distances for hydrographic surveying were formerly estimated by firing a gun and then measuring the time required for the sound to travel to another ship and be echoed back to the point of firing. Ocean depths are determined with depth finders which use echoes of sound from the ocean bottom. Sonar equipment makes use of supersonic signals echoed off the hulls of submarines to determine underwater distances.

It has been discovered during the past few decades that the use of either light waves, electromagnetic waves, infrared, or even lasers offer much more precise methods of measuring distance. Although it is true that some of these waves are affected by changes in temperature, pressure, and humidity, the effects are small and can be accurately corrected. Under normal conditions the corrections amount to no more than a few centimeters in several miles. Numerous portable electronic devices making use of these wave phenomena have been developed that permit the measurement of distance with tremendous precision.

These devices have not completely replaced chaining or taping, but they are commonly used by almost all surveyors. Their costs are at a level where the average surveyor must use them to remain economically competitive. For those who prefer to lease, arrangements may be made with various companies.

Electronic distance-measuring instruments have several important advantages over other methods of measurement (Fig. 3-7). They are very useful in measuring distances that are difficult of access, for example, across lakes and rivers, busy highways, standing farm crops, canyons, and so on. For long distances (say, several thousand feet or more), the time required is in minutes, not hours as would be required for a typical taping party. Two people, easily trained, can do the work better and faster than the conventional four-person taping crew.

Electronic distance-measuring devices will probably not completely replace the steel tape, at least in the foreseeable future. The steel tape will likely be used for short distances for a long time to come because of its convenience and economy. For this reason the surveyor of today and tomorrow should become pro-

Figure 3-7 Model ME5000 distance meter. (Courtesy of Kern Instruments, Inc.)

ficient with the steel tape, even though it is highly probable that he or she will eventually use the tape only a small percentage of the time. Otherwise, the chances of making blunders and large errors will be magnified when the tape is used.

With EDMIs the time required for a light wave to be sent along the required path (or a microwave to be sent along the path and reflected back to the starting position) is measured. From this information the distance may be determined as described in Chapter 5, which is devoted entirely to EDMI measurements.

3-7 SUMMARY OF MEASUREMENT METHODS

Table 3-1 presents a brief summary of the various methods for measuring distances. There is a great variation in the precision obtainable with these different

TABLE 3-1

Method	Precision	Uses
Pacing	1/50 to 1/200	Reconnaissance and rough planning
Odometer	1/200	Reconnaissance and rough planning
Subtense bar	1/1000 to 1/5000	Seldom used in United States and then only when taping is not feasible because of terrain and when electronic distance-measuring devices are not available
Stadia	1/250 to 1/1000	Very commonly used for mapping, rough surveys, and for checking more precise work
Ordinary taping	1/1000 to 1/5000	Ordinary land surveys and building construction
Precision taping	1/10,000 to 1/30,000	Excellent land surveys, precise construction work, and city surveys
Base-line taping	1/100,000 to 1/1,000,000	Formerly used for precise geodetic work performed by the National Geodetic Survey
Electronic distance measurement	$\pm 0.04' \pm 1/300,000$	In the past use has been primarily for precise government geodetic work but is today commonly used for land development, land surveys, and precise construction work

methods and the surveyor will select one that is appropriate for the purposes of the particular survey.

3-8 EQUIPMENT REQUIRED FOR TAPING

A brief discussion of the various types of equipment normally used for taping is presented in this section, and the next two sections are devoted to the actual use of the tapes. A taping party should have at least one 100-ft steel tape, two range poles, a set of 11 chaining pins, a 50-ft woven tape, two plumb bobs, and a hand level. These items are discussed briefly in the following paragraphs.

Steel Tapes

Steel tapes are most commonly 100 ft long, approximately $\frac{5}{16}$ in. wide, 0.025 in. thick, and they weigh 2 or 3 lb. They are either carried on a reel or done up in 5-ft loops to form a figure 8, from which they are thrown into a convenient circle approximately 8 in. in diameter. These tapes are quite strong as long as they are kept straight, but if they are tightened when they have loops or kinks in them, they will break very easily. If a tape gets wet, it should be wiped with a dry cloth and then again with an oily cloth.

The ends of steel tapes are made with heavy brass loops that provide a place to attach leather thongs or tension handles, enabling the user to tighten or tension them firmly. (If the leather thong is missing or breaks, and if a tension handle is not available, one of the chaining pins can be inserted through the loop and used as a handle for tightening the tape.)

Surveyor's tapes are marked at the 1-ft points from 0 to 100 ft. Older tapes have the last foot at each end divided into tenths of a foot, but the newer tapes have an extra foot beyond 0 which is subdivided. Tapes with the extra foot are called *add* tapes; those without the extra foot are called *cut* tapes (see Fig. 3-10). Several variations are available: for example, tapes divided into feet, tenths, and hundredths for their entire lengths, or tapes with the 0- and 100-ft points about $\frac{1}{2}$ ft from the ends, and so on. Needless to say, the surveyor must be completely familiar with the divisions of the tape and its 0- and 100-ft marks before he or she does any measuring. Cut tapes have been used for many decades and they are thus considered to be satisfactory by many surveyors. However, their use seems occasionally to lead to math mistakes, where the end reading is added instead of being subtracted. Mistakes seem to be less frequent when add tapes are used, and as a consequence, add tapes have become the common type manufactured.

Tapes can be obtained in various lengths other than 100 ft. The 300- and 500-ft lengths are probably the most popular. The longer tapes, which usually consist of $\frac{1}{8}$-in.-wide wire bands, are divided only at the 5-ft points, to reduce costs. They are quite useful for rapid, precise measurements of long distances on level ground. The use of long tapes permits a considerable reduction in the time required for marking at tape ends, and it also virtually eliminates the accidental errors that occur while marking.

Although the 100-ft tapes have been used by surveyors for quite a few decades, it seems likely that they will, in the next few years, begin to be replaced by metric steel tapes. Probably these tapes will be available in lengths of 20, 30, and 50 m, and others. It is highly possible that the 30m (98.4-ft) length will become the common metric tape because its length is so close to that of the present 100-ft tape. Perhaps the metric tapes will be divided into decimeters throughout their lengths, with the end decimeter divided into millimeters.

For very precise taping, the *Invar* tape, which is manufactured from a nickel–steel alloy, can be used. It has the considerable advantage that its variations in length caused by temperature changes are $\frac{1}{30}$ or even less of those of the usual steel tape. Very precise taping can be accomplished with Invar tapes even on hot sunny days, but these tapes are perhaps 10 times as expensive as standard tapes and are easily bent and damaged. To avoid kinking, they should be kept on reels with large diameters. They are used only for the most precise operations, such as geodetic work and for checking the length of regular steel tapes. A newer tape called the *Lovar* tape has properties and costs somewhere in between those of regular steel tapes and Invar tapes. Some tapes are made with a graduated thermometer scale which corresponds to temperature expansions and contractions. With different temperatures the surveyor will use a different terminal mark on the tape, thus automatically correcting for the temperature change.

Fiberglass Tapes

In recent years fiberglass tapes made of thousands of glass fibers coated with polyvinyl chloride have been introduced on the market. These less expensive and more durable tapes are available in 50-ft, 100-ft, and other lengths. They are strong and flexible and will not change lengths appreciably with changes in temperature and moisture. They may also be used with little hazard in the vicinity of electrical equipment. When tension forces of 5 lb or less are applied, tension corrections are probably not necessary, but for values greater than 5 lb, length corrections are necessary.

Range Poles

Range poles are used for sighting points, for marking ground points, and for lining up tapemen in order to keep them going in the right direction. They are usually from 6 to 10 ft in length and are painted with alternate bands of red and white to make them more easily seen. The bands are each 1 ft in length and the rods can therefore be used for rough measurements. They are manufactured from wood, fiberglass, or metal. The sectional steel tubing type is perhaps the most convenient one because it can easily be transported from one job to another.

Taping Pins

Taping pins are used for marking the ends of tapes or intermediate points while taping. They are easy to lose and are generally painted with alternating red and white bands. If the paint wears off, they can be repainted any convenient bright color or they can have strips of cloth tied to them which can readily be seen. The pins are carried on a wire loop which can conveniently be carried by a tapeman, perhaps by placing the loop around his or her belt.

Plumb Bobs

Plumb bobs for surveying are usually made of brass (to limit possible interference with compass readings) and weigh from 6 to 18 oz. They have sharp replaceable points and a device at the top to which plumb-bob strings may be tied. Very commonly, plumb bobs are fastened to a *gammon reel*. This is a device that provides easy up-and-down adjustment of the plumb bob, instant rewinding of the plumb-bob string, and a sighting target.

Woven Tapes

Woven tapes (Fig. 3-8) are most commonly 50 ft in length with graduation marks at 0.25-in. intervals. They can be either nonmetallic or metallic. Nonmetallic tapes are woven with very strong synthetic yarns and are covered with a special plastic coating that is not affected by water. Metallic tapes are made with a water-repellant fabric into which fine brass, bronze, or copper wires are placed in the lengthwise direction. These wires strengthen the tapes and provide considerable resistance to stretching. (Because of the metallic wires, they should not be used near electrical units. For such situations nonmetallic woven or fiberglass tapes

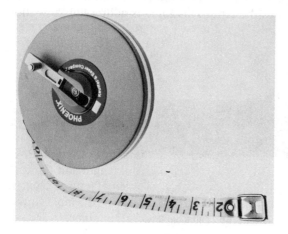

Figure 3-8   Woven tape. (Courtesy of Keuffel & Esser—A Kratos Company.)

should be used.) Nevertheless, since all woven tapes are subject to some stretching and shrinkage, they are not suitable for precise measurement. Despite this disadvantage, woven tapes are often useful and should be a part of a surveying party's standard equipment. They are commonly used for finding existing points, locating details for maps, and measuring in situations where steel tapes might easily be broken (as along highways) or when small errors in distance are not too important. Their lengths should be checked periodically or standardized with steel tapes.

Hand Levels

The hand level is widely used for taping and for the rough determination of elevations. It consists of a metal sighting tube on which is mounted a bubble tube (Fig. 3-9). If the bubble is centered while sighting through the tube, the line of sight is horizontal. Actually, the bubble tube is located on top of the instrument and its image is reflected by means of a 45° mirror or prism inside the tube so that its user can see the bubble at the same time as the terrain. The hand level is very useful to the surveyor for holding the steel tape horizontal, as described in Section 3-10.

Spring Balances

When a steel tape is tightened, it will stretch. The resulting increase in length may be determined with the formula presented in Section 4-7. For average taping the tension applied can be estimated sufficiently to obtain desired precisions, but for very precise taping a *spring balance* or *tension handle* is necessary. The usual spring balance can be read up to 30 lb in $\frac{1}{2}$-lb increments or up to 15 kg in $\frac{1}{4}$-kg increments. If a tape is held above the ground, it will sag, with a resulting shortening of the distance between its ends. This effect may be counteracted by increasing the tension applied to the tape. This topic is addressed in Section 4-7.

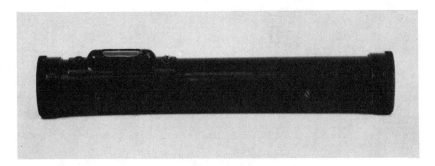

Figure 3-9   Old hand level.

Clamping Handles

Leather thongs are usually placed through the loops provided at the end of the tapes. With these thongs or with spring balances attached to the same loops, the tapes may be tensioned to desired values. When only partial lengths of tapes are used, it is somewhat difficult to pull the tape tightly. For such cases *clamping handles* are available. These have a scissors-type grip which enables one to hold the tape tightly without damaging it.

3-9 TAPING OVER LEVEL GROUND

If taping is done on fairly smooth and level ground where there is little underbrush, the tape can rest on the ground. The taping party consists of the head tapeman and the rear tapeman. The head tapeman leaves one taping pin with the rear tapeman for counting purposes and perhaps to mark the starting point. The head tapeman takes the zero end of the tape and walks down the line toward the other end.

When the 100-ft end of the tape reaches the rear tapeman, the rear tapeman calls "tape" or "chain" to stop the head tapeman. The rear tapeman holds the 100-ft mark at the starting point and aligns the head tapeman (using hand and perhaps voice signals) on the range pole which has been set behind the ending point. Ordinarily, this "eyeball" alignment of the tape is satisfactory, but use of the transit is safer and will result in better precision. Sometimes there are places along a line where the tapemen cannot see the end point and there may be positions where they cannot see the signals of the instrumentman. For such cases it is necessary to set intermediate line points before the taping can be started.

It is necessary to pull the tape firmly (see Sections 4-6 and 4-7). This can be done by wrapping the leather thong at the end of the tape around the hand, or by holding a taping pin that has been slipped through the eye at the end of the tape, or by using a clamp. When the rear tapeman has the 100-ft mark at the starting point and has satisfactorily aligned the head tapeman, he or she calls "all right" or some other such signal. The head tapeman pulls the tape tightly and sticks a taping pin in the ground at right angles to the tape and sloping at 20 to 30° from the vertical. If the measurement is done on pavement, a scratch can be made at the proper point or a taping pin can be taped down to the pavement or the point may be marked with a colored lumber crayon, called *keel*.

The rear tapeman picks up a taping pin and the head tapeman pulls the tape down the line, and the process is repeated for the next 100 ft. It will be noticed that the number of hundreds of feet which have been measured at any time equals the number of taping pins that the rear tapeman has in his or her possession. After 1000 ft has been measured, the head tapeman will have used his eleventh pin and he calls "tally" or some equivalent word so that the rear tapeman will return the taping pins and they can start on the next 1000 ft.

When the end of the line is reached, the distance from the last taping pin to the end point will normally be a fractional part of the tape. For older tapes, the first foot of the tape (from 0 to 1 ft) is usually divided into tenths, as shown in Fig. 3-10. The head tapeman holds this part of the tape over the end point while the rear tapeman moves the tape backward or forward until he has a full foot mark at the taping pin.

The rear tapeman reads and calls out the foot mark, say 72 ft, and the head tapeman reads from the tape end the number of tenths and perhaps estimates to the nearest hundredth, say 0.46, and calls this out. This value is subtracted from 72 ft to give 71.54 ft and the number of hundreds of feet measured before is added. These numbers and the subtraction should be called out so that the math can be checked by each partner.

For the newer steel tapes with the extra divided foot, the procedure is almost identical except that the rear tapeman would, for the example just described, hold the 71-ft mark at the taping pin in the ground. He or she would call out 71 and the head tapeman would read and call out plus 54 hundredths, giving the same total of 71.54 ft.

A comment seems warranted here about practical significant figures as they apply to taping. If ordinary taping is being done and the total distance obtained for this line is 2771.34 ft, the 4 at the end is ridiculous and the distance should

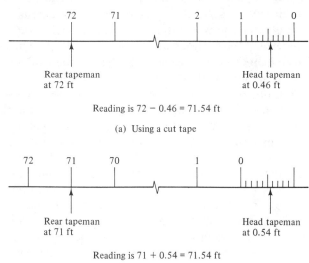

Reading is 72 − 0.46 = 71.54 ft

(a) Using a cut tape

Reading is 71 + 0.54 = 71.54 ft

(b) Using an add tape

Figure 3-10

be recorded as 2771.3 or even 2771 ft because the work is just not done that precisely.

3-10 TAPING ALONG SLOPING GROUND OR OVER UNDERBRUSH

When sloping distances are to be measured, there are three taping methods that can be used. The tape (1) may be held horizontally; or (2) it may be held along the slope, the slope determined, and a correction made to obtain the horizontal distance; or (3) the sloping distance may be taped, a vertical angle measured for each slope, and the horizontal distance later computed. The latter method is sometimes referred to as *dynamic taping*. Descriptions of each of these methods follow.

Holding the Tape Horizontally

Ideally, the tape should be supported for its full length on level ground or pavement. Unfortunately, such convenient conditions are often not available because the terrain being measured may be rough and covered with underbrush. For sloping, uneven ground or areas with much underbrush, taping is handled in a similar manner to taping over level ground. The tape is held horizontally, but one or both tapemen must use a plumb bob, as shown in Fig. 3-11.

If taping is being done uphill, the rear tapeman will have to hold his or her plumb bob over the last taping pin, while the head tapeman may be able to hold his or her end on the ground [Fig. 3-11(a)]. If they are moving downhill, the rear tapeman may be able to hold his or her end on the ground while the head tapeman has to use a plumb bob. As a result, taping downhill is preferable to taping uphill. If the measurement is over uneven ground or ground where there is considerable underbrush, both tapemen may very well have to use plumb bobs as they hold their respective ends of the tape above the ground [Fig. 3-11(b)].

Considerable practice is required for a person to be able to do precise taping in rolling or hilly country. Although for many surveys the tapemen may estimate

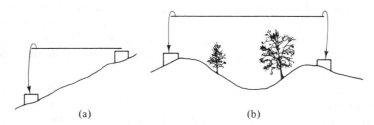

(a)   (b)

Figure 3-11   Holding the tape horizontally.

the horizontal by eye, it pays to use a hand level for this purpose. Where there are steep slopes, it is difficult to estimate by eye when the tape is horizontal because the common tendency is for the downhill person to hold his or her end much too low, causing significant error. If a precision of better than approximately 1/2500 or 1/3000 is desired in rolling country, holding the tape horizontally by estimation will not be sufficient.

Another problem in holding the tape above the ground is the error caused by sagging of the tape (see Section 4-6). Note that both of these errors (tape not horizontal and sag) will cause the surveyor to get too much distance. In other words, either it takes more tape lengths to cover a certain distance or the surveyor does not move forward a full 100 ft each time he or she uses the tape.

If the slope is less than approximately 5 ft per 100 ft (a height above the ground at which the average tapeman can comfortably hold the tape), the tapemen can measure a full 100-ft tape length at a time. If they are taping downhill, the head tapeman holds the plumb-bob string at the 0-end of the tape with the plumb bob a few inches above the ground. When the rear tapeman is ready at his or her end, the head tapeman is lined up on the distant point, and when the tape is horizontal and pulled to the desired tension, the head tapeman lets the plumb bob fall to the ground and sets a taping pin at that point.

For slopes greater than approximately 5 ft per 100 ft, the tapeman will be able to hold horizontally only parts of the tape at a time. Holding the tape more than 5 ft above the ground is difficult, and wind can make it more so. If the tape is held at heights of 5 ft or less above the ground both forearms can be braced against the body and the tape can easily be pulled firmly without swaying and jerking.

Assuming that they are proceeding downhill, the head tapeman pulls the tape along the line for its full length and then, leaving the tape on the ground, returns as far along the tape as necessary for them to hold horizontally the part of the tape from his or her point to the rear tapeman.

The head tapeman holds the plumb bob string over a whole foot mark, and when the tape is stretched, lined, and horizontal, lets the plumb bob fall and sets a taping pin. He or she holds the intermediate foot mark on the tape until the rear tapeman arrives, at which time he or she hands the tape to the rear tapeman with the foot mark that he or she has been holding. This careful procedure is followed because it is so easy for the head tapeman to forget which foot he or she was holding if he or she drops it and walks ahead. The tapemen repeat this process for as much more of the tape as they can hold horizontally until they reach the 0-end of the tape.

This process of measuring with sections of the tape is referred to as *breaking tape* or *breaking chain*. If the head tapeman follows the customary procedure of leaving a taping pin at each of the positions that he or she occupies when breaking tape, counting the number of hundreds of feet taped (as represented by the number of pins in the possession of the rear tapeman) would be confusing. Therefore, at each of the intermediate points the head tapeman sets a pin in the ground and then takes one pin from the rear tapeman. Instead of breaking tape, some surveyors find it convenient to record the partial tape lengths in their notes.

It is probably wise for a beginning surveyor to measure a few distances on slopes of different percentages holding the tape horizontal and then again with the tape along the slopes with no corrections made. These measurements should give him or her a feeling for the magnitude of slope errors.

Taping on Slopes and Making Slope Corrections

Occasionally, it may be more convenient or more efficient to tape along sloping ground with the tape held inclined along the slope. This procedure has long been common for underground mine surveys but to a much lesser extent for surface surveys. Slope taping is quicker than horizontal taping and is considerably more precise because it eliminates plumbing with its consequent accidental errors. Taping along slopes is sometimes useful when the surveyor is working along fairly smooth slopes or when he wants to improve precision. Nevertheless, the method is generally not used because of the problem of correcting slope distances to horizontal values. This is particularly true in rough terrain where slopes are constantly varying and the problem of determining the magnitude of the slopes is difficult.

In some cases it may be impossible to hold the entire tape (or even a small part of it) horizontally. This may occur when taping is being done across a ravine (see Fig. 3-12) or some other obstacle where one tapeman is much lower than the other one and where it is not feasible to "break tape." Here it may be practical to hold both ends of the tape on the ground.

Once the slope distances are determined, they are corrected by means of the formula developed in Section 4-5. To be able to use this formula it is necessary to obtain the values of the slope of the tape for each tape measurement. It is usually convenient to do this by determining elevation difference at the ends of the tape.

The Abney hand level or clinometer (Fig. 3-13) has an arc with which vertical angles can be read from 0 to 90°. With a vernier (see Chapter 7) angle readings can be taken to approximately the nearest 10′. In addition slope values (ratio of vertical distances to horizontal distances) from 1:1 to 1:10 can be read. These devices are just about obsolete.

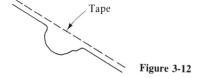

Figure 3-12

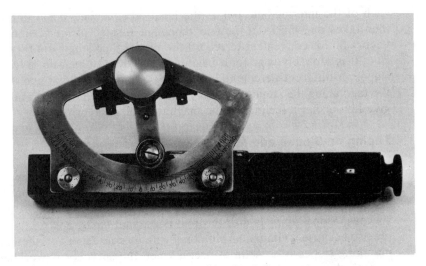

Figure 3-13   Hand level and clinometer.

Dynamic Taping

With this method, which is very similar to the slope taping method, slope distances are measured. Then a transit or theodolite is set up at each taping point and the vertical angle is measured to the next taping point and the horizontal distance computed.

CHAPTER 4

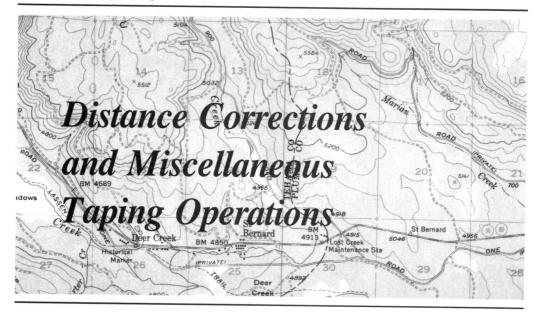

Distance Corrections and Miscellaneous Taping Operations

4-1 INTRODUCTION

To the reader it may seem that the author has too much to say about taping when almost all distance measurements are today made with electronic distance-measuring instruments. This feeling may be even stronger after you study the following discussions pertaining to the corrections of taping measurements for temperature or sag or other items and after you struggle through the detailed information presented concerning errors and mistakes. *The author, however, feels very strongly that if you learn this information concerning tapes, you will understand so much better the entire measuring process, regardless of the surveying operation involved or the equipment used.*

4-2 TYPES OF CORRECTIONS

The five major areas in which the surveyor may need to apply corrections either in measuring or in laying out lines with a tape are as follows:

1. Incorrect tape length or standardization error
2. Temperature variations
3. Slope
4. Sag
5. Incorrect tension

The next few sections of this chapter are devoted to a discussion of these corrections.

4-3 INCORRECT TAPE LENGTH OR STANDARDIZATION ERROR

A very important topic in surveying is the standardization of equipment or, that is, the comparison of the equipment (whether it is a tape, an electronic distance-measuring instrument, a theodolite, or whatever) against a standard. In other words, has the equipment been damaged or shaken out of adjustment, have repairs or weather changes affected it, and so on? If so, we will need to adjust the equipment or make mathematical corrections to compensate for the resulting errors.

It is said that in ancient Egypt the workers on the pyramids were required to compare their cubit sticks against the standard or royal cubic stick each full moon. Those failing to do so were subject to death. Such a practice undoubtedly brought forth the best efforts from their personnel in the area of standardization.

Although steel tapes are manufactured to very precise lengths, with use they become kinked, worn, and imperfectly repaired after breaks. The net result is that tapes may vary by quite a few hundredths of a foot from their desired lengths. Therefore, it is wise to check periodically against a standard. There are several ways in which this might be done. For instance, some surveying offices keep one standardized tape (perhaps an Invar type) that is used only for checking the lengths of their other tapes. Some companies take a tape that has been standardized at 100.00 ft and use it to place marks 100.00 ft apart on a concrete curb, sidewalk, or pavement. The marks are frequently used for "standardizing" their tapes. They feel that the length between the marked points will not change appreciably as temperatures vary because of the mass of the concrete and the friction of the earth.

These practices are advisable for surveyors who have extensive practices and they yield very satisfactory results for surveys where ordinary precision is desired, but they are probably not sufficient for extremely precise work. For such work, tapes can be mailed to the National Bureau of Standards in Gaithersburg, Maryland. For a rather large fee, they will determine the length of a tape for specific tension and support conditions. They issue a certificate for each tape, giving its length to the nearest 0.001 ft at 68°F (20°C) for two conditions: one with the tape supported for its full length and subjected to a 12-lb pull, and the other with the tape supported at its ends only and pulled with 20 lb.

Several municipal governments around the United States, various state agencies, and a good many universities will standardize tapes, occasionally free as a service to the public. In addition, the NGS has established approximately 200 base lines at various locations across the country where tapes and EDM equipment can be calibrated by the surveyor. A detailed description of one of these base lines is presented in Chapter 5. The distances between the monuments for the NGS base lines were measured with Invar tapes and/or electronic distance-measuring equipment with accuracies approaching 1/1,000,000. The locations of these base lines can be obtained by writing to the National Geodetic Survey, National Ocean Survey, Rockville, MD 20852. This information is also available for each state from its surveying society.

If a tape proves to be in appreciable error from the standard, the surveyor must correct the measurements by the required amounts. He or she will note carefully whether such a correction is positive or negative, as explained in the following paragraphs.

The important point to grasp in making corrections is that the tape "says zero ft at one end and 100 ft at the other end" even though its correct length is 99.94 ft, 100.10 ft, or some other value. If the surveyor (ignorant of the tape's true length) uses this tape 10 times, he believes that he has measured a distance of 1000 ft, but has really measured 10 times the actual tape length.

In measuring a given distance with a tape that is too long, the surveyor will not obtain a large enough value for the measurement and will have to make a positive correction. In other words, if the tape is too long, it will take fewer tape lengths to measure a distance than would be required for a shorter, correct-length tape. For a tape that is too short, the reverse is true and a negative correction is required. It should be simple enough to remember this rule: *Tape too long, add; tape too short, subtract.*

Examples 4-1, 4-2, and 4-3 illustrate the correction of taped distances caused by incorrect tape lengths. The problem of Example 4-3 is stated backward from the ones of Examples 4-1 and 4-2, and the sign of the correction is therefore reversed.

Example 4-1

A distance is measured with a 100-ft steel tape and is found to be 896.24 ft. Later the tape is standardized and is found to have an actual length of 100.04 ft. What is the correct distance measured?

Solution   The tape is too long and a + correction of 0.04 ft must be made for each tape length as follows:

$$\text{Measured value} \qquad\qquad = 896.24 \text{ ft}$$

$$\text{Total correction} = +(0.04)(8.9624) = \underline{+0.36 \text{ ft}}$$

$$\text{Corrected distance} \qquad\qquad = \mathbf{896.60 \text{ ft}}$$

Alternate Solution   Obviously, the distance measured equals the number of tape lengths times the actual length of the tape. In this case, it took 8.9624 tape lengths to cover the distance and each tape length was 100.04 ft.

$$\text{Distance measured} = (8.9624)(100.04) = 896.60 \text{ ft}$$

Example 4-2

A distance is measured with a 100-ft steel tape and is found to be 2320.30 ft. Later the tape is standardized and is found to have an actual length of 99.97 ft. What is the correct distance measured?

Solution The tape is too short. Therefore, the correction is minus.

$$\text{Measured value} \qquad\qquad = 2320.30 \text{ ft}$$

$$\text{Total correction} = -(0.03)(23.2030) = \underline{\;-0.70 \text{ ft}}$$

$$\text{Corrected distance} \qquad\qquad = \mathbf{2319.60 \text{ ft}}$$

Example 4-3

It is desired to lay off a dimension of 1200.00 ft with a steel tape that has an actual length of 99.95 ft. What field measurement should be made with this tape so that the correct distance is obtained?

Solution This problem is stated exactly opposite to the ones of Examples 4-1 and 4-2. It is obvious that if the tape is used 12 times, the distance measured (12 × 99.95) is less than the 1200 ft desired, and a correction of the number of tape lengths times the error per tape length must be *added*.

$$12 \text{ tape lengths} = 12 \times 100.00 = 1200.00 \text{ ft}$$

$$+12 \times 0.05 \qquad\qquad = \underline{\;+0.60 \text{ ft}}$$

$$\text{Field measurement} \qquad\qquad = 1200.60 \text{ ft}$$

Check The answer can be checked by considering the problem in reverse. Here a distance has been measured as being 1200.60 ft with a tape 99.95 ft long. What actual distance was measured? The solution is as follows:

$$\text{Measured value} \qquad\qquad = 1200.60 \text{ ft}$$

$$\text{Total correction} = -(0.05)(12.006) = \underline{\;-0.60 \text{ ft}}$$

$$\text{Corrected distance} \qquad\qquad = \mathbf{1200.00 \text{ ft}}$$

One final example of this type of corrected measurement is shown with the field notes of Fig. 4-1, where the sides of the traverse previously paced (shown in Fig. 3-1) are measured with a tape of actual length 99.95 ft. In this case each of the sides was taped twice (forward and back), and an average value was obtained before the wrong-length tape correction was applied.

4-4 TEMPERATURE VARIATIONS

Changes in tape lengths caused by temperature variations can be significant even for ordinary surveys. For precise work they are of critical importance. A temperature change of approximately 15°F will cause a change in length of approximately 0.01 ft in a 100-ft tape. If a tape is used at 20°F to lay off a distance of 1 mile and if the distance is checked the following summer with the same tape when the temperature is 100°F, (no temperature correction being made), there will be

TAPING A TRAVERSE					Oct. 10, 1990	J.B. Johnson HT,N
CHATOOGA FARM					Clear, Warm 80°	R.C. Knight RT
Sta.	Fwd.	Back	Avg.	Corr.	Dist.		
A							
B	189.64	189.60	189.62	-0.09	189.53'		
C	175.26	175.28	175.57	-0.09	175.18'		
D	197.87	197.90	197.88	-0.10	197.78'	Traverse sketch same	
E	142.46	142.47	142.46	-0.07	142.39'	as in Fig 3-1	
A	234.71	234.69	234.70	-0.12	234.58'		
Actual Tape Length = 99.95'						

Figure 4-1

a difference in length of 2.75 ft caused by the temperature variation. Such an error alone would be equivalent to a precision of 2.75/5280 = 1/1920 (not so good).

Steel tapes lengthen with rising temperatures and shorten with falling ones. The coefficient of linear expansion for steel tapes is 0.0000065 per degree Fahrenheit. This means that for a 1°F rise in temperature a tape will increase in length by 0.0000065 times its length.

As described in Section 4-3, the standardized length of a tape is determined at 68°F. A tape that is 100.00 ft long at the standard temperature will at 100°F have a length of 100.00 + (32)(0.0000065)(100) = 100.02 ft. The correction of a distance measured at 100°F with this tape can be made as described previously for wrong-length tapes. The correction of a tape for temperature changes can be expressed with the formula

$$C_t = 0.0000065(T - T_s)(L)$$

In this expression C_t is the change in length of the tape, T is the estimated temperature of the tape at the time of measurement, T_s is the standardized temperature, and L is the tape length.

It is clear that a steel tape used on a hot summer day in the bright sunshine will have a much higher temperature than will the surrounding air. Actually, though, partly cloudy summer days will cause the most troublesome variations in length. For a few minutes the sun shines brightly and then it is covered for awhile by clouds, causing the tape to cool quickly, perhaps by as much as 20 or 30°F. Accurate corrections for tape temperature variations are difficult to make because the tape temperature may vary along its length with sun, shade, dampness (in grass, on ground), and so on. It has been shown that variation of a few degrees may make an appreciable variation in the measurement of distance.

For the best precision it is desirable to tape on cloudy days, early in the mornings, or late in the afternoons to minimize temperature variations. Furthermore, the use of Invar tapes, with their very small coefficients of expansion (0.0000001 to 0.0000002), is very helpful for precise work but such measurement has been made obsolete by electronic distance-measuring equipment.

For very precise surveying, tape measurements are recorded and the proper corrections made. On cloudy, hazy days an ordinary thermometer may be used for measuring the air temperature, but on bright sunny days the temperature of the tape itself should be determined. For this purpose plastic thermometers taped to the tapes near the ends (so that their weights do not appreciably affect sag) should be used. (As mentioned previously some tapes have different end marks to be used, depending on the temperature.)

Regular steel tapes have not been used for geodetic work for quite a few decades because of their rather large coefficients of expansion and because of the impossibility of accurately determining their temperature during daytime operations. Until electronic distance-measuring devices were introduced, almost all of the base-line measurements of the National Geodetic Survey during the twentieth century were done with Invar tapes. Today, almost all of their length measurements are made with EDM instruments.

If SI units are being used, the coefficient of linear expansion is 0.000 011 6 per degree Celsius (°C). The correction in length of a metric tape for temperature changes can be expressed by the formula

$$C_t = (0.000\ 011\ 6)(T - T_s)(L)$$

where T_s is the standardized temperature of the tape at manufacture (usually 20°C), T is the temperature of the tape at the time of measurement, and L is the tape length.

It will be remembered that the expressions for temperature conversion are as follows:

$$°C = \tfrac{5}{9}(°F - 32)$$

$$°F = \tfrac{9}{5}(°C) + 32$$

4-5 SLOPE CORRECTIONS

In Fig. 4-2 a tape of length s is stretched along a slope and it is desired to determine the horizontal distance h that has been measured. It is easy for tapemen to apply an approximate correction formula for most slopes. The expression, derived below, is satisfactory for most measurements, but for slopes of greater than approximately 10 to 15%, an exact trigonometric expression or the Pythagorean theorem should be used. When a 100-ft slope distance is measured, the use of this approximate expression will cause an error of 0.0013 ft for a 10% slope and a 0.0064-ft error for a 15% slope.

It is desired to write an expression for the correction C shown in Fig. 4-2. This value, which equals $s - h$ in the figure, is written in a more practical form by using the Pythagorean theorem as follows:

$$s^2 = h^2 + v^2$$
$$v^2 = s^2 - h^2$$

from which

$$v^2 = (s - h)(s + h)$$
$$s - h = \frac{v^2}{s + h}$$

and since

$$C = s - h$$
$$C = \frac{v^2}{s + h}$$

For the typical 100-ft tape, s equals 100 ft and h varies from 100 ft by a very small value. For practical purposes, therefore, h can also be assumed to equal

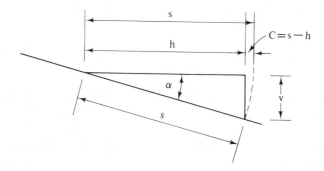

Figure 4-2

100 ft when the slope correction expression is applied. It is written for 100-ft tapes in the form

$$C = \frac{v^2}{200}$$

In taping it is normally convenient to measure a full tape length at a time. Therefore, in measuring along a slope it is often convenient for the head tapeman to calculate (probably in his or her head) the correction and set the taping pin that distance beyond the end of the tape so that they will have measured 100 ft horizontally. For a 6-ft vertical elevation difference

$$C = \frac{(6)^2}{200} = 0.18 \text{ ft}$$

For tape lengths other than 100 ft, the correction expression can be written as

$$C = \frac{v^2}{2s}$$

Sometimes for a long constant slope the tape is held on the ground and the correction is made for the entire length. Such a situation is illustrated in Example 4-4. The reader should carefully note that the correction formula was derived for a single tape length. For a distance of more than one tape length, the total correction will equal the number of tape lengths times the correction per tape length.

Example 4-4

A distance was measured on an 8% slope and found to be 2620.30 ft. What is the horizontal distance measured?

Solution

$$\text{Correction per tape length} = -\frac{(8)^2}{(2)(100)} = -0.32 \text{ ft}$$

$$\text{Total correction} = (26.2030)(-0.32) = -8.38 \text{ ft}$$

$$\text{Horizontal distance} = 2620.30 - 8.38 = \textbf{2611.92 ft}$$

4-6 SAG

When a steel tape is supported only at its ends, it will sag into a curved shape known as the *catenary*. The obvious result is that the horizontal distance between its ends is less than when the tape is supported for its entire length.

To determine the difference in the length measured with a fully supported tape and one supported only at its end or at certain intervals, the following approximate expression may be used:

$$C_s = -\frac{w^2 L^3}{24 P_1^2} = -\frac{W^2 L}{24 P_1^2}$$

where C_s = correction in feet

  w = weight of tape in pounds per foot

  L = unsupported length of tape in feet

  $W = wL$ = total weight of tape between supports

  P_1 = total tension in pounds applied to the tape

This expression, although approximate, is sufficiently accurate for many surveying purposes. It is applicable to horizontal taping or to tapes held along slopes of not more than approximately 10°. *The effect of stretching of the metal in the tape caused by tension is neglected here, but it is considered in Section 4-7.*

To minimize sag errors, it is possible to use this formula and apply the appropriate corrections to the observed distance. Another and more practical procedure for ordinary surveying is to increase the pull or tension on the tape in order to attempt to compensate for the effect of sag. For very precise work, the tape is either supported at sufficient intervals to make sag effects negligible or it is standardized for the pull and manner of support to be used in the field. Examples 4-5 to 4-7 illustrate the application of the sag correction formula. The reader should particularly note in these examples the great reduction in the length correction due to sag when the tape tension is increased. In addition, the greater the tensile force applied to a tape, the longer the tape itself will become. Such stretching is neglected in the next three examples but is considered in the next section of this chapter.

Example 4-5

A steel tape weighing 2 lb is 100.000 ft long when supported continuously on a floor and pulled with a tensile force of 10 lb.

(a) If the tape is lifted from the floor and held at its ends only with the same pull, what is the distance between its ends?

(b) Repeat the problem if the pull is increased to 30 lb.

Neglect stretching of the tape due to tension.

Solution

(a) C_s = correction in length = $-\dfrac{W^2 L}{24 P_1^2}$

$$= -\frac{(2)^2(100)}{(24)(10)^2} = -0.167 \text{ ft}$$

Corrected distance = $100.000 - 0.167 =$ **99.833 ft**

(b) $C_s = -\dfrac{(2)^2(100)}{(24)(30)^2} = -0.019$ ft

Corrected distance $= 100.000 - 0.019 =$ **99.981 ft**

Example 4-6

If the tape of Example 4-5 is supported at its ends and at midpoint as shown in Fig. 4-3, and has a 10-lb pull, what is the corrected distance between its ends? Neglect stretching of the tape.

Solution   The correction is made for each of the 50-ft spans and the weight of the tape is 1 lb for each span.

$$\text{Correction for 50-ft span} = -\dfrac{(1)^2(50)}{(24)(10)^2} = -0.021 \text{ ft}$$

$$\text{Total correction} = (2)(-0.021) = -0.042 \text{ ft}$$

$$\text{Corrected distance} = 100.000 - 0.042 = \textbf{99.958 ft}$$

This expression for the correction of tape lengths caused by sag is applicable to situations where either standard English units of distance or SI units are used, as long as the weights and tension values (W and P_1) are applied in consistent units. Example 4-7 illustrates the shortening of a 30-m tape due to sag when the tape is held above the ground.

Example 4-7

A steel tape weighing 0.910 kg is 30.000 m long when supported continuously across a floor and pulled with a tensile force of 5 kg. What is the sag correction when the tape is held above the ground, supported at its ends only and pulled with a tensile force of 8 kg? Neglect stretching of tape.

Solution

$$C_s = -\dfrac{(0.910)^2(30.000)}{(24)(8)^2} = -0.016 \text{ m}$$

Corrected distance $= 30.000 - 0.016 =$ **29.984 m**

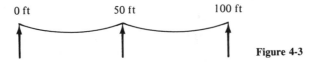

0 ft   50 ft   100 ft

Figure 4-3

4-7 TENSION CORRECTIONS

Variations in Tension

A steel tape stretches when it is pulled, and if the pull is greater than that for which it was standardized, the tape will be too long. If insufficient tension is applied, the tape will be too short. A 100-ft steel tape will change in length by approximately 0.01 ft for a 15-lb change in pull. Since variations in pull of this magnitude are improbable, errors caused by tension variations are negligible for all except the most precise chaining. Furthermore, these errors are accidental and tend to some degree to cancel. For precise taping, spring balances are used so that certain prescribed tensile forces can be applied to the tapes. With such balances it is not difficult to apply tensions within $\frac{1}{2}$ lb or closer to desired values. As with all measuring devices, it is necessary periodically to check the tension apparatus against a known standard.

Despite the minor significance of tension errors, a general understanding of them is important to the surveyor and will serve to improve the quality of the work. The actual elongation of a tape in tension equals the tensile stress in psi over the modulus of elasticity of the steel (29,000,000 psi or 2 050 000 kg/cm^2) times the length of the tape. In the following expression the elongation of the tape in feet is represented by C_p, P_1 is the pull on the tape, A is the cross-sectional area in square inches, L is the length in feet, and E is the modulus of elasticity of the steel in psi.

$$C_p = \frac{P_1/A}{E} L = \frac{P_1 L}{AE}$$

It will be noted that the tape has been standardized at a certain pull P, and therefore the change in length from the standardized situation is desired and the expression is written as

$$C_p = \frac{(P_1 - P)L}{AE} = \frac{(\Delta P)L}{AE}$$

Normal Tension

If, when the tape is suspended, it is pulled very tightly, there will be an appreciable reduction in sag and some increase in the tape length because of tension. As a matter of fact, there is a theoretical pull for each tape at which the lengthening of the tape caused by tension equals its shortening caused by sag. This value is referred to as the *normal tension*. Its magnitude can be practically measured for a particular tape, or it can be computed theoretically as described in the following paragraphs.

A tape may be placed on a floor or pavement, tensioned at its standardized pull, and have its ends marked on the slab. The tape may then be held in the air above the slab supported at its ends only and pulled until its ends (as marked with plumb bobs) coincide with the marked points on the slab. The pull necessary to make the end points coincide is the normal tension. Its value may be measured with spring balances.

A theoretical method of determining the normal tension is to equate the expression for elongation of the tape caused by tension to the expression for shortening of the tape caused by sag. The resulting expression can be solved for P_1, the normal tension:

$$P_1 = \frac{0.204 W \sqrt{AE}}{\sqrt{P_1 - P}}$$

P_1 occurs on both sides of the equation, but its value for a particular tape may be determined by a trial-and-error method. For a normal-weight 100-ft tape, this value will probably be in the range of 20 lb. As described in detail in Section 4-8, most distance measurements are too large because of the cumulative errors

Figure 4-4 Precise distance measurements with Invar tape using spring balance for tension, taping stands or posts, and metal marking strips. (Courtesy of National Geodetic Survey.)

of sag, poor alignment, slope, and so on. As a result, overpulling the tape is a good idea for ordinary surveying because it tends to reduce some of these errors and improve the precision of the work. For such surveys, an estimated pull of approximately 30 lb is often recommended. For very precise surveying, the normal tension is applied to the tape by using accurate spring balances (see Fig. 4-4).

4-8 COMBINED TAPING CORRECTIONS

If corrections must be made for several factors at the same time (e.g., wrong-length tape, slope, temperature), the individual corrections per tape length may be computed separately and added together (taking into account their signs) in order to obtain a combined correction for all. Since each correction will be relatively small, it is assumed that they do not appreciably affect each other and each can be computed independently. Furthermore, the nominal tape length (100 ft) may be used for the calculation. This means that although the tape may be 99.92 ft long at 68°F and a temperature correction is to be made for a 40°F increase in temperature, the increase in tape length can be figured as $(40)(0.0000065)(100)$ without having to use $(40)(0.0000065)(99.92)$.

Examples 4-8 and 4-9 illustrate the application of several corrections to a single distance measurement. The reader should note that the total sag correction used in Example 4-9 is slightly in error. Theoretically the total sag correction for this problem should equal 24 times the sag correction for one full tape length plus the computed sag correction for a partial tape length of 42.67 ft. As the difference is usually negligible the author has not done this for this example nor for the problems at the end of this chapter as solved in the solutions manual.

Example 4-8

A distance was measured on a uniform slope of 8% and was found to be 1665.2 ft. No field slope corrections were made. The tape temperature at the time of measurement was 18°F. What is the correct horizontal distance measured if the tape is 100.06 ft long at 68°F?

Solution   The corrections per tape length are computed, added together, and then multiplied by the number of tape lengths.

Slope correction/tape length $-\dfrac{v^2}{200} = -\dfrac{(8)^2}{200}$ $= -0.3200$ ft

Temp. correction/tape length $- (50)(0.0000065)(100) = -0.0325$ ft

Standardization error/tape length $= \underline{+0.0600 \text{ ft}}$

Total correction/tape length $= -0.2925$ ft

Correction for entire distance $= (16.652)(-0.2925) = -4.87$ ft

Corrected distance $= 1665.2 - 4.87$ $= \textbf{1660.3 ft}$

Example 4-9

A tape weighing 2 lb ($A = 0.006$ sq in.) has a length of 100.020 ft at 68°F when it is supported for its entire length and subjected to a pull of 12 lb. To measure a certain distance this tape is held horizontally, supported at its ends only and pulled with a force of 30 lb. The distance obtained was 2442.67 ft. If the tape temperature at the time of measurement was 108°F, what was the correct distance measured?

Solution   The corrections are again determined for a single tape length and then for the entire measured distance.

Standardization error/tape length $= +0.0200$ ft

Temp. correction/tape length $= +(40)(0.0000065)(100)$ $= +0.0260$ ft

Sag correction/tape length $= -\dfrac{w^2 L}{24 P_1^2} = -\dfrac{(2)^2(100)}{(24)(30)^2}$ $= -0.0185$ ft

Tension correction/tape length $= \dfrac{\Delta P L}{AE} = +\dfrac{(30-12)(100)}{(0.006)(29 \times 10^6)} = \underline{+0.0103 \text{ ft}}$

Total correction/tape length $= +0.0378$ ft

Correction for entire distance $= (24.4267)(+0.0378)$ $= +0.92$ ft

Corrected distance $= 2442.67 + 0.92$ $= \mathbf{2443.59 \text{ ft}}$

4-9 COMMON MISTAKES MADE IN TAPING

Some of the most common mistakes made in taping are described in this section, and a method of eliminating each is suggested.

Reading Tape Wrong.   A frequent mistake made by tapemen is reading the wrong number on the tape, for example, reading a 6 instead of a 9 or a 9 instead of a 6. As tapes become older these mistakes become more frequent because the numbers on the tape become worn. These blunders can be eliminated if tapemen develop the simple habit of looking at the adjacent numbers on the tape when readings are taken.

Recording Numbers.   Occasionally, the recorder will misunderstand a measurement that is called out to him or her. To prevent this kind of mistake, the recorder can repeat the values aloud, including the decimals, as he or she records them.

Missing a Tape Length.   It is not very difficult to lose or gain a tape length in measuring long distances. The careful use of taping pins, described in Section 3-9, should prevent this mistake. In addition, the surveyor can many times eliminate such mistakes by cultivating the habit of estimating distances by eye or by pacing or better, by taking stadia readings whenever possible.

Mistaking End Point of Tape.   Some tapes are manufactured with the 0- and 100-ft points at the very ends of the tapes. Other tapes have them at a little distance from the ends. Clearly, tapemen should not make mistakes like these if they have taken the time to examine the tape before they begin to take measurements.

Making 1-Foot Mistakes.   When a fractional part of a tape is being used at the end of a line, it is possible to make a 1-ft mistake. Mistakes like these can be prevented by carefully following the procedure described for such measurements in Section 3-9. Also helpful are the habits of calling out the numbers and checking the adjacent numbers on the tape.

4-10 ERRORS IN TAPING

In the following paragraphs we discuss briefly the common taping errors. As these errors are studied it is important to notice that the effect of most of them is to cause the surveyor to get too much distance. If the tape is not properly aligned, not horizontal, or sags too much, if a strong wind is blowing the tape to one side, or if the tape has shortened on a cold day, it takes more tape lengths to cover the distance.

Alignment of Tape.   A good rear tapeman can align the head tapeman with sufficient accuracy for most surveys, although it is more accurate to use the transit to keep the tape on line. In some cases it is necessary to use a transit when establishing new lines or when the tapemen are unable to see the ending point because of the roughness of the terrain. For the latter case it may be necessary to set up intermediate points on the line to guide the tapemen.

It is probable that most surveyors spend too much time improving their alignment, at least in proportion to the time they spend trying to reduce other more important errors. In taping a 100-ft distance the tape would have to be 1.414 ft out of line to cause an error of 0.01 ft. From this value it can be seen that for ordinary distances alignment errors should not be appreciable. As a matter of fact, experienced tapemen should have no difficulty in keeping their alignment well within a foot of the correct line by eye, particularly when the lines are only a few hundred feet or less in length. They should be able to keep the poles lined up within at least 0.3 or 0.4 ft, which would cause an error of less than 0.001 ft for each tape length (a negligible error for most steel tape measurements).

Accidental Taping Errors.   Because of human imperfections tapemen cannot read tapes perfectly, cannot plumb perfectly, and cannot set taping pins perfectly. They will place the pins a little too far forward or a little too far back. These errors are accidental in nature and will tend to cancel each other somewhat. Generally, errors caused by setting pins and reading the tape are minor, but errors caused by plumbing may be very important. Their magnitudes can be reduced by increasing the care with which the work is done or by taping along slopes and applying slope corrections to avoid plumbing.

Tape Not Horizontal.   If tapes are not held in the horizontal position, an error results which causes the surveyor to obtain distances that are too large. These errors are cumulative and can be quite large when surveying is done in hilly country. Here the surveyor must be very careful.

If a surveyor deliberately holds the tape along a slope, he or she can correct the measurement with the slope correction expression

$$C = \frac{v^2}{2s}$$

which was presented in Section 4-5. It might be noticed that if one end of a 100-ft tape is 1.414 ft above or below the other end, an error of

$$\frac{(1.414)^2}{200} = 0.01 \text{ ft}$$

is made. From this expression it can be seen that the error varies as the square of the elevation difference. If the elevation difference is doubled, the error quadruples. For a 2.828-ft elevation difference, the error made is

$$\frac{(2.828)^2}{200} = 0.04 \text{ ft}$$

Incorrect Tape Length.   These important errors were discussed in Section 4-3 and must be given careful attention if good work is to be done. For a given tape of incorrect length, the errors are cumulative and can add up to sizable values.

Temperature Variations.   Corrections for variations in tape temperature were discussed in Section 4-4. Errors in taping caused by temperature changes are usually thought of as being cumulative for a single day. They may, however, be accidental under unusual circumstances with changing temperatures during the day and also with different temperatures at the same time in different parts of the tape. It is probably wiser to limit tape variations instead of trying to correct for them no matter how large they may be. Taping on cloudy days, early in the morning, or late in the afternoon or using Invar tapes are effective means of limiting length changes caused by temperature variations.

Sag.   Sag effects (discussed in Section 4-6) cause the surveyor to obtain excessive distances. Most surveyors attempt to reduce these errors by overpulling their tapes with a force that will stretch them sufficiently to counterbalance the sag effects. A rule of thumb used by many for 100-ft tapes is to apply an estimated pull of approximately 30 lb. This practice is satisfactory for surveys of ordinary precision, but it is not adequate for those of high precision because the amount of pull required varies for different tapes, different support conditions, and so on.

It is also difficult to estimate by hand the force being applied. A better method is to use a spring valance for applying a definite tension to a tape, the tension required having been calculated or determined by a standardized test to equal the normal tension of the tape.

Miscellaneous Errors.   Some of the miscellaneous errors that affect the precision of taping are (1) wind blowing plumb bobs; (2) wind blowing tape to one side, causing the same effect as sag; and (3) taping pins not set exactly where plumb bobs touch ground.

4-11 MAGNITUDE OF ERRORS

To get some feeling for the effects of common taping errors, consider Table 4-1. In this table various sources of errors are listed together with the variations that would be necessary for each error to have a magnitude of ± 0.01 ft when a 100-ft distance is measured with a 100-ft tape.

For the discussion to follow it is assumed that a 100-ft steel tape is used to measure a distance of 100 ft. The ground is gently sloping, and the entire length of the tape can be held horizontal at one time. In Table 4-2 several accidental errors and their magnitudes are given. At the bottom of the table the random error theory of Chapter 2 is used to estimate the magnitude of the most probable total error.

TABLE 4-1   ERRORS OF ± 0.01 FT IN 100-FT MEASUREMENTS

Source of error	Magnitude of error
Incorrect tape length*	0.01 ft
Temperature variation*	15°F
Tension or pull variation*	15 lb
Sag*	7.5 in. sag at center line; remainder of tape supported throughout
Alignment*	1.4 ft at one end
Tape not level*	1.4 ft difference in elevation
Plumbing	0.01 ft
Marking	0.01 ft
Reading tape	0.01 ft

Source: J. F. Dracup and C. F. Kelly, *Horizontal Control As Applied to Local Surveying Needs* (Falls Church, Va.: American Congress on Surveying and Mapping, 1973), p. 16.

* May be greatly minimized if sufficient field data are obtained and appropriate mathematical corrections are applied.

TABLE 4-2

Source of error	Magnitude of error (ft)	Magnitude of error squared
Incorrect tape length	±0.005	0.000025
Temperature variation (10°F)	±0.009	0.000081
Tension or pull variation (5 lb)	±0.003	0.000009
Plumbing (0.005 ft)	±0.005	0.000025
Marking (0.001 ft)	±0.001	0.000001
Reading tape (0.001 ft)	±0.001	0.000001
		Σ = 0.000142

$$\text{Most probable error} = \sqrt{0.000142} = \pm 0.0119 \text{ ft}$$

$$\text{Corresponding precision} = \frac{0.0119}{100} = \frac{1}{8403}$$

Source: J. F. Dracup and C. F. Kelly, *Horizontal Control As Applied to Local Surveying Needs* (Falls Church, Va.: American Congress on Surveying and Mapping, 1973), p. 16.

4-12 SUGGESTIONS FOR GOOD TAPING

If the surveyor studies the errors and mistakes that are made in taping, he or she should be able to develop a few rules of thumb that will appreciably improve the precision of the work. Following is a set of rules that have proved helpful in the field:

1. Tapemen should develop the habit of estimating by eye the distances they are measuring because it will enable them to avoid most major blunders.
2. When reading a foot mark at an intermediate point on a tape, a tapeman should glance at the adjacent foot marks to be sure that he or she has read the mark correctly.
3. All points that the tapeman establishes should be checked. This is particularly true when the plumb bob has been used to set a point such as a tack in a stake.
4. It is easier to tape downhill whenever feasible. The rear tapeman can hold the tape end on the ground at the last point instead of having to hold a plumb bob over the point while the head tapeman is pulling against him or her, as would be the case in taping uphill.
5. If time permits (often it may not), distances should be taped twice, once forward and once back. Taping in the two different directions should prevent repeating the same mistakes.
6. Tapemen should assume stable positions when pulling the tape. This usually means feet widespread, leather thong wrapped around the hand (or use of

taping clamp), standing to one side of the tape with arms in close to the body, and applying pull to the tape by leaning against it.

7. Since most taping gives distances that are too large, the surveyor can improve his or her work for ordinary surveys by pulling the tape very firmly, by estimating the smaller number when a reading seems to be halfway between two values, and perhaps even by setting taping pins slightly to the forward side.

4-13 MISCELLANEOUS TAPING OPERATIONS

In addition to the direct determination of the distance between two points, there are several miscellaneous operations that can be performed with a steel tape. These operations, which are presented in the paragraphs to follow, are probably of academic interest only, because with modern equipment these tasks can be handled so much more quickly and with so much better precision. The use of the total station instruments discussed in Chapter 15 particularly make obsolete the discussion here concerning chaining past obstacles.

Angle Measurement

It is normally much wiser to measure angles with a transit or theodolite, but on some occasions such an instrument may not be available and a tape can be used to determine angle values within 5 to 10' of arc. In Fig. 4-5 a convenient *horizontal* distance is measured down each line (say 100 ft), and the chord distance between points *a* and *b* is measured *horizontally*.

From trigonometry the following equation can be used to determine the magnitude of the angle α:

$$\sin \frac{1}{2} \alpha = \frac{\text{chord length}}{200}$$

Laying Off Right Angles

Occasionally, a tape may be used for laying off a right angle (again assuming that a transit or theodolite is not readily available). This can be done quickly and precisely with a steel tape by working with the proportions of a 3–4–5 triangle,

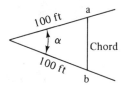

Figure 4-5

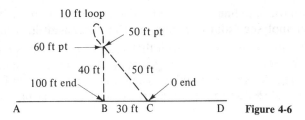

Figure 4-6

as shown in Fig. 4-6. It is assumed in this figure that a perpendicular is to be established at a point B on the straight line AD.

A distance of 30 ft is laid off from B to C, as shown in the figure. Then the 100-ft end of the tape is held at B and the tape is laid out so that a 10-ft loop is made from the 60-ft to the 50-ft marks and the 0-end is at C. When the tape is taut, the 60-ft and 50-ft marks will fall on a perpendicular to line AD passing through point B.

Erecting a Perpendicular from a Given Point to an Existing Line

Another use of steel tapes is the erection of a line from a given point perpendicular to an existing line. See Fig. 4-7, in which it is desired to erect a perpendicular line from point B to line AD. One method of doing this is to make use of the procedure just described for laying off right angles.

The point at which the perpendicular will hit line AD is estimated (C' in the figure). A right angle is laid off, giving the line $B'C'$. This location is obviously in error, but it can be quickly adjusted by measuring the distance from B' to B and then measuring over the same distance from C' to C. Point C should then fall on the desired perpendicular BC. The right-angle procedure should again be used to check the perpendicularity of line BC.

Taping Obstructed Lines

The surveyor is continually faced with the problem of having something obstructing his or her work: for example, trees, buildings, and lakes. Several methods of running straight lines where obstructions occur are described in the following paragraphs.

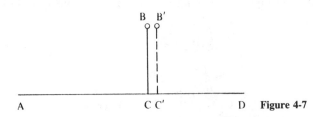

Figure 4-7

1. *Random lines.* It is desired to establish a straight line from *A* to *B*, but the surveyor cannot see from one point to the other and does not know the precise direction between them (see Fig. 4-8). A common method of handling the problem is for the surveyor to run convenient random straight lines working from the first point until he or she reaches the second one. In other words, he or she starts at point *A* and runs a straight line in what seems to be the easiest way to proceed (i.e., fewest obstructions, reasonable slopes, etc.) in the general direction of *B*. After some distance he sets a point (*x* in the figure) and proceeds to run another convenient straight line to *y*, then to *z*, and so on, until he or she reaches *B*. It is necessary for him or her to tape each of these distances and measure the angles involved. By a quick and convenient method called *latitudes and departures* (described in Chapter 12), he or she can compute the length and direction of the straight line *AB* and if necessary return to the field and establish the line on the ground.

Figure 4-8

2. *Offsetting.* It is assumed that the surveyor is running a straight line. He or she has already established points *A*, *B*, and *C*, but beyond *C* there is a tree along the line (see Fig. 4-9). One method of handling the problem is to offset a convenient distance along the existing line, as at *B* and *C*. The distances *BB'* and *CC'* are made with sufficient length to pass the tree (say 1 or 2 ft), and then a straight line is run through *B'* and *C'* past the tree until points *D'* and *E'* are established. The offset distances *D'D* and *E'E* can then be established in order to get back on the original line. The surveyor, however, may choose to keep running the offset line until he or she encounters another tree, at which time the surveyor can offset back to the original line. It should be obvious that the distances *BC* and *DE* should be fairly large and that the offset distances should be very short; otherwise, the tape will not be too desirable for laying off the distance because of the problem of laying off the 90° angle with the tape.

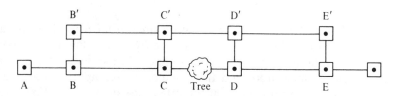

Figure 4-9

3. *Equilateral triangles*. A convenient method for getting around large obstacles such as lakes or large buildings involves using an equilateral triangle, as illustrated in Fig. 4-10. A 60° angle is established at *A* and a sufficient distance is chained (*AB*) to pass the obstacle. Then a 60° angle is laid off at *B* and the distance *BC* is taped equal to *AB*. Point *C* will be on the original line and the survey can continue. The distance desired, *AC*, will equal *AB* or *BC*.

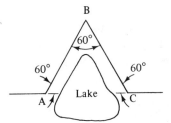

Figure 4-10

4. *Other triangles*. The student can easily think of several other precise methods of taping past an obstacle. For instance, in Fig. 4-11 a convenient angle α is laid off at *B* and a convenient distance is taped. Then, as shown at *C*, the angle 2α is turned and the distance *l* is measured to point *D*. This puts the surveyor back on the original line *AE*. By measuring the angle α shown at *D*, the surveyor will be able to move on toward point *E*. From trigonometry, the distance *BD* equals 2*l* cos α.

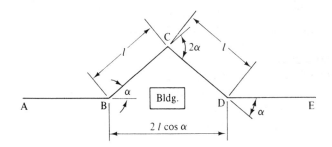

Figure 4-11

5. *Triangulation*. To obtain the distance from *B* to *D* across a large body of water (see Fig. 4-12), the distance *BC* can be laid off in a convenient direction, perhaps perpendicular to line *AB*, the distance taped, the angles at *B* and *C* measured, and the distance *BD* computed. For checking, one can repeat the process from the other side of the lake. For better accuracy, it is desirable for the distance *BC* to equal at least one-half of the unknown

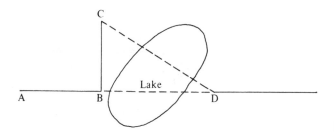

Figure 4-12

distance *BD*. This type of problem is discussed further and illustrated numerically in Chapter 20.

4-14 TAPING PRECISION

Different kinds of surveys require different accuracies and many surveys must meet definite specifications. For instance, many states, cities, and towns have passed laws requiring certain minimum accuracy standards that must be met for work within their boundaries for various kinds of work. For instance, a common set of values might include the minimum values 1/5000 for rural land surveys and 1/10,000 for property within city boundaries.

The presentation of definite values that should be obtained for good, average, and poor taping is difficult because what is good under one set of conditions may be poor for another set of conditions. For example, a precision of 1/2500 is poor when taping is done along a level road, but it may very well be satisfactory when the work is done through heavy underbrush in mountainous terrain.

Below are presented some supposedly reasonable precision values that should be expected under ordinary taping conditions. The average value is probably sufficient for most preliminary surveys and the good value is desirable for most other surveys.

> Poor   1/2500
>
> Average   1/5000
>
> Good   1/10,000

Taping can be done with precision much higher than 1/5000 if careful attention is paid to the reduction of the errors discussed previously in this chapter. Thus by carefully controlling tape tensions, precisely measuring tape temperatures, applying corrections, using hand levels to keep tapes horizontal, or taping along slopes and minimizing the use of plumb bobs but necessitating the measurement of the slopes and the application of the appropriate corrections, the surveyor will be able to tape with precisions of 1/10,000 and better.

PROBLEMS

For Problems 4-1 to 4-5, distances were measured with tapes assumed to be 100.00 ft long. Later the tapes were standardized and found to have different tape lengths. Determine the correct distances measured in each case.

	Recorded distance (ft)	Correct tape length (ft)
4-1.	1343.29	99.96
4-2.	2461.42	100.05
4-3.	2139.88	100.03
4-4.	717.54	99.94
4-5.	3696.02	99.97

(*Ans.:* 1342.75 ft)

(*Ans.:* 2140.52 ft)

(*Ans.:* 3694.91 ft)

For Problems 4-6 to 4-8, distances were measured with tapes assumed to be 30.00 m long. Later the tapes were standardized and found to have different lengths. Determine the correct distances measured in each case.

	Recorded distance (m)	Correct tape length (m)
4-6.	534.67	29.98
4-7.	848.83	30.03
4-8.	1642.10	29.96

(*Ans.:* 849.68 m)

4-9. The actual distance between two marks used at a university for standardizing tapes is 100.03 ft. When a certain tape was held along this line, the surveyor, thinking that the distance between the marks was 100.00 ft, observed that the tape was 99.95 ft long. What is the correct length of the tape?

(*Ans.:* 99.98 ft)

4-10. Repeat Problem 4.9 for values of 99.97, 100.00, and 100.03 ft, respectively.

4-11. Repeat Problem 4.9 for values of 29.96, 30.00, and 30.03 m, respectively.

(*Ans.:* 29.99 m)

For Problems 4-12 to 4-17, it is desired to lay off certain horizontal distances for building layouts. The lengths of the tapes used in Problems 4-12 to 4-15 are not 100.00 ft and not 30.00 m for Problems 4-16 and 4-17. Determine the field dimensions (or actual tape readings) that should be used with the incorrect tape lengths so that the correct dimensions are obtained.

	Desired dimension	Correct tape length
4-12.	600.00 ft × 150.00 ft	100.06 ft
4-13.	440.00 ft × 280.00 ft	100.05 ft
4-14.	380.00 ft × 820.00 ft	99.96 ft
4-15.	213.20 ft × 322.40 ft	99.93 ft
4-16.	120.00 m × 180.00 m	30.03 m
4-17.	220.00 m × 420.50 m	29.96 m

(*Ans.:* 439.78 ft × 279.86 ft)

(*Ans.:* 213.35 ft × 322.63 ft)

(*Ans.:* 220.29 m × 421.06 m)

4-18. It is desired to lay off a horizontal distance equal to 704.40 ft with a tape that is 49.90 ft long (not 50.00 as indicated on the scale). What should the recorded distance be?

4-19. A 50-ft woven tape is used to set the corners for a building. If the tape is actually 50.12 ft long, what should the recorded distances be if the building is to be 182.00 ft by 97.50 ft?

(*Ans.:* 181.56 ft × 97.27 ft)

4-20. A distance is measured through rough country and found to be 3476.2 ft. If on the average a plumb bob is used every 50 ft with a probable error of ±0.02 ft, what is the probable total error in the whole distance?

4-21. Rework Problem 4-20 if the plumb bob is used every 30 ft on the average and the distance measured is 1792.3 ft.

(*Ans.:* ±0.15 ft)

For Problems 4-22 to 4-25, distances were measured with 100-ft steel tapes and their average temperatures estimated. From these values and the standardized tape lengths (100.000 ft at 68°F), determine the correct distances measured.

	Recorded distance (ft)	Average tape temperature at time of measurement (°F)
4-22.	3154.99	18
4-23.	843.86	108
4-24.	1766.52	98
4-25.	2201.60	8

(*Ans.:* 844.08 ft)

(*Ans.:* 2200.74 ft)

4-26. A steel tape that has a length of 100.000 ft at 68°F is to be used to lay off a building with the dimensions 450.00 ft by 620.00 ft.
   (a) What should be the tape readings if the tape temperature is 18°F at the time of the measurement?
   (b) Repeat part (a) if the tape temperature is 108°F.

4-27. A 1-acre lot (43,560 ft^2) is to be staked out on level ground with the dimensions 204.00 ft by 213.53 ft. The standardized length of the tape at 68°F is 99.92 ft. If the tape temperature is 28°F, what dimensions should the survey party use to lay out this lot?

(*Ans.:* 204.22 ft × 213.76 ft)

4-28. Repeat Problem 4-27 if the tape temperature is 92°F.

For Problems 4-29 to 4-31, distances were measured with 30-m steel tapes and their average temperatures estimated. From these values and the standardized tape length (30.000 m at 20°C), determine the correct distances measured.

	Recorded distance (m)	Average tape temperature at time of measurement (°C)	
4-29.	316.23	40	(*Ans.:* 316.30 m)
4-30.	842.96	8	
4-31.	1 264.32	32	(*Ans.:* 1 264.50 m)

4-32. (a) A distance was measured at a temperature of 8°F and was found to be 3349.66 ft. If the tape has a standardized length of 99.96 ft at 68°F, what is the correct distance measured?

(b) If the same distance were measured at a temperature of 108°F, what would be the probable value obtained, neglecting other errors?

4-33. A distance was measured with a 100-ft steel tape at a temperature of 98°F and found to be 2498.32 ft. The next winter the distance was remeasured with the same tape at a temperature of 18°F and was found to be 2493.73 ft. What part of the discrepancy between the two measurements should be caused by the temperature difference?

(*Ans.:* 1.30 ft)

4-34. (a) A distance is measured as 2223.40 ft when the tape temperature is 28°F. If the same distance is measured again with the same tape at a temperature of 98°F, what distance should be expected, neglecting other errors?

(b) If the tape is 100.00 ft long at 68°F, what is the "true" distance measured?

4-35. A 100.00-ft tape is used to measure an inclined distance and the value determined is 6920.20 ft. If the slope is 6%, what is the correct horizontal distance obtained using the slope correction formula?

(*Ans.:* 6907.74 ft)

4-36. A 200.00-ft tape is used to measure an inclined distance and the value determined is 3682.50 ft. If the slope is 8%, what is the correct horizontal distance? Use an appropriate slope correction formula.

4-37. Repeat Problem 4-36 if the measured value is 4832.40 ft and the slope is 5%.

(*Ans.:* 4826.36 ft)

4-38. A 30.00 m tape is used to measure an inclined distance and the value determined is

1642.32 m. If the slope is 6%, what is the correct horizontal distance? Use an appropriate slope correction formula.

4-39. A tape weighing 2 lb (A = 0.006 sq in.) has a length of 100.00 ft under a pull of 12 lb when supported for its full length. The tape is used to measure a line holding the tape on the ground with a pull of 35 lb. If the value obtained is 1929.61 ft, what should the correct distance be?

(Ans.: 1929.87 ft)

4-40. A tape weighing 2 lb (A = 0.006 sq in.) has a length of 100.00 ft under a pull of 12 lb when supported for its full length. This tape is used to measure a line holding the tape on the ground with a pull of 40 lb. If the value obtained is 4006.33 ft, what should the correct distance be?

4-41. A 2-lb steel tape (with a cross-sectional area of 0.006 sq in.) is 100.00 ft long when it is supported for its full length and subjected to a pull of 12 lb. What is the length of this tape when it is supported only at its ends and subjected to a pull of 40 lb?

(Ans.: 100.01 ft)

4-42. Repeat Problem 4-39 if the tape is supported at its ends and center line.

4-43. Rework Problem 4-40 if the tape is supported at its ends only and the pull is 20 lb.

(Ans.: 4004.84 ft)

4-44. Rework Problem 4-40 if the tape is supported at its ends and center line and if the pull is 30 lb.

4-45. A 30-m steel tape weighs 0.030 3 kg/m. Determine the sag effect if a tension of 10 kg is applied and the tape is supported at its ends only.

(Ans.: −0.010 m)

4.46. A tape weighing 0.910 kg and with a cross-sectional area of 3.87 mm² has a length of 30.00 m under a pull of 5 kg when supported for its full length. This tape is used to measure a line holding the tape on the ground with a tension of 12 kg. If the value obtained is 854.28 m and E = 2 050 000 kg/cm², what is the correct distance measured?

4-47. A tape weighing 0.910 kg and with a cross-sectional area of 3.87 mm² has a length of 30.00 m under a pull of 5 kg when supported for its full length. This tape is used to measure a distance with its ends only supported and E = 2 050 000 kg/cm² with a pull of 10 kg. If the value obtained is 1168.32 m, what should the correct distance be?

(Ans.: 1167.99 m)

4.48. A 100-ft tape weighs 2 lb, has a cross-sectional area equal to 0.006 sq in., and is 100.00 ft long at 68°F while subjected to a 12-lb pull and supported for its full length. A distance is measured in the field with this tape and found to be 698.32 ft. If the pull on the tape is 20 lb and if the tape is supported only at its ends, what is the correct distance measured?

In Problems 4-49 to 4-52, a 2-lb steel tape is 99.95 ft long at 68°F while supported for its full length with a 12-lb pull. It has a cross-sectional area of 0.006 sq in. and E = 29 × 10⁶ psi. For these problems determine the correct value for each length assuming that the tape is supported only at its ends.

	Recorded distance (ft)	Tape temperature (°F)	Pull on tape (lb)	Slope (%)
4-49.	1542.91	98	35	4
4-50.	2846.36	108	30	6
4-51.	1400.28	18	40	5
4-52.	922.22	28	30	5

(*Ans.:* 1541.20 ft)

(*Ans.:* 1397.45 ft)

In Problems 4-53 to 4-55, a 0.910-kg steel tape is 30.03 m long at 20°C while supported for its full length with a 5-kg pull. It has a cross-sectional area of 3.87 mm² and $E = 2\,050\,000$ kg/cm². For these problems determine the correct value for each distance.

	Recorded distance (m)	Tape temperature (°C)	Pull on tape (kg)	Slope (%)
4-53.	873.32	10	10	4
4-54.	1469.84	26	12	6
4-55.	616.19	22	14	5

(*Ans.:* 873.15 m)

(*Ans.:* 616.01 m)

4-56. A 2-lb steel tape ($A = 0.006$ sq in.) has a length of 100.00 ft at a temperature of 68°F when suppported for its entire length and subject to an applied total tension of 10 lb. With this tape a distance is measured as being 100.00 ft long under the following conditions:

Supported at its ends only
Total applied tension 30 lb
Alignment off by 2.00 ft (bent around tree on line at center line)
Tape not horizontal by 4.00 ft
Average tape temperature 18°F
What is the correct distance measured?

CHAPTER FIVE

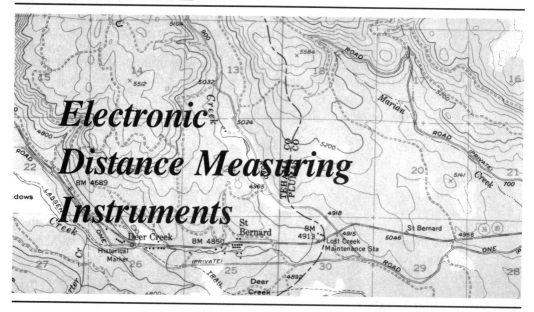

Electronic Distance Measuring Instruments

5-1 INTRODUCTION

Over the last 100 years there has been a constant but gradual improvement in the quality of surveying equipment. Not only have better tapes been manufactured but better instruments for measuring elevations and directions have been produced. Within the past few decades the pace of improvement has quickened tremendously as the electronic distance-measuring instruments (EDMIs) and the total station instruments have come of age.

Now there are available to the surveyor devices with which he or she can measure short distances of a few feet or long distances of many miles with extraordinary speed and precision. These devices save time and money and reduce the sizes of conventional survey parties. Furthermore, they may be used with the same facility where there are obstacles intervening, such as lakes, canyons, standing crops, swampy and timbered terrain, hostile landowners, or heavy traffic. Electronic distance-measuring devices have revolutionized distance measurement not only for geodetic surveys but also for ordinary land surveys.

Another important point is that these instruments automatically display direct readout measurements, with the result that mistakes are greatly reduced. At the push of a button the operator may have the value shown in feet or in meters as desired. Normally, the distances displayed are slope distances, but in many of the newer instruments vertical angles are measured and horizontal distances computed and displayed.

For the surveyor to fully understand the electronic measurement of distances, he or she would have to have a good background in physics and electronics. Fortunately, however, a person can readily use electronic distance-measuring equipment without really understanding very much about the physical phenomena involved. Although the equipment is complex, its actual operation is automatic and requires less skill than is needed with traditional instruments.

Electronic distance-measuring devices were not much used until the 1960s. The equipment was heavy, expensive, and required rather sophisticated maintenance. Nevertheless, their tremendous advantages—the instant and precise measurement of distances up to 10 miles or more over forbidding terrain—inspired much research, which has led to far better models. As a result, the equipment has become lighter, cheaper, and more maintenance-free. Generally, the manufacturers do not recommend that the owners of EDMIs service the units. They feel that more often than not, worse problems will be created, resulting in even higher repair costs. It may be very practical, however, to call the company when an instrument problem occurs and describe the difficulty. They may be able to explain the necessary adjustment over the phone without the necessity of returning the equipment to the factory.

5-2 BASIC EDMI THEORY

The basic theory of EDMIs is that distance equals time multiplied by velocity. Thus if the velocity of a radio or light wave and the time required for it to go from one point to another are known, the distance between the two points can be calculated. A very important relationship in the transmission of sound, radio, microwave, and light waves is given by the following expression:

$$V = f\lambda$$

where V is the velocity of propagation, f is the modulating frequency or number of cycles per unit of time of the energy, and λ is the wavelength or the distance traveled in one cycle. The unit of frequency is the hertz (Hz), which is equal to 1 cycle per second. A frequency of 10 MHz (or 10 megahertz) equals 10 million cycles per second. The wavelength λ is normally expressed in meters.

Both radio waves and light waves are electromagnetic. They have identical velocities in a vacuum (or space) equal to 299 792.458 ± 0.001 km/s. These velocities are reduced in the atmosphere, however, because of density variations between atmospheric layers. In addition, the velocities of radio and light waves are different from each other in the atmosphere. These are the velocities that need to be used in the calculations, not laboratory values, since surveying measurements are of course made in the atmosphere. The ratio of the velocity V_0 of an electromagnetic wave in a vacuum divided by its velocity V in the atmosphere is called the *index of refraction* (η) and has a typical value of about 1.0003:

$$\eta = \frac{V_0}{V}$$

The velocity of the waves in the atmosphere is affected by the air's density. As a wave moves through the innumerable layers of the atmosphere, it changes its velocity and direction because of refraction. To determine the value of the index of refraction, it is necessary to measure the temperature, humidity, and barometric pressure and then to substitute these values into an appropriate equation. Various publications have presented equations for computing the index. Although there are slight variations in the form of the equations and the results obtained, the differences are insignificant. The part of these equations involving relative humidity can be neglected for the electro-optical instruments (described in the next section), as its effect is negligible. An index of refraction of about 1.0003 has been applied to the internal circuitry of the instruments. If the meteorological conditions existing at the time of the measurement yield an index appreciably different from this value, a correction should be made.

With EDMIs the frequency of the electromagnetic waves can be controlled precisely, but the velocity of those waves varies with temperature, pressure, and humidity. To make accurate measurements it is necessary to determine the meteorological conditions and make appropriate corrections. Usually, these conditions are determined at each end of a line and averaged. For very large distances it may be desirable to determine these conditions at intermediate points as well.

As we have seen, a wave is transmitted from a transmitter to a reflector and returned. The total distance traveled is thus equal to two times the slope distance. When the wave returns it is detected and compared at the transmitter with the phase (a definite change with time, controlled by the frequency) of the original signal in order that the travel time can be determined.

To calculate the distance within an accuracy of ± 0.5 ft would require that the time interval be measured to the nearest billionth of a second. Even with current equipment, such a time measurement is extraordinarily difficult to make. Yet the distance should be measured closer than ± 0.5 ft, with the result that the time measurement problem is worse than stated. Fortunately, however, it is possible to solve the problem by making a very accurate phase measurement of the signal.

As stated previously, distance is measured with EDMIs by determining the time required for an electromagnetic wave to travel from a transmitter at one end of a line to a reflector at the other end and back to the transmitter. This measurement is made indirectly by transmitting the wave to the reflector, and on its return it is detected and converted into an electric signal. This signal is a replica of the original waveform. It is, however, delayed in phase by an amount that is proportional to the time required for the wave to travel over the return path. Since the velocity of propagation of the wave can be determined accurately, the distance can be calculated from the integral number of phase rotations plus the fraction of phase delay.[1]

[1] M. O. Schmidt and K.W. Wong, *Fundamentals of Surveying*, 3rd ed. (Boston: PWS Engineering, 1985), pp. 230–233.

5-3 TYPES OF EDMIs

EDMIs are classified as being electro-optical instruments or microwave instruments. The distinction between them is based upon the wavelengths of the electromagnetic energy which they transmit. The electro-optical instruments transmit light in short wavelengths of about 0.4 to 1.2 μm (micrometers or microns). This light is visible or just above the visible range of the spectrum. Microwave instruments transmit long wavelengths somewhere between 10 and 100 μm.

Figure 5-1 shows in one sketch the relative frequencies and wavelengths for various forms of electromagnetic radiation including gamma rays, lasers, radar, and others. The light-wave systems (including lasers and infrared) have a transmitter at one end of the line to be measured and a reflector at the other end. The reflector consists of one or more corner prisms, which are discussed in the next section. For short distances of a few hundred feet, bicycle reflectors or reflector tape is satisfactory.

Almost all short-range EDMIs in use today for measurements up to a few miles are of the infrared type, although there are infrared instruments available that can be used for measurements up to and over 10 miles under normal conditions. There are laser EDMIs available which are useful for short-range measurements as well as for much longer distances. One advantage that lasers have over the infrared type is that they are visible, and that makes them very useful for situations where sighting is difficult. With some laser equipment it is possible to sight the EDMI to a point, place a red dot of light on the point with the laser, press a button, and measure the distance to the dot. This is very useful for mea-

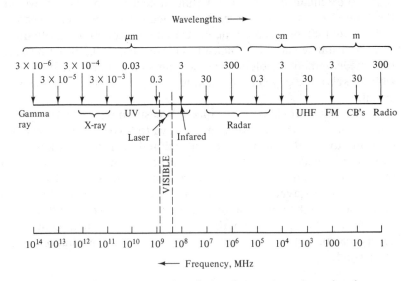

Figure 5-1 Electromagnetic radiation: frequencies and wavelengths.

Figure 5-2 DM-80 electronic distance module. (Courtesy of Cubic Precision.)

suring to points of difficult access, such as towers, church steeples, bottoms of holes in the ground, and so on.

With microwave systems two instruments are necessary: a transmitting system and a receiving system. The beam is transmitted from one end of the line, received at the other end, and is returned to the master instrument. Microwave instruments have the advantages that the waves penetrate through fog or rain and that there is a speech link between the two instruments. However, they are more affected by humidity than are the light-wave instruments. Another problem with microwave systems is the wider beam induced. This can cause some difficulties when surveys are being made inside buildings, in underground situations, or near water surfaces.

5-4 BASIC TERMS

Below are presented a few brief definitions that the reader will encounter in the sections to follow.

An *electronic distance-measuring device* is an instrument that transmits a carrier signal of electromagnetic energy from its position to a receiver located at another position. The signal is returned from the receiver to the instrument such that two times the distance between the two positions can be measured.

Visible light is generally defined as that part of the electromagnetic spectrum to which the eye is sensitive. It has a wavelength in the range of 0.4 to 0.7 μm.

Infrared light has frequencies below the visible portion of the spectrum; they lie between light and radio waves with wavelengths from 0.7 to 1.2 μm. Nevertheless, infrared light is generally put in the lightwave category because the distance calculations are made by the same technique.

An *electro-optical* instrument is one that transmits modulated light, either visible or infrared. It consists of a measuring unit and a reflector. The reflector (Figure 5-3) consists of several so-called retrodirective glass cube corner prisms mounted on a tripod. The sides of the prisms are perpendicular to each other within very close tolerances. Due to the perpendicular sides, the prisms will reflect the light rays back in the same direction as the incoming light rays, hence the term *retrodirective*. Figure 5-4 illustrates this situation.

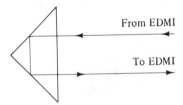

Figure 5-3 Reflection of visible or infrared light by a reflector.

Figure 5-4 Prism retroreflectors. (Courtesy of Topcon Instrument Corporation of America.)

The trihedral prisms mounted on a tripod will reflect light back to the transmitting unit even though the reflector is out of perpendicularity with the light wave by as much as 20°. The number of prisms used depends on the distance to be measured and on visibility conditions.

The distance-measuring capability of an electro-optical instrument can be increased by increasing the number of prisms used. In general, the distance that can be measured is doubled if the number of reflectors is squared. If a cluster of nine prisms is used instead of a cluster of three, the distance measured can probably be doubled. Using more than about 12 or 15 prisms does not help very much for most instruments. If more than 12 or 15 prisms are necessary another EDMI with a greater range capability is probably needed. The distance-measuring capability of a particular EDMI is affected not only by the number of prisms used but also by their quality and cleanliness.

A *laser* is one of several devices that produces a very powerful single-color beam of light. The word *laser* is an acronym for "light amplification by stimulated emission of radiation." Low-intensity light waves are generated by the device and amplified into a very intense beam that spreads only slightly even over long distances. The waves produced fall into the visible or into the infrared frequencies of the electromagnetic spectrum. Personnel must be informed about the necessary precautions to the eyes when working with lasers and they must strictly adhere to these safety requirements.

A *microwave* is an electromagnetic radiation that has a long wavelength and a low frequency and lies in the region between infrared and shortwave radio. The microwaves used in distance measurements have wavelengths from 10 to 100 μm.

Radar with low frequencies and long wavelengths makes use of the special properties of electromagnetic waves. A radar set transmits these waves, which are reflected or bounced back from objects in their path, in the same manner that sounds bounce back from solid objects as echoes. The term *radar* was taken from the first letters of the words "*ra*dio *d*etecting *a*nd *r*anging."

The *tribrach* or base is a detachable part of an EDMI, theodolite, or total station instrument. It contains three leveling screws (which are enclosed and dustproof) and a circular or bull's-eye level. A special locking device is used to hold the instrument head and the tribrach together.

5-5 THE FIRST EDMIs

Radar was developed in the 1930s and was used very successfully in combat during World War II. Distances could be measured almost instantaneously with fairly good accuracy and used to guide aircraft to targets or used by defenders to locate attacking aircraft. As a result of radar's successful use, it is not surprising that geodesists began to study the possibility of applying radar principles to distance measurements in the years immediately following the war.

The first electronic distance-measuring device developed for surveyors was

Figure 5-5 DM 550 electro-optical distance meter. (Courtesy of Kern Instruments, Inc.)

the *Geodimeter* (the name being derived from the words "*geo*detic *di*stance *meter*"). It was conceived by a Swedish geodesist, Erik Bergstrand, around 1941, reduced to a workable device in 1947, and put on the market in 1952. Distance measurement with the original Geodimeters was based on the accurate measurement of the time required for a visible light beam to travel from the instrument to a reflector (which acts as a mirror) and back to the instrument.

About a decade after the Geodimeter was developed, microwave distance-measuring devices were introduced. These included the Tellurometer, the Electrotape, and others. While the Geodimeter made use of light waves, the latter instruments measured the time required for microwaves to travel from the transmitter to a receiver at the other end of the line and back.

The Tellurometer was developed in 1954 by the Telecommunications Research Laboratory of the South African Council for Scientific and Industrial Research. T. L. Wadley conducted the work and his device was introduced in the United States in 1957. A short electromagnetic wave (a refined kind of radar) is sent from one unit, called the master, to another, called the remote instrument, where it is picked up and sent back to the master and the received signal compared with the transmitted signal.

The first EDMI produced in the United States for the surveying profession was the *Electrotape*. It was produced by the Cubic Corporation of San Diego in

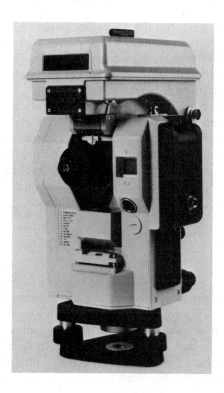

Figure 5-6 The Eldi 10 electronic distance meter is shown connected to the ETh 3 electronic theodolite. (Courtesy of Carl Zeiss, Inc., Thornwood, NY 10594.)

1958 and could be used to measure distances as large as 30 miles. The time required (again by a phase comparison technique) for a microwave beam traveling at 186,000 miles/sec to travel to and from the other end of the line was measured.

The light-wave and microwave devices permitted superior work for medium- and long-range distances up to 30 miles and soon became indispensable for organizations that needed to perform long-distance measurements with great precision: for example, the National Geodetic Survey. But they did not really benefit the average American surveyor because they were very expensive, and they did not provide sufficient precision for measurements under 1500 ft—yet that is where most of the average surveyor's work falls.

During the 1960s a solid-state device for generating infrared radiation permitted the development of relatively low-cost electronic distance-measuring devices that could be used for short-distance measurements of only a few feet and for longer distances up to several miles. The much higher frequency of infrared produces much narrower beams than those of radio microwaves, with the result that there are several advantages. More precise measurement can be made for short distances; there is less atmospheric interference and less trouble caused by reflection by objects along the path (as was the case with the wider beams), and reflection is permitted by a passive mirrorlike reflector.

The infrared EDMIs operate with almost twice the wavelengths of the instruments that utilize visible light, but the precisions obtained are approximately equal. The infrared diodes, however, have lower power transmission, with the result that ranges are usually limited to maximum distances of several miles. The infrared EDMI units are light and compact and yield precision equal to that of very careful taping even for short distances.

Nearly all of the EDMIs manufactured in recent years have used as their carrier source either infrared or red laser light. Red laser and infrared carriers have a great advantage over the tungsten light used for earlier units. A filter can be used at the instrument which will permit only transmitted light to be passed. This means that the instruments can be used in daytime and direct sunlight and other reflected stray light (representing noise) can be prevented from entering the instruments.

5-6 TIME-PULSE INSTRUMENTS

In the last few years EDMIs have been developed that use a time-pulse signal for determining distances. The time required for a signal to travel to and from an object is required. These amazing instruments can be used with or without prisms. If prisms are used, they can be used to measure distances up to several miles. If prisms are not used, they can measure distances up to 500 or 1000 ft, depending on light conditions. Notice that all objects are reflective, and if something moves onto the beam (such as a car or a tree limb) the distance to that object will be determined and not the intended one.

When prisms are not used, these EDMIs can be used to obtain distances to topographic features that have vertical components, such as buildings, bridges, stockpiles of materials, and so on. Best results are obtained if the item being observed has a smooth light-colored surface perpendicular to the beam. For such cases distances can be measured with a standard deviation of ± 5 to 10 mm for distances up to 500 or 600 ft. For medium-dark to dark surfaces and to corners, edges, and inclined surfaces, the maximum distances are not as high and the accuracies are poorer.

Just think of all the measurements that can be taken with one of these instruments without having the rodmen climb all over tanks, buildings, stockpiles of materials, and other items to hold the reflectors. It can also be used to locate shorelines for hydrographic surveys, the inside walls of tunnels, and so on. Furthermore, if the equipment is interfaced with an electronic theodolite and data collector, areas and volumes can be computed, and cross-sections and profiles plotted.[2]

[2] B. F. Kavanagh and S. J. G. Bird, *Surveying Principles and Applications*, 2nd ed. (Englewood Cliffs, N.J.: Prentice-Hall, Inc., 1989), p. 243.

5-7 STEPS NECESSARY TO MEASURE WITH EDMIs

EDMIs are so completely automated that their use can be learned very quickly. To measure a distance with an EDMI, it is necessary to set up the instrument and the reflectors, to sight on the reflectors, and finally to measure and record the value obtained. These steps are briefly described in this section.

1. *Setup*. The EDMI is set up at one end of the line to be measured, centered over the point with the optical plummet (a device described in section 5-8), and leveled. The prisms are set over the other end point on a prism pole or tripod. The telescope is sighted toward the prisms and the power turned on.

2. *Sighting*. The tangent screws are used to point the instrument toward the reflector until the maximum strength returning signal is indicated on a signal scale.

3. *Measuring*. The measurement is accomplished simply by pushing a button. The user may measure in feet or in meters, as desired. The display will be to two decimal points if the measurement is in feet, or to three decimal points if in meters. If the measurements are recorded in a field book, it is not a bad idea to take an extra measurement using different units (feet or meters). Such a habit may enable one to pick up mistakes in recording, such as transposing numbers.

4. *Recording*. The values obtained are recorded in the field book or they may be recorded in an electronic data collector. Total station instruments (discussed in Chapter 15) automatically record the measurements taken.

5-8 ERRORS IN EDMI MEASUREMENTS

Although EDMIs are splendidly manufactured and their use can lead to very accurate surveys, it is still necessary to take precautionary measures to minimize errors. The source of errors in EDMI work is the same as for other surveying work: personal, natural, and instrumental.

Personal Errors

Personal errors are caused by such items as not setting the instruments or reflectors exactly over the points and not measuring instrument heights and weather conditions perfectly.

For precise distance measurement with an EDMI device, it is necessary to center the instrument and reflector accurately over the end points of the line. This topic of centering is discussed in detail in Chapter 10. Plumb bobs hanging on strings from the centers of instruments have long been used by surveyors for centering over points. They are still in constant use and have been employed by surveyors for countless centuries. For very precise work, however, the *optical plummet* is much to be preferred over the plumb bob. This is a special telescope device with which the surveyor can sight vertically from the center of the instrument to the point below. With the optical plummet, errors in centering can be greatly reduced (usually, to a fraction of a millimeter). Its advantage over the plumb bob is multiplied when there is appreciable wind. The axis of the optical plummet must be checked periodically under lab conditions.

Natural Errors

The natural errors present in EDMI measurement are caused by variations in temperature, humidity, and pressure. Some EDMIs automatically correct for atmospheric variables, for others it is necessary to "dial in" corrections into the instruments, and for others it is necessary to make mathematical corrections. For instruments requiring adjustments, manufacturers provide tables, charts, and explanations in their direction manuals regarding how the corrections are to be made. For the microwave instruments it is necessary to correct for temperature, humidity, and pressure, whereas for the electro-optical instruments humidity can be neglected. (The humidity effect on microwaves is more than 100 times its effect on light waves.) Meteorological data should be obtained at each end of a line and sometimes at intermediate points and averaged if a higher precision is desired. On hot sunny days it is desirable to protect both the EDMI equipment and any meteorological equipment with umbrellas.

A new, clean optical reflector probably does not scatter the light by more than about 2 arc-seconds. After a few days or weeks of use, however, the presence of dirt scatters the light, with the result that the divergence may be as large as 8 or more arc-seconds.

Cleaning of the prisms is very involved and seldom can be done adequately by the average user. Soap and water are not satisfactory and the units should not be scrubbed, since they may be scratched. A very lengthy cleaning process can be used involving first the blowing off of the dust, followed by repeated applications of a light solvent such as acetone with clean, unmedicated cotton balls. To clean a retro-reflector adequately probably takes 45 minutes to an hour and a few hundred cotton balls. As a result the average surveyor very seldom if ever cleans the reflectors.

Snow, fog, rain, and dust affect the visibility factor for EDMIs and drastically reduce the distances that can be measured. The distance that can be measured with a particular instrument is sometimes affected by the shimmering phenomenon when sights are taken near the earth's surface. Obviously, keeping sights as high as possible above the ground will reduce this factor. It is also a good practice to keep electro-optical instruments pointed away from the sun, due to background radiation effects.

Whenever possible, surveyors using microwave equipment should try to avoid high-voltage power lines, microwave towers, and so on. Should they have to work in the vicinity of any of these structures, such information should be noted carefully in the field records since it may later be helpful should specifications not be met.

Instrumental Errors

Instrumental errors are usually quite small if the equipment has been carefully adjusted and calibrated. Each EDMI has a built-in error, which varies from model to model. In general, the more expensive the equipment, the smaller the error.

When a measurement is made with an EDMI, the beam goes from the electrical center of the instrument to the effective center of the reflector and then back to the electrical center of the instrument. However, the electrical center of an EDMI does not coincide exactly with a plumb line through the center of the tribrach, nor does the effective center of the reflector coincide exactly with the point over which it is placed. The difference between the electrical center of the EDMI and the plumb line is determined by the manufacturer and adjustments are made at manufacture to compensate for these errors.

The exact location of the effective center of the reflector is not very easy to obtain because of the fact that light travels through the glass prisms more slowly than it does through air. The effective center is actually located behind the prisms

Figure 5-7 Geodimeter 220 electro-optical distance meter. (Courtesty of Geotronics of North America, Inc.)

and these distances need to be subtracted from measured values. This error, which is a constant, may be compensated for at manufacture or it may be done in the field. It is to be realized that on some occasions reflectors made by different manufacturers (and thus having different constants) may be used with the EDMI. If an EDMI is supposed to be used with a prism with a 30-mm constant but is being used with a reflector with a 40-mm constant, it may be necessary to dial the 40 mm into the instrument.

5-9 CALIBRATION OF EDMI EQUIPMENT

It is important to check EDMI measurements periodically against the length of a National Geodetic Survey (NGS) base line or other accurate standard. From the differences in the values an *instrument constant* can be determined. This constant, which is a systematic error, enables the surveyor to make corrections to future measurements. Although a constant is furnished with the equipment, it is subject to change. It is as though we have an incorrect-length tape and have to make a numerical correction. With the value so determined, a correction is applied to each subsequent measurement. In addition, a record of the results and dates when the checks were made should be kept in the surveyor's files in case of future legal disputes involving equipment accuracy.

Unfortunately, there seems to be a rather large percentage of practicing surveyors who think that EDMIs can continually be used accurately, without the necessity of calibration. However, just as other measuring devices, these instruments must be standardized periodically. Both electro-optical and microwave equipment should be checked against an accurate base line at frequent intervals. The electronically obtained distances should be determined while taking into consideration differences in elevations, meteorological data, and so on.

Most modern theodolites and automatic levels remain in good calibration for quite a few years when subjected to normal usage. Sadly, this is not the case for EDMIs (aging of the electronic components is one reason) and they must be calibrated at least every few months, even if they are very carefully used. Recognizing the need for frequent calibration of these instruments, in 1974 the NGS began setting up calibration base lines around the United States.[3]

The NGS compiles and publishes a calibration of base lines for each state showing location, elevations, horizontal distances, and other pertinent data. Copies may be obtained by writing to NGS, National Ocean Survey, Rockville, MD 20852. Copies are also on hand at each state's surveying society offices. A detailed description of one of these base lines and its location are presented in Fig. 5-8.

If a base line is not available, two points can be set up (as much as 5 miles apart for microwave equipment) and the distances between them measured. A

[3] M. O. Schmidt and K.W. Wong, *Fundamentals of Surveying*, 3rd ed. (Boston: PWS Engineering, 1985), pp. 243–244.

US DEPARTMENT OF COMMERCE - NOAA
NOS - NATIONAL GEODETIC SURVEY
ROCKVILLE MD 20852 - MAY 27, 1981

CALIBRATION BASE LINE DATA
BASE LINE DESIGNATION: CLEMSON
PROJECT ACCESSION NUMBER: G16441

QUAD: N340824
SOUTH CAROLINA
PICKENS COUNTY

LIST OF ADJUSTED DISTANCES (APRIL 22, 1981)

FROM STATION N	ELEV.(M)	TO STATION N	ELEV.(M)	ADJ. DIST.(M) HORIZONTAL	ADJ. DIST.(M) MARK - MARK	STD. ERROR(MM)
0	228.600	150	229.071	149.9999	150.0006	.2
0	228.600	430	231.412	429.9949	430.0041	.4
0	228.600	1070	241.844	1069.9287	1070.0106	.6
150	229.071	430	231.412	279.9950	280.0048	.4
150	229.071	1070	241.844	919.9287	920.0174	.5
430	231.412	1070	241.844	639.9336	640.0186	.4

DESCRIPTION OF CLEMSON BASE LINE
YEAR MEASURED: 1981
CHIEF OF PARTY: WJR

THE BASE LINE IS LOCATED ABOUT 4.3 KM (2.7 MI) SOUTHEAST OF CLEMSON AND 2.9 KM (1.8 MI) WEST OF PENDLETON ALONG THE
RIGHT-OF-WAY ON THE WEST SIDE OF UNITED STATES HIGHWAY 76 WHERE IT CROSSES THE ANDERSON-PICKENS COUNTY LINE.
OWNERSHIP--MR. GEORGE WEATHERS, SOUTH CAROLINA HIGHWAY DEPARTMENT, PRE-CONSTRUCTION ENGINEER, POST OFFICE BOX 191,
COLUMBIA, SOUTH CAROLINA 29202, TELEPHONE 803-758-3414.

TO REACH THE BASE LINE FROM THE SOUTH CAROLINA STATE HIGHWAY 93 OVERPASS ON UNITED STATES HIGHWAY 76 EAST OF CLEMSON, GO
SOUTH ON HIGHWAY 76 FOR 2.6 KM (1.65 MI) TO NEW HOPE ROAD ON THE RIGHT AND THE 1070 METER POINT IN THE SOUTHWEST ANGLE OF
THE INTERSECTION. TO REACH THE OTHER MARKS AND 0 METER POINT, CONTINUE SOUTH ON HIGHWAY 76 FOR 0.64 KM (0.4 MI) TO THE
430 METER POINT ON THE RIGHT, CONTINUE SOUTH FOR 0.32 KM (0.2 MI) TO A SIDE ROAD RIGHT AND THE 150 METER POINT IN THE
SOUTHWEST ANGLE OF THE INTERSECTION, AND CONTINUE 0.16 KM (0.1 MI) SOUTH TO THE 0 METER POINT ON THE RIGHT ABOUT 0.9 M
(3.0 FT) LOWER THAN THE HIGHWAY AND 21.5 M (70.5 FT) SOUTH OF THE ANDERSON COUNTY LINE.

THE 0 METER POINT IS A STANDARD NATIONAL GEODETIC SURVEY DISK STAMPED 0 1980, SET INTO THE TOP OF A ROUND CONCRETE
MONUMENT 38 CM (15 IN) IN DIAMETER FLUSH WITH THE GROUND, LOCATED 40.9 M SOUTHEAST OF TELEPHONE JUNCTION BOX NUMBER 6,
21.5 M SOUTH OF ANDERSON COUNTY LINE SIGN, 21.5 M EAST OF WEST EDGE OF THE WOODS, 3.65 M WEST OF WEST EDGE OF HIGHWAY 76,
AND 1.15 M SOUTHEAST OF A METAL WITNESS POST.

THE BASE LINE IS A NORTH-SOUTH LINE WITH THE 0 METER POINT ON THE SOUTH END. IT IS MADE UP OF THE 0, 150, 430, AND 1070
METER POINTS WITH A POINT FOR THE CALIBRATION OF 100 FOOT TAPES SET SOUTH OF THE 0 METER POINT. ALL OF THE MARKS ARE SET
ON A LINE PARALLEL TO THE HIGHWAY AND IN THE DITCH ON THE WEST SIDE OF THE ROAD. THIS BASE LINE IS NOT CONNECTED TO THE
NATIONAL NOR THE LOCAL CONTROL NETWORKS.

THIS BASE LINE WAS ESTABLISHED IN CONJUNCTION WITH THE STATE OF SOUTH CAROLINA. FOR FURTHER INFORMATION, CONTACT THE
DIRECTOR, SOUTH CAROLINA GEODETIC SURVEY, SOUTH CAROLINA DIVISION OF RESEARCH AND STATISTICAL SERVICES, OFFICE OF
GEOGRAPHIC STATISTICS, 915 MAIN STREET, SUITE 203, COLUMBIA, SOUTH CAROLINA 29201. TELEPHONE 803-758-3604.

Figure 5-8

point can be set in between the other two and the two segmental distances measured (see Fig. 5-9). The sum of those two values should be compared with the overall length. Should the three points not be in a line, angles will be needed to compute the components of the two segments to compare with the overall straight-line distance between the end points. The instrument constant can be calculated as follows—noting that the constant will be present in each of the three measurements:

$$\text{instrument constant} = AC - AB - BC$$

It is also desirable to check barometers, thermometers, and psychrometers approximately once a month or more often if they are subject to heavy use. These checks can usually be made with equipment that is available at most airports.

A   B   C   **Figure 5-9**   Calibrating EDM equipment.

5-10 ACCURACIES OF EDMIs

The manufacturers of EDMIs usually list their accuracies as a standard deviation. (Remember: It is anticipated that 68.3% of measurements of a quantity will have an error equal to or less than the standard deviation.) The manufacturers give values that consist of a fixed or constant *instrumental error* which is independent of distance, plus a *measuring error* in parts per million (ppm) that varies with the distance being measured.

EDMIs are listed as having accuracies in the range from about ±3mm instrumental error ± a proportional part error of 1 ppm up to ±(10 mm + 10 ppm). The first of these errors is of little significance for long distances but may be very significant for short distances of 100 or 200 ft or less. On the other hand, the proportional part error is of little significance for short or long distances. It can be seen that for short distances EDMI equipment may on occasion not provide measurements as precise as those obtained by taping.

5-11 COMPUTATION OF HORIZONTAL DISTANCES FROM SLOPE DISTANCES

All EDMI equipment is used to measure slope distances. For most models the values obtained are corrected for the appropriate meteorological and instrumental corrections and then reduced to horizontal components. It is possible at the same time to determine the vertical components (or differences in elevations) for the slope distance. If the distance involved is rather short and/or the precision required is not extremely high, the horizontal distance equals the slope distance times the cosine of the vertical angle α:

$$h = s \cos \alpha$$

For greater distances and higher precision requirements, earth's curvature and atmospheric refraction will need to be considered. With many of the newer instruments, however, the computations are made automatically. As with taping along slopes, the horizontal values may be computed by making corrections with the slope correction formula (described in Section 4-5), by using the Pythagorean theorem, or by applying trigonometry. If the slope is quite steep, say greater than 10 to 15%, the slope correction formula (which is only approximate) should not be used.

To compute horizontal distances it is necessary either to determine the elevations at the ends of the line or to measure vertical angles at one or both ends. Example 5-1 illustrates the simple calculations involved when the elevations are known.

Example 5-1

A slope distance of 1654.32 ft was measured between two points with an EDMI. It is assumed that the atmospheric and instrumental corrections have been made. If the difference in elevation between the two points is 183.36 ft and if the heights of the EDMI and its reflector above the ground are equal, determine the horizontal distance between the two points using:

(a) The slope correction expression.

(b) The Pythagorean theorem.

Solution

(a) Using the slope correction expression:

$$C = \frac{v^2}{2s} = \frac{(183.86)^2}{(2)(1654.32)} = 10.22 \, \text{ft}$$

$$h = 1654.32 - 10.22 = \textbf{1644.10 ft}$$

(b) Using the Pythagorean equation (right triangles):

$$h = \sqrt{(1654.32)^2 - (183.86)^2} = \textbf{1644.07 ft}$$

For Example 5-1 it was assumed that the distance measured was parallel to the ground: that is, the heights of the EDMI and the reflector above the end points were equal. (Adjustable-length prism poles are available that enable the surveyor to adjust the height of the prism so that it equals the height of the instrument above the ground.) If these values are not equal, that fact must be accounted for in the calculations. Such a situation is shown in Fig. 5-10, where h_E and h_R represent the heights of the instrument and the reflector, respectively, d is the difference in elevation between the instrument and reflector, s is the slope distance determined, and h is the horizontal component desired. From this figure the value of d can be determined as follows:

$$d = \text{elevation } A + h_E - (\text{elevation } B + h_R)$$

Once d is determined, the horizontal distance can be calculated as described in Example 5-1. If the Pythagorean theorem is used, the results can be expressed in formula fashion as follows:

$$h = \sqrt{s^2 - [(\text{elev. } A + h_E) - (\text{elev. } B + h_R)]^2}$$

Such a problem is illustrated in Example 5-2.

Example 5-2

A slope distance of 1836.42 ft was measured between two points with an EDMI. It is assumed that the necessary atmospherical and instrumental corrections have been made. The EDMI is centered over one point with a ground elevation of 632.52 ft, while the reflector is located over the end point, which has an elevation of 506.46 ft. If the heights of the EDMI and the reflector above these points are 4.80 ft and 5.20 ft, respectively, determine the horizontal distance between the two points.

Solution

$$h = \sqrt{(1836.42)^2 - [(632.52 + 4.80) - (506.46 + 5.20)]^2} = \mathbf{1832.12 \ ft}$$

If, instead of measuring elevations and instrument heights, vertical angles are measured, the horizontal component may be calculated directly with trigonometry. For instance, if the vertical angle α' shown in Fig. 5-11 between the horizontal and a line from the EDMI to the center of the reflector is measured, the horizontal distance will equal $s \cos \alpha'$. If, however, a theodolite is used to measure the true vertical angle between the two points A and B, represented by α in Fig. 5-11, it will be necessary to calculate the value of $\Delta\alpha$ and use it to correct the angle α. In the figure, $\alpha' = \alpha - \Delta\alpha$. The value of $\Delta\alpha$ in seconds can be

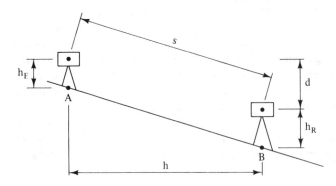

Figure 5-10

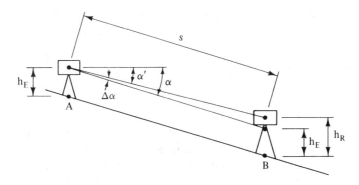

Figure 5-11

calculated with the following expression:

$$\Delta\alpha = \frac{(h_R - h_E)\cos\alpha}{s\,\sin 1''}$$

Some EDMIs have the capability of vertical angle measurement, but with others it is necessary to set up a transit or theodolite to measure the angles. As a result, the heights of the EDMI, the reflector, the theodolite, and even the target may sometimes be different.

5-12 TRAINING OF PERSONNEL

Measurements obtained with EDMIs may be extremely accurate or very inaccurate. The difference can usually be attributed to the amount of training (or lack of it) provided to personnel. The procedures used for field measurements vary somewhat with equipment from different manufacturers. As a result, such procedures are not described here. Each manufacturer provides an equipment manual in which operating instructions are given. The instructions are quite simple and little training is needed before the units can be used.

Despite the simplicity of EDMI operation, however, it is advisable to have personnel trained as thoroughly as possible to get optimum results. No matter how fine and expensive the instruments used for a survey, they are of little value if they are not used knowledgeably. If a company spends $5000 or $10,000 or more for EDMI equipment but will not spend an extra few hundred dollars for detailed equipment training, the result will probably be poor economy.

A rather common practice among instrument companies is for one of their representatives to provide a few hours or even a day of training when an instrument is delivered. Such a short period is probably not sufficient to obtain the best results and to protect the large investment in equipment. Some manufacturers offer a week-long training course. Participation in such a course is normally a wise investment and will yield long-range dividends. In addition, surveyors should consider attending workshops or short courses on EDMI theory and practice given by the National Geodetic Survey and by manufacturers and universities.

5-13 CONSTRUCTION LAYOUT

In construction work, a steel tape is normally used for laying out distances, since it is so convenient to handle. This is particularly true for the average construction job where distances are relatively small. When rather long lines are involved, EDMI devices can be used advantageously. Today, EDMIs are available that can be operated in a high-speed tracking mode with which the prisms can be tracked when they are moved at speeds up to and above 10 ft/sec. As the prisms are moved back and forth, a constantly adjusted distance is obtained. You can see that such equipment is particularly useful for layout and construction work.

It is to be remembered that horizontal distances are needed and slope corrections may have to be made. Field layout work can be expedited appreciably if an EDMI is used which has an angle-measuring capability. Many such devices can measure vertical angles and compute and display horizontal distances as well as vertical distances or elevation differences.

5-14 EDMI EQUIPMENT AVAILABLE

Because EDMI equipment is changing rapidly year by year, it is not feasible to list and describe all the different instruments available. In 1990, more than 30

TABLE 5-1 SOME EDMIs AVAILABLE

Company	Instrument no.	Type
Cubic Precision	Auto Ranger II-X, DM-80, DM-81	Infrared
	Rangemaster III, Ranger V-A	Visible laser
Geodimeter, Inc.	114, 210, 216, 220, 6000	Infrared
Geo-Fennell	Pulsar 50	Infrared
IR Inc.	Stinger, Stinger 2500	Infrared
Kern Instruments Inc.	DM104, DM150, DM504, DM550	Infrared
Lietz Co.	RED Mini 2, RED 2A, RED 2L	Infrared
Nikon Inc.	ND-20, ND-21, ND-26	Infrared
Pentax Instruments	MD-14, MD-20	Infrared
Topcon Instrument Corp. of America	DM-S2, DM-S3, DM-A2, DM-A3	Infrared
Wild Herrbrugg Instruments, Inc.	DI 5S Distomat, DI 1000, DI 1000L, DI 2000	Infrared
	DI 3000	Infrared, time-pulse

Source: P. W. McDonnell, "P.O.B. 1987 EDMI Survey," (Wayne, Mich.; *P.O.B. Magazine,* Oct.–Nov. 1987, vol. 13, no. 1, pp. 22–32.

commonly used models were on the market. To learn more about this array of equipment, the reader may want to obtain brochures from the various manufacturers or contact their representatives. For short-range instruments with which maximum distances up to about 2 miles can be measured, prices range from about $3000 to $10,000. For instruments with greater ranges—up to 15 or 20 miles—prices generally will go above $20,000, and for those with ranges up to 40 or 50 miles or more, prices may exceed $30,000.

Although limited to shorter distances than microwave devices, light-wave instruments are instruments in common use today. Table 5-1 presents a list of many of the EDMIs now sold in the United States. It should be noted that all the instruments listed use infrared, with the exception of two of Cubic Precision's instruments, which use visible lasers.

5-15 SUMMARY

EDMIs enable us to measure short or long distances quickly and accurately over all types of terrain. Survey or traverse points can quickly be selected without having to pick those to which we can conveniently tape. If the instruments do not convert slope distances to horizontal components, we have to make the conversions. Furthermore, it becomes necessary to consider earth's curvature and atmospheric refraction in determining horizontal components if elevation differences at the ends of a line are more than a few hundred feet and/or if the required accuracy is $>1/50,000$. We particularly need to remember that all instruments get out of adjustment and thus need to be checked against a standard at frequent intervals.

CHAPTER SIX

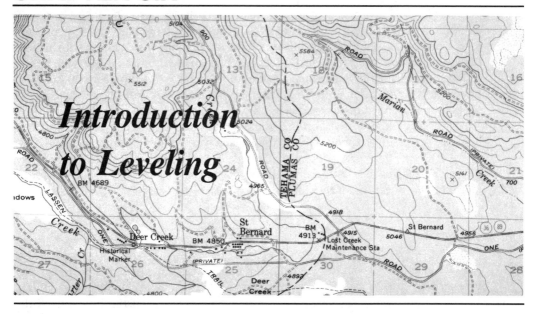

Introduction to Leveling

6-1 IMPORTANCE OF LEVELING

The determination of elevations with a surveying instrument, which is known as *leveling*, is a comparatively simple but extremely important process. The significance of relative elevations cannot be exaggerated. They are so important that one cannot imagine a construction project in which they are not critical. From terracing on a farm or the building of a simple wall to the construction of drainage projects or the largest buildings and bridges, the control of elevations is of the greatest importance.

6-2 BASIC DEFINITIONS

Below are presented a few introductory definitions that are necessary for the understanding of the material to follow. In this chapter and the next, additional definitions are presented as needed for the discussion of leveling. Several of these terms are illustrated in Fig. 6-1.

A *vertical line* is a line parallel to the direction of gravity. At a particular point it is the direction assumed by a plumb-bob string if the plumb bob is allowed to swing freely. Because of the earth's curvature, plumb-bob lines at points some distance apart are not parallel, but in plane surveying they are assumed to be.

A *level surface* is a surface of constant elevation that is perpendicular to a plumb line at every point. It is best represented by the shape that a large body of still water would take if it were unaffected by tides.

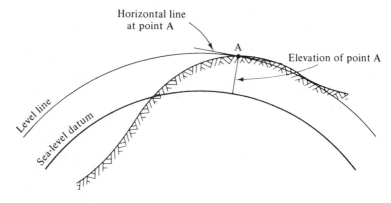

Figure 6-1

The *elevation* of a particular point is the vertical distance above or below a reference level surface (normally, sea level).

A *level line* is a curved line in a level surface all points of which are of equal elevation. Every element of the line is perpendicular to gravity.

A *horizontal line* is a straight line tangent to a level line at one point.

6-3 REFERENCE ELEVATIONS OR DATUMS

For a large percentage of surveys, it is reasonable to use some convenient point as a reference or datum with respect to which elevations of other points can be determined. For instance, the surface of a body of water in the vicinity can be assigned a convenient elevation. Any value can be assigned to the datum, for example, 100 ft or 1000 ft, but the assigned value is usually sufficiently large so that no nearby points will have negative elevations.

In the past many assumed reference elevations were used in the United States, even for surveys of major importance. An assumed value may have been given to the top of a hill, or the surface of one of the Great Lakes, or the low-water mark of a river, or any other convenient point. These different datums created confusion and the current availability of the *sea-level datum* is a great improvement.

In the United States the sea-level datum is the value of mean sea level determined by averaging the hourly elevations of the sea over a long period of time, usually 19 years. This datum, which is mean sea level, is defined as the position the ocean would take if all tides and currents were eliminated. Its position is rising very slowly—probably because of the gradual melting of the polar ice caps and perhaps because of the erosion of ground surfaces. The change, however, is so slow that it does not prevent the surveying profession from using sea level as a datum. Over the last century the level of the world's oceans has risen by approximately 6 in. (15 cm).

In 1878, at Sandy Hook, New Jersey, the National Geodetic Survey (then the U.S. Coast and Geodetic Survey) began working on a transcontinental system of precise levels. In 1929, the agency made an adjustment of all first-order leveling in the United States and Canada, and established a datum for elevations throughout the country. This datum is referred to as the *National Geodetic Vertical Datum of 1929* (NGVD 29) or just the *sea-level datum of 1929*. A readjustment of NGVD 29 is now being completed and will be referred to as the North American Vertical Datum of 1988 (NAVD 88). The datum is based on the adjustment of several hundred thousand miles of level lines and 500,000 bench marks in the United States and Canada. It will probably not be completed until 1991.

Mean sea level varies from one part of the country to another. For instance, if we determine the elevation of a point from the Atlantic coast or the Gulf coast and then do it again from the Pacific coast, we would find the one from the Pacific to be approximately 2 ft higher. Furthermore, working from different points along the same coast line will cause some slight discrepancies.

The National Geodetic Survey establishes first- and second-order vertical control stations about 1 km apart in interrelated square grids that are 50 to 100 km on a side. Other governmental agencies (federal, state, county, and city) provide vertical control of lower orders. Information concerning first- and second-order control can be obtained by contacting the Director, NGS Information Center, National Ocean Survey, NOAA, Rockville, MD 20852, telephone (301) 443-8611.

One of the most important tasks of the National Geodetic Survey and the U.S. Geological Survey is the establishment of a network of known elevations throughout the country. As a result of their work, eventually there will be no area in the country that will be far from a point whose elevation in relation to sea level is not known.

The National Geodetic and U.S. Geological Surveys have set up throughout the U.S. monuments whose elevations have been precisely determined. A monument, called a *bench mark*, is usually of concrete and has a brass disk embedded in it, as illustrated in Fig. 6-2. For many years the practice was to record the elevations on the monuments. This, however, is no longer the case because the elevations of these monuments may be changed by frost action, earthquakes, vandalism, and so on. The elevations are carefully checked at intervals and the surveyor who needs the elevation of a particular monument should contact the agency that set it, even though a value is given on the monument. More than half a million of these monuments have been placed in the United States and its possessions.

6-4 METHODS OF LEVELING

There are three general methods of leveling: trigonometric, barometric, and spirit. Although the surveyor is concerned almost entirely with spirit leveling, a brief description of each type is given below.

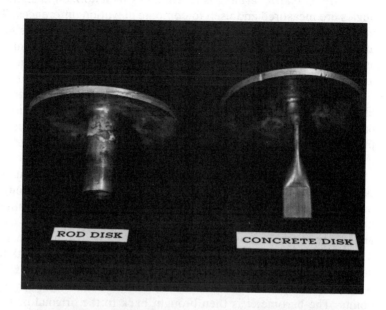

Figure 6-2   National Geodetic Survey brass disk marks used beginning in 1972 are designed to set in concrete or clamped to rods. (Courtesy of National Geodetic Survey.)

Trigonometric leveling is leveling in which horizontal distances and vertical angles are measured and used to compute elevation differences. This method can be used for inaccessible points such as mountain peaks, offshore construction, and so on. For purposes of mapping, rough leveling, or preliminary surveys, the stadia method, which is actually a variation of trigonometric leveling, is very useful.

Barometric leveling involves the determination of elevations by measuring changes in air pressure. Although air pressures can be measured with mercurial barometers, they are cumbersome and fragile and are impractical for surveying purposes. Instead, the light and sturdy but less precise aneroid barometers commonly called *altimeters* are used.

Surveying altimeters have been manufacturered with which elevations can be determined within about 2 ft. Such precision is sufficient only for preliminary or reconnaissance work. They do nevertheless offer the advantage that approximate elevations over a large area can quickly be determined. If careful procedures and larger aneroid barometers are used, much better results can be obtained.

Barometer readings at the same elevation vary with local air pressure conditions and are affected by temperature and humidity variations. If one barometer is used, it is adjusted at a known elevation and then readings are taken at other points. The barometer is then brought back to the original or starting point and read again. If its reading is different, it is necessary to distribute the difference around to the other points. A very similar procedure is illustrated in Section 8-2.

It is desirable to use more than one barometer. At least three are desirable for reasonably accurate elevation measurements. Ideally, one barometer is placed at a known elevation higher than that of the desired point, and one is placed at a lower point of known elevation. All of the barometers are read, and from the readings at the known elevations corrections can be made at the point at which the elevation is being determined.

Spirit leveling, also called *direct leveling*, is the usual method of leveling. Vertical distances are measured in relation to a horizontal line, and these values are used to compute the differences in elevations between various points. A spirit level (discussed in Section 6-6) is used to fix the line of sight of the telescope. This line of sight is the assumed horizontal line with respect to which vertical distances are measured. The term *spirit leveling* is often used because the bubble tubes of many of the older levels were filled with alcohol.

6-5 LEVELS

A level consists of a high-powered telescope (20 to 45 diameters) with a spirit level attached to it in such a manner that when its bubble is centered, the line of sight is horizontal. The purposes of the telescope are to fix the direction of the line of sight and to magnify the apparent sizes of objects observed. The invention of the telescope is generally credited to the Dutch optician Hans Lippershey about 1607. In colonial times telescopes were too large for practical surveying and they were not used on surveying instruments until the end of the nineteenth century. Their use tremendously increases the speed and precision with which measurements can be taken. These telescopes have a vertical cross hair for sighting on points and a horizontal cross hair with which readings are made on level rods. In addition, they may have stadia hairs.

The telescope has three main parts: the objective lens, the eyepiece, and the reticle. The *objective lens* is the large lens located at the front or forward end of the telescope. The *eyepiece* is the small lens located at the viewer's end. It is actually a microscope that magnifies and enables the viewer to see clearly the image formed by the objective lens. The cross hairs and stadia hairs form a network of lines that is fastened to a metal ring called the *reticle, reticule,* or *cross-hair ring*. In older instruments the cross hairs were formed from spider webs or fine wires. In newer instruments, however, the cross hairs are formed by etching lines on a glass reticule. A line drawn from the point of intersection of the cross hairs and the optical center of the objective system is called the *line of sight* or *line of collimation* (the word *collimation* meaning lining up or adjusting the line of sight).

Some old telescopes are said to be *external focusing*. Their objective lens is mounted on a sleeve that moves back and forth as the focusing screw is turned. *Internal focusing* is used for modern telescopes. These instruments have a lens that moves back and forth internally between the objective lens and the reticle.

The level tube is an essential part of most surveying instruments. The closed glass tube is precisely ground on the inside surface to make the upper half curved in shape so that the bubble is stabilized. If the tube were not so curved, the bubble would be very erratic. The tube is filled with a sensitive liquid (usually a purified synthetic alcohol) and a small air bubble. The liquid used is stable and nonfreezing for ordinary temperature variations. The bubble rises to the top of the liquid against the curved surface of the tube. The tangent to the circle at that point is horizontal and perpendicular to gravity. The tube is marked with divisions symmetrical about its midpoint, and thus when the bubble is centered, the tangent to the bubble tube is a horizontal line and is parallel to the axis of the telescope. For some very precise instruments the amount of fluid can be increased or decreased, enabling the surveyor to adjust and control the length of the bubble when temperature changes occur.

6-6 TYPES OF LEVELS

Several types of levels are discussed in this section. These include the dumpy level, the automatic or self-leveling level, and the tilting level. In addition, a few comments are presented concerning construction site lasers used for elevation work.

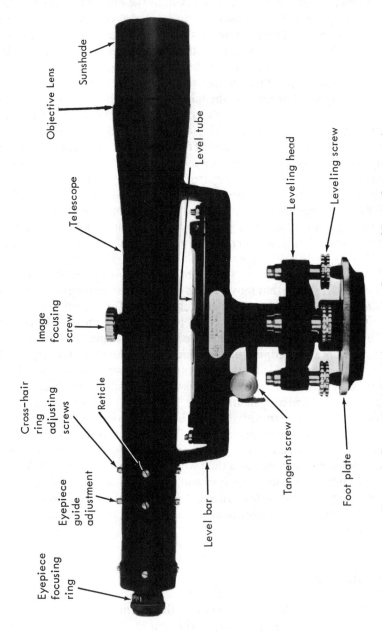

Figure 6-3 Dumpy level (an old instrument). (Courtesy of Berger Instruments.)

Objective Lens

Sunshade

Level tube

Telescope

Leveling head

Leveling screw

Image focusing screw

Cross-hair ring adjusting screws

Reticle

Eyepiece guide adjustment

Tangent screw

Foot plate

Eyepiece focusing ring

Level bar

Dumpy Level

The dumpy level was commonly used in surveying work until the last few decades. Although these excellent, sturdy, and long-lasting devices have very largely been replaced with more modern instruments, they are discussed at some length here to help the student understand leveling.

Originally, the dumpy level had an inverting eyepiece and as a result was shorter (thus the name "dumpy") than its predecessors with the same magnification power. A typical dumpy level with its various parts is shown in Fig. 6-3. Its major components are its *telescope*, *level tube*, and *leveling head*. These and other parts are indicated in the figure.

Automatic or Self-Leveling Level

Automatic levels are the standard instruments used by today's surveyor. This type of level (an example of which is shown in Fig. 6-4) is very easy to set up and to use and is available with almost any desired range of precision. They are usually satisfactory for second-order leveling and may be satisfactory for first-order leveling if an optical micrometer (described in the next section of this chapter) is used. The automatic level has a small circular spirit level called a *bull's-eye* level and three leveling screws. The bubble is approximately centered in the bull's eye and then the instrument itself automatically does the fine leveling. For the best operation of the automatic level it is necessary to center this bubble carefully and keep it in good adjustment, as will be described later.

Figure 6-4 Topcon AT-F4 automatic level, engineer's series. (Courtesy of Topcon Instrument Corporation of America.)

The self-leveling level has a prismatic device called a *compensator* suspended on fine, nonmagnetic wires. When the instrument is approximately centered, the force of gravity on the compensator causes the optical system to swing almost instantaneously into a position such that its line of sight is horizontal.

The automatic level speeds up leveling operations and is particularly useful where the ground is soft and/or when strong winds are blowing because the instrument automatically relevels itself when it is thrown slightly out of level. When the surveyor uses an ordinary level under these adverse conditions, he or she must constantly check the bubble to see that it remains centered.

Tilting Level

A tilting level is one whose telescope can be tilted or rotated about its horizontal axis. The instrument can be leveled quickly and approximately by means of a bull's eye or circular-type level. With the telescope pointed at the level rod, the surveyor rotates a tilting knob that moves the telescope through a small vertical angle until the telescope is level. A tilting level is shown in Fig. 6-5.

The tilting level has a special arrangement of prisms that enables the user to center the bubble by means of a *split* or *coincident bubble*. The two halves of the bubble are actually the half ends of a single bubble and they will coincide when the bubble is centered. A split image of the bubble is seen through a small microscope that is located next to the telescope eyepiece. As the leveling screws are adjusted and the bubble moves in its tube, the images of the two ends of the

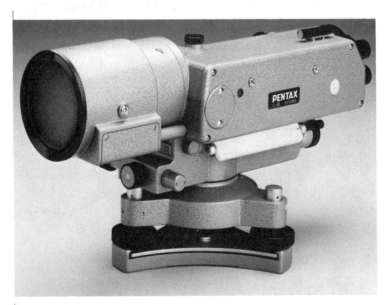

Figure 6-5 Pentax L-10 precision level with 42 magnification telescope, high-sensitivity main level, and tilting screw. (Courtesy of Pentax Corporation.)

bubble move in opposite directions. When the instrument is correctly leveled, the two images will coincide in a continuous U-shaped curve. Manufacturers claim that split bubbles can be centered several times more precisely than the non-coincident type. Tilting levels are useful when a very high degree of precision is required. If they were used for ordinary, everyday surveying operations such as earthwork, the extra time required to bring the bubbles to coincidence probably could not be justified economically.

When the surveyor is using a dumpy level, he or she must check constantly to see if the bubble is centered, as it must be when a reading is taken. Using a tilting level while sighting through the telescope, the surveyor can, at the same time look through a window on one side of the eyepiece and see if the bubble is coincident.

Laser Level

The laser (mentioned in Chapter 5 in relation to EDMI equipment) is effectively used today for several leveling operations. It is commonly used to create a known reference elevation or point from which construction measurements can be taken.

The lasers used for surveying and construction fall into two general classes: single-beam lasers and rotating-beam lasers. A single-beam laser projects a string line that can be seen on a target regardless of lighting conditions. The line may be projected in a vertical, horizontal, or inclined direction. The vertical line provides a very long plumb line, which builders have needed throughout history. Horizontal and inclined lines are very useful for pipelines and tunnels. Section 17-16 describes the use of lasers for pipeline construction.

A rotating-beam laser, which provides a plane of reference over open areas, can be rotated rapidly or slowly or can be stopped and used as a single beam. Today's rotating lasers are self-leveling and self-plumbing, thus providing both horizontal and vertical reference planes. The laser beam will not come on until the instrument is level. If the instrument is bumped out of position, the beam shuts off and will not come back on until it is level again. Thus it is particularly advantageous when severe winds are occurring. The rotating beam can be set to provide a horizontal plane for leveling or a sloped plane, as might be required for the setting of road or parking lot grades. It is precise for distances up to 1000 ft. This means that fewer instrument setups are required, and for construction jobs, it means that the laser can be set up some distance away and out of the way of construction equipment.

The laser is not usually visible to the human eye in bright sunlight, and thus some type of detector is needed. The detector can either be a small hand-held or rod-mounted unit that may be moved up and down the level rod, or it may be an automatic detector. The latter device has an electronic carriage that moves up and down inside the rod and locates the beam.

Lasers can be very useful for staking pipelines and parking lots, for setting

Figure 6-6 CLS Accusweep 731 automatic laser level for general construction. (Courtesy of CLS Industries.)

stakes to control excavation and fills, for topographic surveys, and so on. A detailed discussion of lasers is provided in an article by Tom Liolios.[1]

Transit or Theodolite Used as Level

Although they are primarily used for angle measurement (as described in Chapter 9), they can also be used for leveling. The results are fairly precise but not as good as those obtained with standard levels with their better telescopes and their more sensitive bubble tubes.

Total Stations Used as Levels

As described in Chapter 15 total stations are today commonly used for leveling with splendid results.

[1] Tom Liolios, "Lasers and Construction Surveying," *P.O.B. Magazine*, Aug./Sept. 1981, vol. 6, no. 6, pp. 38–41, and 63.

6-7 LEVEL RODS

There are many kinds of level rods available. Some are in one piece and others (for ease of transporting) are either telescoping or hinged. Level rods are usually made of wood and are graduated from zero at the bottom. Level rods are usually read directly through the telescope by the instrumentman and are commonly called *self-reading rods*. Sometimes a sliding target is placed on the rod as shown in Fig. 6-7. The rodman sets the target at the desired position in accordance with signals from the instrumentman and then makes the reading himself or herself. These rods are commonly called *target rods*, although such a designation is technically incorrect, as the target is only an accessory to the regular self-reading level rod.

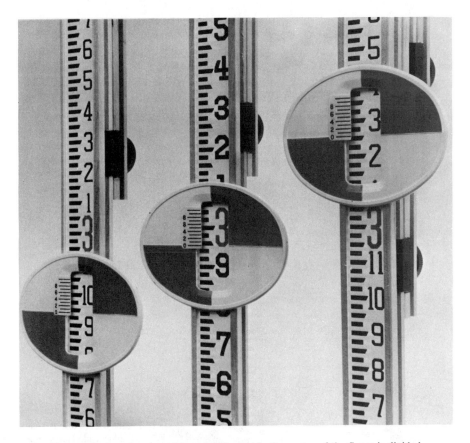

Figure 6-7   Level rods and targets. The rod in the center of the figure is divided into feet, tenths of a foot, and hundredths of a foot, while the ones to the left and right are divided into feet, inches, and eighths of an inch (convenient for construction work). (Courtesy of Berger Instruments.)

Among the several types of level rods available are the *Philadelphia rod*, the *Chicago rod*, and the *Florida rod*. The Philadelphia rod has one major advantage over these other rods, as will be seen in Section 7-7. This is the fact that the rodman can independently check the readings taken by the instrumentman. This rod, the most common one, is made in two sections. It has a rear section that slides on the front section. For readings between 0 and 7 ft, the rear section is not extended; for readings between 7 and 13 ft, it is necessary to extend the rod. When the rod is extended, it is called a *high rod*. The Philadelphia rod is distinctly divided into feet, tenths, and hundredths by means of alternating black and white spaces painted on the rod. Figure 6-7 shows a picture of Philadelphia rods with targets. The scales, or verniers, shown on the targets are discussed in Section 7-5. In Fig. 6-8 are shown a Philadelphia rod and some sample readings

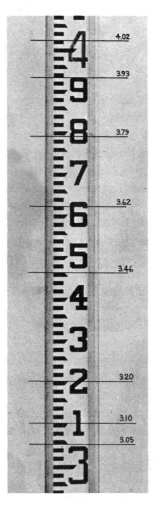

Figure 6-8 Various readings on a Philadelphia rod. Lines represent the horizontal cross hair of the instrument. (Legault/McMaster/Marlette, *Surveying*, © 1956, renewed 1983, p. 124. Reprinted by permission of Prentice Hall, Inc., Englewood Cliffs, New Jersey)

on the rod taken through the telescope at locations where the telescope horizontal cross hair appears to intersect the rod.

The Chicago rod is 12 ft long and is graduated in the same way as the Philadelphia rod, but it consists of three sliding sections. The Florida rod is 10 ft long and is graduated in white and red stripes, each stripe being 0.10 ft wide. Also available for ease of transportation are tapes or ribbons of waterproofed fabric which are marked in the same way that a regular level rod is marked and which can be attached to ordinary wood strips. Once a job is completed, the ribbon can be removed and rolled up. The wood strip can be thrown away. The instrumentman can clearly read these various level rods through the telescope for distances up to 200 or 300 ft, but for greater distances he or she must use a target. A target is a small red and white piece of metal (see Figs. 6-7 and 7-7) attached to the rod. The target has a vernier (see Section 7-5) that enables the rodman to take readings to thousandths of a foot.

When the Philadelphia rod is used as a target rod and the readings are 7 ft or less, the target is moved up and down until the horizontal cross hair bisects the target or, that is, coincides with the line dividing the red and white colors on the target. At this time the target is clamped to the rod and the reading is taken using the vernier on the target.

If the reading is higher than 7 ft, the target is clamped so that its center line is located at the 7-ft mark on the rod and the rear part of the rod is raised (and that moves the target with it) until the level horizontal hair bisects the target. Then the back and front strips of the level rod are clamped together and the reading taken on the rear face of the sliding part of the rod. The graduations on the rear face of the rod are numbered down from 7 to 13 ft. If the rod is extended 2 ft, the reading on the back will be $7 + 2 = 9$ ft. As the back section is pushed upward it runs under an index scale and vernier, which enables the rodman to estimate the reading on the rod's back side to thousandths of a foot.

Readings taken with a self-reading rod are almost as precise as those taken with a target rod, particularly if they are estimated to, say, the nearest 0.002 ft. The target rod, however, is quite useful in some situations: for long sights, in heavy woods in rather dark places, during strong winds, and so on.

It used to be fairly common to use a target and vernier to make rod readings. This time-consuming procedure is today rather obsolete, however, and a reasonably economical device called an *optical micrometer* is available which enables the surveyor to make much more accurate readings. Micrometers can usually be read to about $\frac{1}{100}$th of the least division on the level rod. For the usual level rod with 0.01 ft divisions this means $(\frac{1}{100})(0.01) = 0.0001$ ft.

The device is mounted in front of the telescope. It has a micrometer screw which can be used to turn a parallel-plate prism through a range equal to the least division of the rod. The surveyor sets up the instrument and turns the micrometer screw until the line of sight falls on a full rod division. This division is read and then is added to the micrometer reading (which is the proportional distance the horizontal hair was above the full rod division).

6-8 SETTING UP THE LEVEL

Before setting up the level the instrumentman should give some thought to where he or she must stand to make the sights. In other words, he or she will consider how to place the tripod legs so that he or she can stand comfortably between them for the layout of the work that he or she has in mind, as shown in Fig. 6-9.

The tripod is desirably placed in solid ground, where the instrument will not settle as it most certainly will in muddy or swampy areas. For such locations it may be necessary to provide some special support for the instrument, such as stakes or a platform. The tripod legs should be well spread apart and adjusted so that the footplate under the leveling screws is approximately level. The instrumentman walks around the instrument and pushes each leg firmly into the ground. On hillsides it is usually convenient to place one leg uphill and two downhill, as shown in Fig. 10-3. Such a setup provides better stability.

There are two types of tripods available: the extension-leg types and the fixed-leg types. The fixed-leg tripods are more rigid and provide more stability during measurements. On the other hand, extension-leg tripods are more easily transported in vehicles and they provide more flexibiltiy in setting up.

Four-Screw Instruments

A four-screw instrument is attached to the tripod by means of a threaded base. After the instrument has been leveled as much as possible by adjusting the tripod legs, the telescope is turned over a pair of opposite leveling screws if a four-screw instrument is being used. In centering the bubble, opposite screws are used and the screws are turned in toward the center, ⟩⟨ , or in the opposite directions, ⟨⟩ . In each case, as the screws are turned, the bubble will follow the motion of the left thumb.

The bubble is roughly centered and then the telescope is turned over the other pair of leveling screws and the bubble is again roughly centered. The telescope is turned back over the first pair and the bubble is again roughly centered, and so on. This process is repeated a few more times with increasing care until the bubble is centered with the telescope turned over either pair of screws. If the level is properly adjusted, the bubble should remain centered when the telescope is turned in any direction. It is to be expected that there will be a slight malad-

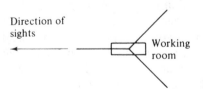

Figure 6-9 Planning an instrument setup.

justment of the instrument that will result in a slight movement of the bubble; however, the precision of the work should not be adversely affected if the bubble is centered each time a rod reading is taken.

If, when leveling a four-screw instrument, one screw is turned more rapidly than the other, the screws may bind or the leveling head may become unstable. If this should happen, we should loosen all the screws slightly and then turn one or more of them until proper bearing is achieved and then proceed with the leveling. Little time or effort will have been lost.

Three-Screw Instruments

A three-screw instrument is attached to the tripod with a threaded bolt that sticks up from the top of the tripod into the leveling base of the instrument. The first step in leveling a three-screw instrument is to turn the telescope until the bubble tube is parallel to two of the screws (see 1 and 2 in Fig. 6-10). The bubble is centered by turning these two screws in opposite directions. Next, the telescope is turned so that the bubble tube is perpendicular to a line through screws 1 and 2. The bubble is centered by turning screw 3 (see Fig. 6-11). These steps are repeated until the bubble stays centered when the telescope is turned back and forth.

Two-Screw Instruments

Some newer levels have only two leveling screws. They are constructed similar to the three-screw levels, but one of the screws is replaced with a fixed point. This means that when the instruments are set up and leveled at a given point, they will be at a fixed elevation. If the instrument is moved slightly and releveled, it will return to the same elevation. To level the two-screw instrument, the telescope is turned until it is parallel to a line from the fixed point to one of the leveling screws. The instrument is leveled and then the telescope is turned to the other leveling screw; this instrument is again leveled and the process is repeated a few more times as needed.

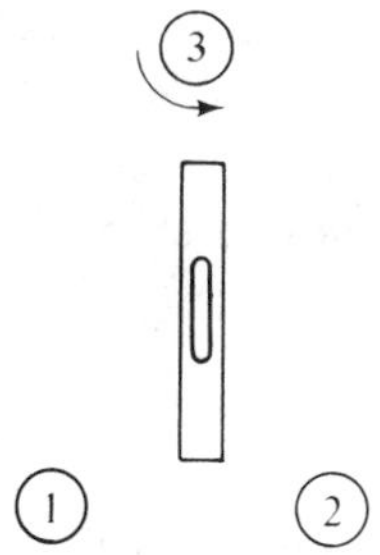

Figure 6-10   Leveling a three-screw instrument.

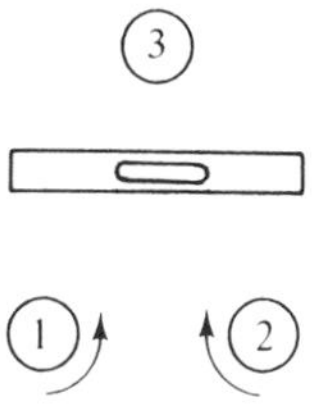

Figure 6-11   Leveling a three-screw instrument.

6-9 SENSITIVITY OF BUBBLE TUBES

The divisions on bubble tubes were at one time commonly spaced at $\frac{1}{10}$-in. intervals, but today they are usually spaced at 2-mm intervals. The student often wants to know how much the rod readings will be affected if the bubble is off center by one, two, or more divisions on the bubble tube when the readings are made. One way for him or her to find out is to place the level rod a certain distance from the level (say 100 ft) and take readings with the bubble centered; then with it one division off center; two divisions off center; and so on.

The *sensitivity* of the bubble tube may be expressed in terms of the radius of curvature of the bubble tube (see Fig. 6-12). Obviously, if the radius is large, a small movement of the telescope vertically will result in a large movement of the bubble. An instrument with a large radius of curvature is said to be sensitive. For very precise leveling very sensitive levels need to be used. The centering of the bubble for such instruments takes more time and therefore less sensitive instruments (which are more quickly centered) may be more practical for surveys of lesser precision.

Another way of expressing the sensitivity of the bubble tube is to give the angle through which the axis of the bubble tube must be rotated (usually given in seconds of arc) to cause the bubble to move by one division on the scale (Fig. 6-12). If the movement of one division on the scale corresponds to 10″ of rotation, the bubble is said to be a 10″ bubble. For very precise levels, 0.25″ or 0.5″ bubbles or even more sensitive ones may be used. For ordinary construction leveling, 10″ or 20″ instruments may be satisfactory.

The average dumpy level has about a 20″ bubble and the radius of curvature is about 68 ft. For first-order leveling the levels used may have 2″ bubbles with a 680-ft radius of curvature.

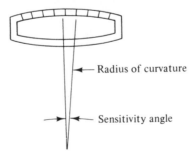

Figure 6-12 Sensitivity of bubble tubes.

6-10 CARE OF EQUIPMENT

Although surveying equipment is very precisely and delicately manufactured, it can be very durable when properly used and maintained. In fact, these instruments may very well last a lifetime in the hands of a careful surveyor, but a few seconds of careless treatment can result in severe or irreparable damage. Because of the long-lasting qualities of surveying equipment, the surveyor will occasionally have to use some rather antique levels and transits. This is not a disadvantage because a good old instrument will still be quite precise. Used transits and levels decline very little in value as the years go by. As a matter of fact, their selling prices in dollars often remain stable for several decades. However, finding parts for older instruments is sometimes a problem.

Several suggestions are made in the following paragraphs concerning the care of the instruments used for leveling.

The Level

Before the level is removed from its box, the tripod should be set up in a firm position. The user should observe exactly how the level is held in the box so that after use he or she can return it to exactly the same position. After the instrument is taken from its box, it should be handled by its base when it is not on the tripod. It should be carefully attached to the tripod. The surveyor must not permit anything to interrupt him or her until this task is completed. Many levels have been severely damaged when careless surveyors began to attach them to their tripods but then allowed their attention to be diverted before tightening was completed.

If possible, levels should not be set up on smooth hard surfaces, such as building floors, unless the tripod points can be either set in indentations in the floor or firmly held in place by other means, perhaps by triangular frames made for that purpose. Particular care should be exercised when the instruments are being carried inside buildings to avoid damage from possible collisions with doors, walls, or columns. In such locations, therefore, the level should be carried in the arms instead of on the shoulder. When working outside where it is normal to carry the instrument on the shoulder, the clamps should be left loose to allow the instrument to turn if it hits limbs or bushes. If the level is to be moved for long distances or over very rough terrain, it should be placed in its carrying case.

A level should never be left unattended unless it is in a very protected location. If it is turned over by wind, cattle, children, or cars, the results will probably be disastrous. Some surveyors paint their tripod legs with bright colors. Such a scheme is particularly wise when work is being done near heavy traffic. Another reason for not leaving instruments unprotected is thieves. Surveyors' levels, transits, EDMIs, and total stations can be sold quickly and easily for good prices. In addition, the surveyor should protect instruments as much as possible from moisture. Waterproof hoods are desirable in case of sudden rain. If the level does get wet, all but the lens should be gently wiped dry. Moisture on the lens is usually permitted to evaporate because the lens can so easily be scratched while being dried. The lens should never be touched with anything other than a camel's-hair brush, or less desirably, a soft silk handkerchief. If the objective lens or the eyepiece lens becomes so dusty that it interferes with vision, it may be cleaned with a camel's-hair brush or with lens paper.

Leveling Screws

Perhaps the most common injury to levels by the novice surveyor is caused by applying too much pressure to the leveling screws. If the instrument is in proper condition, these screws should turn easily and it should never be necessary to use more than the fingertips for turning them.

To emphasize the "light touch" that should be used with leveling screws, many surveying instructors require that students carry out the following test: A leveling screw is loosened and a piece of paper is slipped underneath. Then the screw is tightened just enough to hold the paper in place against a slight tug of the hand. The instructor then explains that leveling screws should never be turned with any greater force than is required to hold the paper. This demonstration should make the student conscious of the need for care of these screws, which can become bound or even stripped if too much pressure is applied. If the threads are stripped, they usually have to be shipped back to the manufacturer for repair.

If the leveling screws do not turn easily, they may be cleaned with a solvent, such as gasoline, and the interior threads may be *very* lightly oiled with a light watch oil. When the level is taken indoors for storing or outdoors for use, its screws and clamps should be loosened because severe temperature changes may cause severe damage.

Level Rod

The level rod should never be dragged on the ground or through water, grass, or mud, and its metal base should never be allowed to strike rocks, pavement, or other hard objects; such use will gradually wear away the metal base and will thereby cause leveling errors due to the change in length of the rod itself. The Philadelphia rod must not be carried over the shoulder when it is fully extended, or leaned against trees or walls—it is just too flexible. It should be placed flat on the ground with the numbers facing upward.

CHAPTER SEVEN

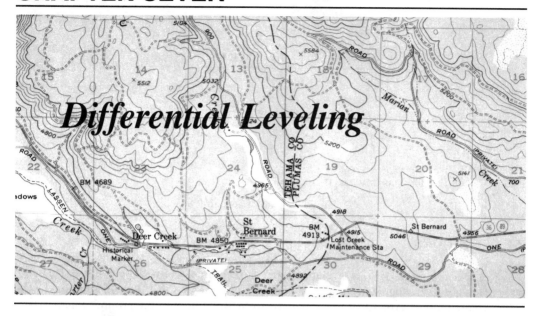

Differential Leveling

7-1 THEORY OF SPIRIT LEVELING

For an introductory description of direct or spirit leveling it is assumed that the surveyor has set up the instrument and has leveled it carefully. He or she then sights on the level rod held by the rodman on some point of known elevation (this sight is called a *backsight*, or BS). If the backsight reading is added to the elevation of the known point, we will have the *height of the instrument* (HI), that is, the elevation of the line of sight of the telescope.

To illustrate this procedure, refer to Fig. 7-1, in which the HI is seen to equal 100.00 + 6.32 = 106.32 ft. If the HI is known, the telescope may be used to determine the elevation of other points in the vicinity by placing the level rod on each point whose elevation is desired and by taking a reading on the rod for each point. Since the elevation of the point where the line of sight of the telescope intersects the rod is known (the HI), the rod reading called a *foresight* (FS) may be subtracted from the HI in order to obtain the elevation of the point in question. In Fig. 7-1 a FS reading of 3.10 ft was taken on the rod. The elevation at the bottom of the rod at this new position is thus 106.32 − 3.10 = 103.22 ft.

The level may be moved to another area by using temporary points called *turning points* (TPs). The telescope is sighted on the rod held at a convenient turning point and a foresight is taken. This establishes the elevation of the point. Then the level can be moved beyond the TP and set up at a convenient location. A *backsight* is taken on the rod held at the turning point and the HI for the new location is established. This process can be repeated over and over for long distances.

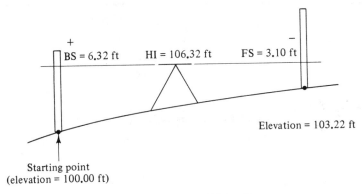

BS = 6.32 ft   HI = 106.32 ft   FS = 3.10 ft

Elevation = 103.22 ft

Starting point
(elevation = 100.00 ft)

Figure 7-1   Direct or spirit leveling.

7-2 DEFINITIONS

A *bench mark* (BM) is a relatively permanent point of known elevation. It should easily be recognized and found and should be set fairly low in relation to the surrounding ground. It may be a concrete monument in the ground, a nail driven into a tree, an × mark in a concrete foundation, a bolt on a fire hydrant, or a similar object that is not likely to move. Bench marks that are to be permanent should be supported by structures that have completely settled and which extend below the frostline. The foundation of an old building usually meets these requirements very well. Parts of structures that must resist significant lateral forces such as retaining walls make rather poor bench marks. Such structures may move for years due to lateral earth pressure. Similar discussions can be made for the supporting foundations of poles and towers. Especially careful records should be made of bench marks because they may frequently be reused during the life of the job or for future work in the vicinity. They should be so completely and carefully described in the notes that another surveyor unfamiliar with the area can find and use them, perhaps years later.

A *turning point* (TP) is a temporary point whose elevation is determined during the process of leveling. It may be any convenient point on the ground, but it is usually wise to use a readily identifiable point such as a rock, a stake driven into the ground, or a mark on the pavement, so that the level rod can be removed and put back in the same location as many times as required. It is essential that solid objects be used for turning points. Never should points be selected on soft or wet ground or grass, which will give or settle under the pressure of the rod. Should satisfactory natural objects not be available, a metal turning point pin, a wood stake, or the head of an ax or hatchet may be used if they are driven firmly into the ground.

A *backsight* (BS) is a sight taken to the level rod held on a point of known elevation (either a BM or a TP) to determine the height of the instrument (HI).

Backsights are also frequently referred to as *plus sights* because they are added to the elevations of points being sighted on to determine the height of the instrument.

A *foresight* (FS) is a sight taken to any point to determine its elevation. Foresights are often called *minus sights* because they are subtracted from HIs to obtain the elevations of points. *Notice that for any position of the instrument where the HI is known, any number of foresights may be taken to obtain elevations of other points in the area.* The only limitations on the number of sights are the length of the level rod and the power of the telescope on the level. Normally, one cannot sight on the level rod held at a point whose elevation is greater than that of the HI. There are a few exceptions, such as when it is desired to determine the elevation of the underside of a bridge or the top of a mine shaft or tunnel. For such cases the level rod is inverted and held up against the point whose elevation is sought. Then the backsights are negative and the foresights are positive.

7-3 DIFFERENTIAL LEVELING DESCRIBED

Differential leveling, which is the process of determining the difference in elevation between two points, is illustrated in Fig. 7-2, in which a line of levels is run from BM_1 to BM_2. The instrumentman sets up the level at a convenient point and backsights on the level rod held on BM_1. This gives the HI. The rodman moves to a convenient point (TP_1 in the figure) in the direction of BM_2. The instrumentman takes a foresight on the rod, thus enabling him or her to compute the elevation of TP_1. The level is then moved to a convenient location beyond TP_1 and a backsight taken on TP_1. This gives the new HI. The rodman moves forward to a new location (TP_2), and so on. This procedure is continued until the elevation of BM_2 is determined.

It is very important in differential leveling to keep the lengths of the backsights and foresights approximately equal for each setup of the instrument. Such a practice will result in greatly reduced errors for cases where the instruments

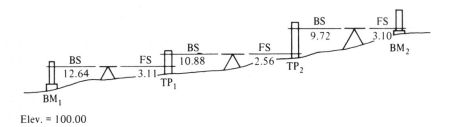

Figure 7-2

are out of adjustment and also for errors due to atmospheric refraction and the earth's curvature errors. The easiest way to obtain roughly equal distances is by pacing, but a better way is by the use of stadia measurements. These topics are discussed further in Sections 7-4 and 9-3.

The usual form for recording differential leveling notes is presented in Fig. 7-3 for the readings that were shown in Fig. 7-2. *The student would be wise to study these notes very carefully before attempting leveling; otherwise, he or she may become confused in recording the readings.* In studying these notes, he or she should particularly notice the math check. Since the backsights are positive and the foresights are negative, the surveyor should total them separately. The difference between these two totals must equal the difference between the initial and final elevations or else a math blunder has been made in the field book. *Since the math check is easy to make, there is no excuse for omitting it. For this reason, level notes are considered incomplete unless the check is made and shown in the notes.*

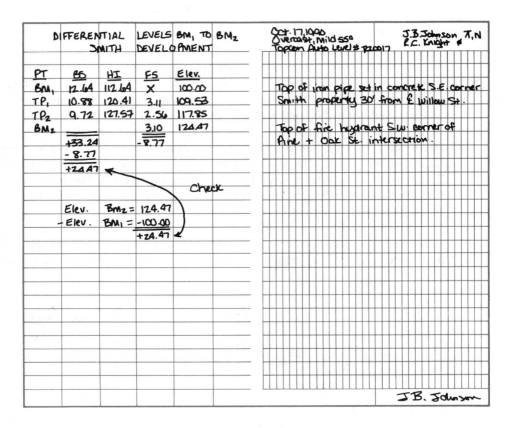

Figure 7-3 Differential level notes.

7-4 EARTH'S CURVATURE AND ATMOSPHERIC REFRACTION

Up to this point the discussion of leveling has assumed that when the instrument has been leveled, its line of sight is a level line or a line of equal elevation. This is obviously not so and, in fact, the line of sight is perpendicular to a plumb line at one point only—at the instrument location. The student may then decide that the line of sight is a horizontal line, but this is not correct either.

When rays of light pass through air strata of different densities, they are refracted or bent downward. This means that to see an object on the ground some distance away, a person actually has to look above it. The amount of refraction is dependent on temperatures, pressures, and relative humidities. It is greatest when the line of sight is near the ground or near bodies of water where temperature differences are large and therefore where large variations in air densities occur.

As the amount of refraction is difficult to determine, an average value is usually used. This value is approximately 0.093 ft in 1 mile, (about one-seventh of the effect of the earth's curvature) and varies directly as the square of the horizontal distance.

Variations from the 0.093-ft value are not usually significant for the relatively short sight distances used in differential leveling—but they may have to be considered for very precise leveling and for extreme conditions for ordinary leveling. Refraction has been known to be as large as 0.10 ft in a 200-ft sight distance.

Because of the earth's curvature, a horizontal line departs from a level line by 0.667 ft in 1 mile, also varying as the square of the horizontal distance. The effects of the earth's curvature and atmospheric refraction are represented in Fig. 7-4.

The combination of the earth's curvature and atmospheric refraction causes the telescope's line of sight to vary from a level line by approximately 0.667 minus 0.093 or 0.574 ft in 1 mile, varying as the square of the horizontal distance in miles. This may be represented in formula form as follows, where C is the departure of a telescope line of sight from a level line and M is the horizontal distance in miles:

$$C = 0.574M^2$$

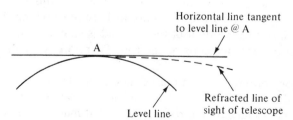

Horizontal line tangent to level line @ A

A

Refracted line of sight of telescope

Level line

Figure 7-4

For a telescope reading on a rod 100 ft away, the reading would be in error by

$$(0.574)\left(\frac{100}{5280}\right)^2 = 0.000206 \text{ ft}$$

Similarly, a reading on a rod at a 300-ft distance would be in error by 0.00185 ft. At a 1000-ft distance the error would be 0.0206 ft.

If the backsight distance were exactly equal to the foresight distance for each setup of the instrument, it could be seen that errors caused by atmospheric refraction and the earth's curvature would cancel each other. Each of the readings would be too large by the same amount and since the same error would be added with the backsight and subtracted with the foresight, the net result would be to cancel them. *Keeping BS and FS distances approximately equal may be more important for minimizing errors due to instrument inadjustment than for atmospheric refraction and earth's curvature.*

For surveys of ordinary precision, it is reasonable to neglect the effect of the earth's curvature and atmospheric refraction. The instrumentman, however, might like to, by eye, make BS and FS distances approximately equal. For precise leveling, it is necessary to use more care in equalizing the distances. Pacing or even stadia may be used.

For SI units the correction for earth's curvature and atmospheric refraction is given by the following expression, in which C is in meters and k is the distance in kilometers.

$$C = 0.0675 \, k^2$$

7-5 VERNIERS

A vernier is a device used for making readings on a divided scale closer than the smallest divisions on the scale. The vernier, which was invented by the Frenchman Pierre Vernier in 1620, is a short auxiliary scale attached to or moved along the divided scale.

Most of the targets used on level rods have verniers on them with which rod readings can be made to the nearest 0.001 ft. Figure 7-5 illustrates the construction and the reading of level rod verniers. The numbers on this particular rod are for 3.1 ft and 3.2 ft and thus the divisions in between are the 0.01-ft divisions. The vernier is shown to the right of the rod and is so constructed that 10 divisions on the vernier cover 9 divisions on the rod. Therefore, each division on the vernier is 0.009 ft.

In this figure the 0 or bottom mark on the vernier coincides with the 3.10-ft mark on the rod. It will be noted that the next division of the vernier falls $\frac{1}{10}$ of a division short of the next mark on the rod (or its 3.11-ft mark). The second division on the vernier falls $\frac{2}{10}$ of a division short of the 3.12-ft mark on the rod.

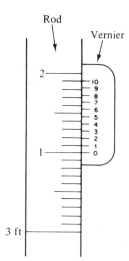

Figure 7-5

If the bottom of the vernier is moved up until the first division on the vernier coincides with the first division on the rod or the 3.11 ft mark, the bottom of the vernier will be located at 3.101 ft on the rod. In the same manner, if the vernier is moved up until the second division on the vernier coincides with the second mark on the main scale (3.12 ft), the bottom of the vernier will be located at 3.102 ft on the rod.

To read a level rod vernier, the bottom of the vernier (which is the center of the target) is lined up with the horizontal cross hair and the reading is determined by counting the number of vernier divisions until a division on the vernier co-incides with a division on the rod. This reading is added to the last division on the rod below the bottom of the vernier. In Fig. 7-6 the rod reading at the bottom of the vernier is between 3.12 and 3.13 ft. The sixth division up on the vernier coincides with a division on the rod, so the rod reading is 3.12 plus the vernier reading 6 equals 3.126 ft. This also could be read as $3.18 - (6)(0.009) = 3.126$ ft.

For the level rod and vernier described here, the smallest subdivision that can be read equals one-tenth of the scale division on the rod. Ten divisions on the vernier cover nine divisions on the rod, and the smallest subdivision that can be read with the vernier is

$$\frac{0.01}{10} = 0.001 \text{ ft}$$

If another vernier is used for which 20 divisions of the vernier cover 19 divisions of the rod, the smallest subdivision that can be read is $\frac{1}{20}$ of the scale division, or

$$\frac{0.01}{20} = 0.0005 \text{ ft}$$

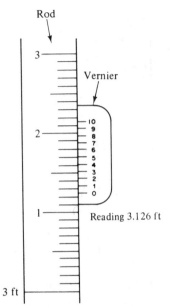

Figure 7-6

From this information an expression can be written for the smallest subdivision that can be read with a particular vernier. Letting n be the number of vernier divisions, s the smallest division on the main scale, and D the smallest subdivision that can be read, the following expression can be written:

$$D = \frac{s}{n}$$

Historically in surveying, one of the principal uses of verniers was for angle measurements, and for this reason the subject of verniers is continued in Chapter 10, which deals with such measurements.

7-6 LEVEL ROD TARGETS

For long sights or for situations in which readings to the nearest 0.001 ft are desired, a level rod target may be used. Targets are small circular or elliptical pieces of metal approximately 5 in. in diameter painted red and white in alternate quadrants. They are clamped to the rods. As shown in Fig. 7-7, a vernier is part of the target.

The target is moved up or down as directed by the instrumentman until it appears to be bisected by the cross hair. At this point the horizontal cross hair

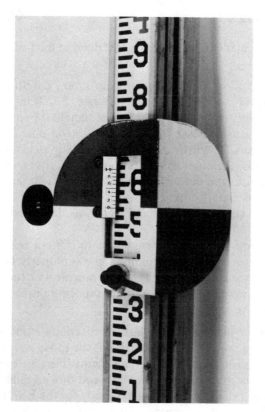

Figure 7-7 Level rod target.

coincides with the bottom of the vernier. The rodman takes the reading to the nearest 0.001 ft and this value is approximately checked by the instrumentman. An important item to remember about target readings is that although they are read to the nearest 0.001 ft, they can be no more accurate than the accuracy obtained in setting the target. In other words, for ordinary leveling "do not attach too much significance to readings taken to the nearest 0.001 ft." Although more accurate than those taken to the nearest 0.01 ft, the readings are probably not accurate to the third place. The reason is that the instrumentman just cannot signal the rodman to set the target on the rod as precisely as he or she can set the telescope cross hairs on the rod.

The precision of ordinary leveling can be increased somewhat by the laborious use of targets as described in the preceding paragraph. *A very simple, quick and practical way in which precision can be appreciably improved is by merely limiting the lengths of sights, perhaps to 100 ft or less, and by having the instrumentman estimate the rod readings through the telescope to the nearest 0.002 ft.*

7-7 COMMON LEVELING MISTAKES

The most common mistakes made in leveling are described in the following paragraphs.

Misreading the Rod.   Unless the instrumentman is very careful, he or she may occasionally read the rod incorrectly; as, for instance, 3.72 ft instead of 4.72 ft. This mistake most frequently occurs when the line of sight to the rod is partially obstructed by leaves, limbs, grass, rises in the ground, and so on. There are several ways to prevent such mistakes. The instrumentman should always carefully note the foot marks above or below the point where the horizontal cross hair intersects the rod. If those red foot marks are not visible, he or she may ask the rodman to raise the rod slowly until a foot mark can be seen. To do this he or she may either call "raise for red" or give an appropriate hand signal, as described in Section 7-12.

An excellent procedure for the instrumentman is to call out readings as he or she takes them. The rodman, while still holding the rod properly can point to the reading with a pencil. Obviously, the pencil should coincide with the horizontal cross hair if the reading was taken correctly. Another procedure is to use a target and have both persons take readings.

Moving Turning Points.   A careless rodman causes serious leveling mistakes if he or she moves the turning points. The rodman holds the rod at one point while the instrumentman takes the foresight reading, and then while the level is being moved to a new position, the rodman puts the level rod down while he or she does something else. If, when the instrumentman is ready for the BS, the rodman holds the rod at some other point, a serious mistake can be made because the new location may have an entirely different elevation. Obviously, a good rodman prevents mistakes like this by using well-defined turning points or by clearly marking them with crayon (keel) if on pavement or by driving a stake or ax head into the earth.

Field Note Mistakes.   To prevent the recording of incorrect values, the instrumentman should call out the readings as he or she reads and records them. This is particularly effective if the rodman is checking the readings with a pencil or with a target. To prevent addition or subtraction mistakes in level notes, the math check described in Section 7-3 should be carefully followed.

Mistakes with Extended Rod.   When readings are taken on the extended portion of the level rod, it is absolutely necessary to have the two parts adjusted properly. If they are not, mistakes will be made.

7-8 LEVELING ERRORS

A brief description of the most common leveling errors and suggested methods of minimizing them are presented in the next several paragraphs.

Level Rod Not Vertical.   When sighting on the rod, the instrumentman can see if the rod is leaning to one side or the other by means of the vertical cross hair in the telescope and, if necessary, can signal the rodman to straighten up. This rodman cannot, however, usually tell if the rod is leaning a little toward or away from the instrument. If the student thinks about this for a while, he or she will see that the smallest possible rod reading will occur when the rod is vertical. Many surveyors, therefore, have their rodman slowly "wave" the rod toward and away from the instrument and then record the smallest reading observed through the telescope. A method used by some surveyors for ordinary leveling is to require each rodman to hold the level rod so that it touches his or her nose and belt buckle, but this practice is not as satisfactory as waving the rod. Some level rods, particularly those used for precise work, are equipped with individual circular levels that allow the rodman to plumb the rod by merely centering the bubble. Other rods are equipped with conventional bubble tubes. The use of rod levels or other methods of plumbing is preferable to waving the rod. Waving the ordinary flat-bottomed rod can cause small errors due to the rotation of the rod about its edges instead of about the center of its front face.

Settling of Level Rod.   It is essential to hold the level rod on firm definite points which will not settle and which are readily identifiable, so that the rodman, if called away for some other work, may return to exactly the same spot. If such convenient points are unavailable, it may be necessary to take turning points on the ordinary earth with the resulting possibility of settlement. To minimize this possibility, the rodman should hold the rod on a stake driven in the ground, on the head of an ax stuck in the ground, a metal turning point pin, or on some other similar base.

Mud, Snow, or Ice Accumulation on Base of Rod.   If the rodman is not careful, mud, snow, or ice may stick to the bottom of the rod. This can cause severe errors in leveling. The rod must not be dragged on the ground at any time and the bottom of the rod must be carefully cleaned when there is snow, ice, or mud.

Rod Not Fully Extended on High Rod.   When the rear part of a Philadelphia rod is extended, it is called a high rod. Such an extension is necessary for readings from 7 to 13 ft. Frequently, level rods have been damaged by letting the upper part of the rod slide down so rapidly that the blocks on the two sections are damaged. The result is that the high rod readings may be in error and the rodman must carefully check the rod extension.

Incorrect Rod Length.   If a level rod is of incorrect length (and no rod is of perfect length), the rod readings will be in error. If the length errors occur at the bottom of the rod, they will theoretically be canceled in the differential leveling process. Errors due to a misfit in extending the rod, however, will not cancel if some of the readings are taken above the joint and some of them below. This generally occurs when the surveyor is leveling up or down a slope. For such cases one tends to read on the high part of the rod for downhill shots and on the low part for uphill shots. Rod lengths should be checked periodically with a steel tape.

BS and FS Distances Not Equal.   In Section 7-4 it was shown that if the lengths of backsights and foresights were kept equal for a particular setup, there would theoretically be no error caused by the earth's curvature and atmospheric refraction. For ordinary work it is sufficient to neglect or merely approximate by eye equal distances. For more precise work, it is necessary to pace distances or even use stadia to keep BS and FS distances equal.

Errors due to instrument maladjustment are usually much more important than those due to atmospheric refraction and the earth's curvature. Particularly significant is the error produced if the axis of the bubble tube is not parallel to the line of sight of the telescope (see Section 7-13). However, if BS and FS distances are kept equal, such errors will be greatly minimized.

Bubble Not Centered on Level.   If the bubble is not centered in the level tube when a reading is taken, the readings will be in error. It is surprising how easily this can happen. The instrumentman may sometimes brush against the instrument, the tripod legs may settle in soft ground, or the instrument may not be properly leveled or adjusted, with the result that when the telescope is turned the bubble does not stay centered. All of these factors mean that the instrumentman must be particularly careful. If the bubble is checked before and after readings are taken to be sure that it is centered, these errors will be reduced substantially. (It is to be remembered that automatic levels will relevel themselves when they are thrown slightly out of level.)

Settling of Level.   In soft or swampy ground or asphalt there definitely will be some settling of the tripod. Between the time of the backsight and the foresight readings there will be some settlement, with the result that the foresight reading will be too small. Special care should be taken in selecting the firmest possible places to set up the instrument. In addition, as little time as possible should be taken between the readings (use two rodmen, if possible). A further precaution in minimizing settlement errors is to take the foresight reading first on alternate setups.

Instrument Out of Adjustment.   The adjustment of levels is discussed in Section 7-13 and is very important. The surveyor will with experience learn to make simple checks constantly to see that the instruments are adjusted properly.

Improper Focusing of Telescope (Parallax).   If we look at the speedometer of a car from different angles, we will read different values. This is due to *parallax*. If the indicator and the speedometer scale were located exactly in the same plane, parallax would be eliminated.

Sometimes when we sight through a telescope, we find that if we move an eye a slight distance from one side to the other, there is an apparent movement of the cross hairs on the image or the object seems to move. Again, this is parallax and it can cause appreciable errors unless it is corrected. The surveyor should carefully focus the objective lens until the image and the cross hairs appear to be exactly in the same place, that is, until they do not move in relation to each other when the eye is moved back and forth. The distorting effect of parallax will then be prevented.

Heat Waves.   On hot sunny days, heat waves from the ground, pavement, buildings, pipes, and other objects can seriously reduce the accuracy of work. Sometimes these waves are so intense that they cause large errors in rod readings. They may be so bad in the middle part of the day that work must be stopped until the waves subside. Heat wave errors can be minimized by reducing the lengths of sights. In addition, since the waves are worse near the ground, points should be selected so that the lines of sight are 3 or 4 ft or more above the ground.

Wind.   Occasionally, high winds cause accidental errors because the winds actually shake the instrument so much that it is difficult to keep the bubble centered. These errors can be reduced by using shorter sight distances and by pushing the tripod shoes deeper in the ground and placing the legs farther apart.

7-9 SUGGESTIONS FOR GOOD LEVELING

After reading the lengthy list of mistakes and errors described in the preceding two sections, the novice surveyor may be as confused as the beginning golfer who is trying to remember 15 different things about his or her swing when trying to hit the ball. To perform good leveling, however, remember the following few general rules:

1. Anchor tripod legs firmly.
2. Check to be sure that the bubble tube is centered before and after rod readings.
3. Take as little time as possible between BS and FS readings.
4. For each setup of the level, use BS and FS distances that are approximately equal.
5. Either provide rodmen with level rods that have level tubes (circular, conventional, etc.) with which the rods can be plumbed or have them wave the rods slowly toward and away from the instrument.
6. It is wise to use straight-leg (nonadjustable) tripods.

7-10 COMMENTS ON TELESCOPE READINGS

The instrumentman should learn to keep both eyes open when looking through the telescope. First, it is quite tiring to keep closing one eye all day to take readings. Second, it is convenient to keep one eye on the cross hair and the other eye open to locate the target.

If a person wears ordinary glasses for magnification purposes with no other corrections, it will not be necessary to wear glasses while looking through the telescope. The adjustment of the lens will compensate for the eye trouble.

7-11 PRECISION OF DIFFERENTIAL LEVELING

In this section we present as a guide the approximate errors that should result in differential leveling of different degrees of precision. It is assumed that levels of average condition and in good adjustment are used. *Rough leveling* pertains here to preliminary surveys in which readings are taken only to the nearest 0.1 ft and in which sights of up to 1000 ft may be used. In *average leveling*, rod readings are taken to the nearest 0.01 ft, and BS and FS distances may be approximately balanced by eye, particularly when leveling on long declines or upgrades, and sights up to 500 ft may be used. It is probable that 90% of all leveling falls into this category. In *excellent leveling*, readings are made to the nearest 0.001 ft, BS and FS distances are approximately equalized by pacing, and readings are taken for distances no greater than 300 ft.

The average errors will probably be less than the values given here.[1] In these expressions M is the number of miles leveled and the values resulting from the expressions are in feet.

Rough leveling	$\pm 0.4 \sqrt{M}$
Average	$\pm 0.1 \sqrt{M}$
Excellent	$\pm 0.05 \sqrt{M}$

For instance, if differential leveling is done over a route of 6 miles, the maximum error resulting from surveying of average precision should not exceed $\pm 0.1 \sqrt{6} = \pm 0.24$ ft. These values are given for leveling done under ordinary conditions. Surveyors who frequently work in very hilly parts of the country might have some difficulty in maintaining these degrees of precision. If very rough leveling is done using the hand level, the maximum error can be limited to approximately $3.0 \sqrt{M}$.

Beginning in 1957 federal government agencies classified leveling as being first-, second- and third-order. Since 1975, however, they have used five accuracy grades. These are, in descending order: first-order, second-order, and third-order, with the first two orders subdivided into classes I and II. Table 7-1 is a copy of this classification system. It will be noted in the table that first- and second-order leveling is applicable to the basic control for national geodetic surveys and to the secondary framework of those systems, respectively. Second-order leveling and third-order leveling apply to large and small engineering projects, respectively.

---

[1] This information on maximum probable errors is as given in *Elementary Plane Surveying* by Raymond E. Davis (New York: McGraw-Hill Book Company, 1955), pp. 82–83.

TABLE 7-1(a) STANDARDS OF CLASSIFICATION: VERTICAL CONTROL (FEDERAL GEODETIC CONTROL COMMITTEE)

	First-order		Second-order		Third-order
	Class I	Class II	Class I	Class II	
Recommended uses	Basic framework of the National Network and metropolitan area control Regional crustal movement studies Extensive engineering projects Support for subsidiary surveys		Secondary framework of the National Network and metropolitan area control Local crustal movement studies Large engineering projects Tidal boundary reference Support for lower-order surveys	Densification within the National Network Rapid subsidence studies Local engineering projects Topographic mapping	Small-scale topographic mapping Establishing gradients in mountainous areas Small engineering projects May or may not be adjusted to the National Network
Relative accuracy between directly connected points or bench marks (standard error)	0.5 mm $\sqrt{K}$	0.7 mm $\sqrt{K}$	1.0 mm $\sqrt{K}$	1.3 mm $\sqrt{K}$	2.0 mm $\sqrt{K}$

TABLE 7-1(b) ENGLISH EQUIVALENTS OF VALUES IN TABLE 7-1(a)

Relative accuracy	0.0021 ft $\sqrt{M}$	0.0029 ft $\sqrt{M}$	0.0042 ft $\sqrt{M}$	0.0054 ft $\sqrt{M}$	0.0083 $\sqrt{M}$

Source: Classification, Standards of Accuracy, and General Specifications of Geodetic Control Surveys, February 1974 (Rockville, Md.: U.S. Department of Commerce), p. 3.

Leveling performed by trigonometric or barometric procedures would be classified as fourth-order leveling or worse. In the table the letter K represents the distance between bench marks in kilometers. The English equivalents of these accuracy standards are given in Table 7-1(b). In these latter expressions M represents the distance between bench marks in miles.

7-12 HAND SIGNALS

For all types of surveying it is essential for the various personnel to keep close communication with each other. Very often, calling back and forth is completely impractical because of the distances involved or because of noisy traffic or earth-moving machinery in the area. In the absence of walkie-talkies, therefore, a set of hand signals that is clearly understood by everyone involved is often a necessity.

The instrumentman should remember that he or she has a telescope with which the rodman can be observed; the rodman, however, cannot see the instrumentman nearly so clearly. As a result, the instrumentman must be very careful to give clear signals to the rodman. Following are some commonly used hand signals:

Plumb the Rod. One arm is raised above the head and moved slowly in the direction that the rod should be leaned [Fig. 7-8(a)].

Wave the Rod. The instrumentman holds one arm above his or her head and moves it from side to side [Fig. 7-8(b)].

High Rod. To give the signal for extending the rod, hold arms out horizontally and bring them together over the head [Fig. 7-8(c)].

Raise for Red. Sometimes for very short sights the red foot marks will not fall within the telescope's field of view, and with the "raise for red" signal the instrumentman asks that the rod be raised a little so he or she can determine the correct foot mark. One arm is held straight forward, with palm up, and raised a short distance [Fig. 7-8(d)].

All Right. The arms are extended horizontally and waved up and down [Fig. 7-8(e)].

Pick Up the Instrument. The party chief may give this signal when a new setup of the instrument is desired. The hands are raised quickly from a downward position as though an object is being lifted [Fig. 7-8(f)].

Raise the Target. If one hand is raised above the shoulder with the palm visible, it means to raise the target [Fig. 7-8(g)]. If a large movement is needed, the hand is moved abruptly, but if only a small movement is needed, the hand is moved slowly.

(a) Plumb the rod.   (b) Wave the rod.   (c) High rod.

(d) Raise for red   (e) All right.   (f) Pick up the   (g) Raise the target.
(side view).   instrument.

(h) Lower the target.   (i) Clamp the target.

Figure 7-8

Lower the Target.   Lowering the hand below the waist means to lower the target [Fig. 7-8(h)].

Clamp the Target.   The instrumentman, keeping one arm horizontal, moves his or her hand in vertical circles. This means to clamp the target [Fig. 7-8(i)].

7-13 ADJUSTMENTS OF DUMPY LEVELS

A level is manufactured with great care and precision, but it must be checked periodically to see that it is in proper adjustment. The instrument will get out of adjustment and parts will become worn and loose fitting even though the instrument may be handled very carefully. If it is handled roughly, the problem will be magnified. This section is devoted to those adjustments that can be made by the surveyor in the field. The surveyor should constantly check the instruments to see that they are properly adjusted. It is a good habit to check at least one of the relations discussed below each day that the instrument is used.

In the following paragraphs are listed the relations that should exist in a dumpy level, the procedure that should be used to see if they do exist, and the adjustments necessary if they do not. The checking and adjusting of levels described herein should desirably be carried out under ideal conditions: that is, favorable atmospheric conditions, with levels preferably shaded and located on terrain that permits easy and stable instrument setups. Figure 7-9 illustrates the terms used in the following discussion.

Bubble Tube

When the instrument is leveled, the axis of the bubble tube should be perpendicular to the vertical axis of the instrument.

To check to see if this relation is present, the instrument is carefully leveled over opposite pairs of leveling screws. Then the telescope is turned through 180°, that is, end to end over the same pair of screws. If the bubble moves, the desired relation does not exist. The amount of movement of the bubble equals twice the error involved. The reason for the doubled error is discussed in Section 7-14. This can be corrected by moving the bubble halfway back to center by using the adjusting screws at the ends of the bubble tube. Once this is done, the test should be run again to see that the proper correction has been made.

Cross-Hair Ring

The horizontal cross hair should lie in a plane perpendicular to the vertical axis of the level. If it does not meet this requirement, an error will occur with each reading. That error will tend to cancel if the same location on the cross hair is used each time for making readings. Furthermore, the error will theoretically be zero if all readings are taken at the center of the reticle: where the horizontal and vertical hairs meet.

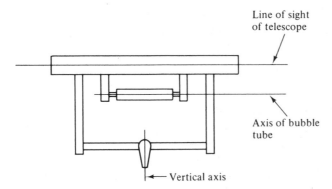

Line of sight of telescope

Axis of bubble tube

Vertical axis

Figure 7-9

To make this test, the level is set up and the telescope sighted on a sharply defined point with one end of the horizontal cross hair. Then the telescope is moved slowly about the vertical axis of the instrument (i.e., the telescope is moved horizontally) to see if the cross hair remains on the point. If it does not, it is necessary to rotate the cross-hair ring. This is done by slightly loosening the four adjacent capstan screws on the ring and turning the ring slightly either with light pressure of the fingers or by tapping the ring lightly with a pencil. Then the test is run again and the process repeated until the desired relation exists.

Line of Sight

The line of sight of the telescope should be parallel to the axis of the bubble tube.

The "two-peg test" is commonly used to see whether or not this relation exists. The instrument is carefully set up at a convenient point. Stake A is driven in the ground, say 150 ft from the instrument, in one direction and stake B is driven in the ground the same measured distance on the other side of the instrument. If the line of sight of the level is not horizontal and a BS is taken on the rod at point A and a FS on the rod at point B, there will be equal errors in each of the readings. However, since, the BS is added and the FS subtracted the errors will theoretically cancel each other and the correct elevation of point B will be determined. In Fig. 7-10 the readings we take are shown as a and b and the correct difference in elevation between the two points is $a - b$.

Next, the level is moved to one of the stakes, say B, and is placed so that the telescope will swing within approximately $\frac{1}{2}$ in. of the level rod held on the stake. The instrument cannot be focused on the rod that close, but if the instrumentman looks through the telescope backward to the rod (the cross hairs will not be visible), he or she will be able to see a very short segment of the rod. By holding a pencil at the center of this segment, he or she can take a very precise reading. This value (labeled c in Fig. 7-11) will be correct even if the line of sight of the telescope is not parallel to the bubble tube axis.

The instrumentman can then take a reading on the rod at point A. If the instrument is properly adjusted, the difference in readings c and d will equal the difference between readings a and b. If reading d does not check, the instrument is out of adjustment. The correct value of d can be calculated and the instrument adjusted with the top and bottom cross-hair ring capstan screws. One is loosened and the other tightened, or vice versa, until the horizontal cross hair is on the desired reading. For example, if $a = 6.42$ and $b = 3.32$, the difference in elevation from A to B is 3.10. If reading c is 5.04, then d should equal 8.14 for an instrument

Figure 7-10

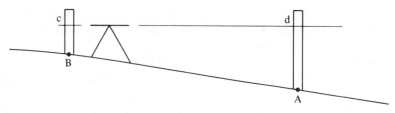

Figure 7-11

in proper adjustment. The effects of the earth's curvature and atmospheric refraction have been neglected for this discussion.

The desired relations and tests have been given in a particular order so that any adjustments made will have the least possible effect on other desired relations. When the series of tests is completed, it will be necessary to run through the series again (unless no adjustments were necessary) to see if the adjustments have affected other relations. For an instrument in extremely poor adjustment, it may be necessary to repeat the series several times.

7-14 PRINCIPLE OF REVERSION

Adjustments of surveying instruments are frequently checked by the principle of reversion. The instrument is inverted or turned 180° in its position with the result that the error in question is doubled and is thus more obvious. As an illustration of this principle, it is assumed that the bubble for a level is carefully centered as described previously. With the bubble centered and the telescope over a pair of leveling screws, the telescope is rotated through 180°. If the bubble does not remain centered, the axis of the bubble tube is not perpendicular to the vertical

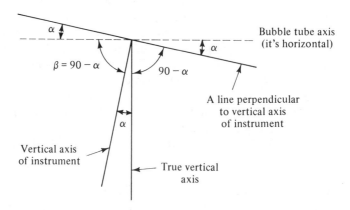

Figure 7-12 Initial position (bubble centered).

axis of the instrument. As described in the paragraphs to follow, the error has been doubled. As a result, the bubble should be moved halfway back to the centered position by raising or lowering one end of the capstan screws at the end of the tube.

If the axis of the bubble tube is not perpendicular to the vertical axis of the instrument, the centering of the bubble will not put the vertical axis of the instrument in a truly vertical position. This can be seen in Fig. 7-12, where it is shown that if the bubble tube axis is not perpendicular to the vertical axis of the instrument by an angle α, the instrument vertical axis will be located an angle α from the correct vertical axis.

In Fig. 7-12 the smaller angle from the vertical axis of the instrument to the bubble tube is designated as β and equals 90 − α. If the telescope is now rotated 180° horizontally, the angle β will also rotate 180°. The axis of the bubble tube will move away from its original horizontal position by an angle 2α as shown in Fig. 7-13. To correct the error, it will be necessary to raise or lower one end of the bubble tube with the capstan screws at the ends of the tube until the bubble is halfway back to the center position. Then the axis of the bubble tube will be correctly perpendicular to the vertical axis of the instrument. If the bubble is centered, it will stay centered as the telescope is rotated around the vertical axis.

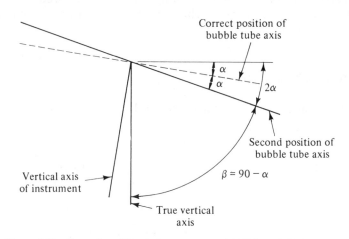

Figure 7-13   Second position (telescope rotated 180° from initial position).

7-15 ADJUSTMENTS OF AUTOMATIC LEVELS

Adjustments for the automatic level are quite similar to those for the dumpy level. The brief discussion that follows applies specifically to the Topcon Auto Level but is quite similar to other automatic levels.

Circular Level Tube

The instrument is set up and leveled with the three leveling screws. The telescope is rotated through 180°. If the bubble moves away from the center of the bull's eye or circular level, it is out of adjustment. The adjustment screws are located by the circular level. The adjustment screw on the side to which the bubble has moved is tightened until the bubble moves halfway back to the center.

Line of Sight (or Collimation) of Level

The two-peg test may be used as it is for the dumpy level. If an adjustment is required, the horizontal cross hair is moved up or down until the correct reading is observed. This is done with the capstan screws, which can be seen when the eyepiece cover is removed.

Cross-Hair Ring

The test is run as for the dumpy level, and if the horizontal cross hair is not horizontal, the reticule glass is rotated with the three fixing screws, which are exposed when the eyepiece cover is removed. They are located next to the capstan screws used for the line-of-sight adjustment. The three screws are slightly loosened, the reticule is very slightly rotated, the screws are retightened, and a check is made again.

PROBLEMS

In Problems 7-1 to 7-3, complete and check the following sets of level notes.

7-1.

Station	BS	HI	FS	Elevation
BM_1	3.49			100.00
TP_1	6.21	103.49	8.02	95.47
TP_2	8.42	101.68	5.96	95.72
TP_3	5.37	104.14	6.44	97.70
BM_2		103.07	3.57	99.50 ✓

(*Ans.:* $BM_2 = 99.50$)

7-2.

Station	BS	HI	FS	Elevation
BM_1	8.124			932.862
TP_1	6.348		1.212	
TP_2	7.126		7.386	
TP_3	4.448		5.626	
TP_4	3.964		5.024	
BM_2			4.742	

7-3.

Station	BS	HI	FS	Elevation
BM₁	5.492			764.324
TP₁	4.884		4.088	
TP₂	3.266		3.126	
BM₂	5.108		2.144	
TP₃	6.322		4.364	
BM₃			6.862	

(*Ans.:* BM₃ = 768.812)

In Problems 7-4 to 7-6, set up and complete differential level notes for the information shown in the accompanying illustrations. Include the customary math checks.

7-4.

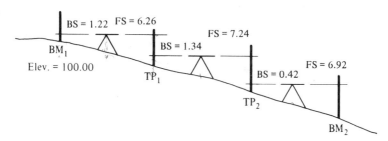

7-5.

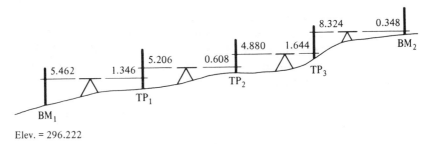

(*Ans.:* BM₂ = 316.148)

7-6.

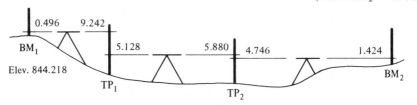

In Problems 7-7 to 7-9, the accompanying illustrations represent plans for differential leveling. The values shown on each line represent the sights taken along those lines. Prepare and complete the necessary field notes for this work (Y = instrument setup).

7.7.

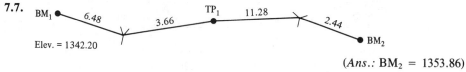

BM₁ 6.48 3.66 TP₁ 11.28 2.44 BM₂

Elev. = 1342.20

(*Ans.*: BM₂ = 1353.86)

7-8.

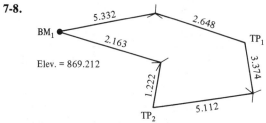

5.332 2.648

BM₁ 2.163 TP₁

Elev. = 869.212 3.374

1.222 5.112

TP₂

7-9.

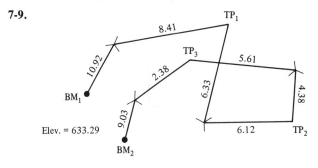

TP₁

8.41

TP₃ 5.61

10.92 2.38

BM₁ 6.33 4.38

Elev. = 633.29 9.03

6.12 TP₂

BM₂

(*Ans.*: BM₂ = 628.13)

In Problems 7-10 to 7-14, the level rod readings are given in the order in which they were taken. In each case the first reading is taken on BM₁ and the last reading is taken on BM₂, the point whose elevation is desired. Set up the differential level notes, including the customary math check. The elevation of BM₁ is given under each problem number.

Problem 7-10	Problem 7-11	Problem 7-12	Problem 7-13	Problem 7-14
100.00	352.62	1466.88	841.182	498.242
4.64	5.12	10.32	3.128	8.442
8.32	9.38	3.49	7.466	3.460
3.61	5.36	9.38	1.090	8.292
9.44	7.29	2.46	8.322	3.112
5.19	5.22	11.51	3.484	8.160
6.67	8.62	5.09	9.642	2.044
7.26	3.19	9.47	3.962	4.882
5.33	10.64	4.82	6.874	3.114
	4.16		2.260	
	5.33		9.822	

(*Ans.*: BM₂ = 334.41)   (*Ans.*: BM₂ = 812.980)

7-15. In running a line of levels from BM_1 (elevation 914.60) to BM_2, the following readings were taken in the order given: 6.12, 7.24, 3.14, 5.68, 4.84, 6.47, 9.62, and 6.81. Set up and complete the level notes, including the math check.

(*Ans.:* $BM_2 = 912.12$)

7-16. In running a line of levels from BM_1 (elevation 342.88) to BM_2, the following readings were taken in the order given: 1.01, 3.06, 4.94, 5.81, 6.72, 2.86, 3.74, 9.09, 8.46, and 1.11. Set up and complete the level notes including the math check.

7-17. A line of levels was run into a mine shaft. All of the points (BMs and TPs) were located in the shaft ceiling and readings were taken by inverting the level rod. Complete the resulting level notes shown, including the math check.

(*Ans.:* $BM_2 = 682.18$)

Station	BS	HI	FS	Elevation
BM_1	8.42			696.89
TP_1	7.92		5.32	
TP_2	8.20		4.26	
TP_3	7.84		4.11	
BM_2			3.98	

7-18. Repeat Problem 7-16, assuming that all points (BMs and TPs) were located in the top of a tunnel and were taken by inverting the level rod.

For Problems 7-19 to 7-21, complete and check the level notes.

7-19.

Station	BS	HI	FS	Elevation
BM_1	4.92			696.36
TP_1	8.11		2.18	
			4.64	
			6.61	
TP_2	6.76		5.33	
BM_2			8.12	

(*Ans.:* 700.52)

7-20.

Station	BS	HI	FS	Elevation
BM_{11}	9.18			1462.84
TP_1	4.44		1.96	
			3.28	
			2.74	
TP_2	5.29		5.91	
			6.48	
			7.33	
TP_3	6.39		9.34	
BM_{12}			8.26	

7-21.

Station	BS	HI	FS	Elevation
BM₁	6.38			1503.26
			8.42	
			5.12	
TP₁	6.86		3.42	
			2.94	
TP₂	1.11		4.06	
BM₂			0.26	

(Ans.: 1509.87)

7-22. Compute the combined effect of the earth's curvature and atmospheric refraction for distances of 100 ft, 200 ft, 500 ft, 2000 ft, and 10 miles.

7-23. A BS of 3.72 ft is taken on a level rod at a 100-ft distance, and a FS of 9.46 ft is taken on the rod held 1000 ft away. (a) What is the error caused by the earth's curvature and atmospheric refraction? (b) What is the correct difference in elevation between the two points?

(Ans.: −0.02 ft, −5.72 ft)

7-24. In differential leveling from BM₁ to BM₂, the BS and FS distances for readings were as follows: BS 300 ft, FS 100 ft, BS 600 ft, FS 100 ft, BS 450 ft, FS 300 ft, BS 400 ft, and FS 200 ft. What is the error in the elevation of BM₂ caused by atmospheric refraction and the earth's curvature?

7-25. What BS or FS distances for an instrument setup will cause an error due to the earth's curvature and atmospheric refraction equal to 0.005 ft? 0.02 ft? 0.10 ft?

(Ans.: 493 ft, 986 ft, 2204 ft)

7-26. Two towers A and B are located on flat ground and their bases have equal elevations above sea level. A person on tower A whose eye level is 60 ft above the ground can just see the top of tower B, which is 110 ft above the ground. How far apart are the towers?

7-27. A man whose eye level is 5.3 ft above the ground is standing by the ocean. He can just see the top of a lighthouse across the water. Neglecting tidal and wave effects, how high is the lighthouse if it's 20 miles away?

(Ans.: 165 ft)

7-28. A surveyor is going to take an 8-mile sight across a lake from the top of one tower to a target on the top of another tower. It is desired to keep the line of sight 10 ft above the lake surface. At what equal heights above the shoreline should the instrument and the target be located?

7-29. Calculate the error involved in the following level rod readings if a 13.0-ft rod is assumed to be 6 in. out of plumb at its top.
(a) A BS of 12.100 ft.

(Ans.: 0.009 ft)

(b) A FS of 3.800 ft.

(Ans.: 0.003 ft)

7-30. A line of levels is run from point A (elevation 642.80 ft) to point B to determine its elevation. The value so obtained is 896.20 ft. If a check of the rod after the work is done reveals that the bottom 0.04 ft has been worn away, what is the correct elevation of point B if there were 15 instrument setups?

7-31. The smallest divisions on a level rod are $\frac{1}{10}$ of a foot. If 20 divisions on the vernier cover 19 divisions on the main scale, what is the least value readable?

(Ans.: 0.005 ft)

7-32. The least divisions on a level rod are tenths of a foot. Describe a vernier that will enable the surveyor to read the rod to the nearest 0.01 ft.

7-33. For a level rod that is graduated to hundredths of a foot, design a target vernier so that it can be read to the nearest 0.002 ft.

(Ans.: 5 divisions on vernier covering 4 divisions on level rod)

7-34. Repeat Problem 7-32 except that the rod to be used will be read to the nearest 0.005 ft.

7-35. The two-peg test is used to check to see if the line of sight of the telescope is parallel to the bubble tube axis. The instrument is set up halfway between points A and B and rod readings on A and B are 6.46 and 7.09 ft, respectively. The level is moved very close to point A and readings of 5.10 ft on A and 5.67 ft on B are taken. What should this last reading on B equal for the instrument to be in proper adjustment?

(Ans.: 5.73 ft)

7-36. A two-peg test is run as described in Problem 7-35. With the instrument halfway between points A and B, the readings on those two points are 1.923 and 2.162 m, respectively. With the level located near point A, the reading on A is 1.524 m, while on B it is 1.802 m. What should the last reading on B equal for the instrument to be in proper adjustment?

CHAPTER EIGHT

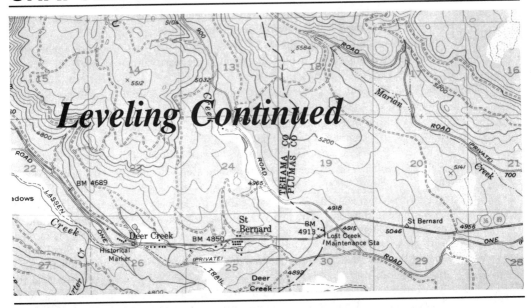

Leveling Continued

8-1 RECIPROCAL LEVELING

A useful method for leveling across wide, deep ravines, wide rivers, or other large bodies of water is called *reciprocal leveling*. We note that long sights are required across these features, and for such sights, errors will be large because of the earth's curvature and atmospheric refraction. In addition, rod reading errors and errors caused by imperfect instrument adjustment (inclination of the line of sight of the instrument) can be appreciable on these shots. Reciprocal leveling is especially devised to reduce these errors.

For this description, reference is made to Figs. 8-1 and 8-2 in which the elevation of point A on one side of a river is known and the elevation of point B on the other side is desired.

The instrument is set up very close to A, as shown in Fig. 8-1, and a BS is taken on a level rod held at A after which a foresight is taken on a level rod at B. Because the distance is large and the errors are appreciable, several FS readings are taken and averaged. The instrument is moved to a point near B, as shown in Fig. 8-2; several BS readings are taken on A and averaged, after which a FS reading is taken on B.

When the instrument is near A, the average FS reading on B (a $-$sight) contains the appreciable error. When the instrument is near B, the average BS reading on A (a $+$sight) is the one that has the appreciable error. If the difference in elevation between points A and B, determined in each of the cases, is averaged, the large errors should theoretically be removed, as shown in the following expressions, where D is the difference in elevation between the two points. In these

152

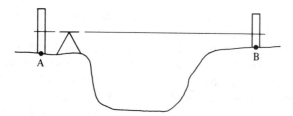

Figure 8-1

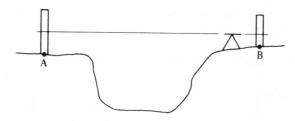

Figure 8-2

expressions it is assumed that the averages of the long sights have already been calculated.

$$D = \frac{[(BS) - (FS_{avg} \text{ including error})] + [(BS_{avg} \text{ including error}) - (FS)]}{2}$$

$$D = \frac{(\text{sum of backsights}) - (\text{sum of foresights})}{2}$$

If two leveling rods are used so that no appreciable time elapses between BS and FS readings, the precision of the work can be improved. In addition, the use of two levels with which simultaneous observations can be made will reduce the effects of atmospheric refraction variations. If two rods are used, it is necessary to use each of them for the same number of backsights as foresights. In this way errors due to incorrect rod lengths will be greatly reduced, as described in Section 7-8.

8-2 ADJUSTMENTS OF LEVEL CIRCUITS

Levels over One Route

If a surveyor runs a line of levels from a bench mark in order to set and establish the elevations of several bench marks some distance away, he or she will have to tie back into the original bench mark (or to some other bench mark). If he or she does not do this, he or she will not be able to check against serious discrepancies in his or her work. If the surveyor were, however, to level back to the starting bench mark, he or she would in all probability (no matter how careful the methods) obtain a value different from the starting one. As a result, he or she would have to adjust proportionately the measured elevations for the new bench marks that he or she set along the route.

The majority of the errors occurring in leveling are accidental; that is, there is just as much chance of getting too large a value with a particular reading as there is of getting too small a value. It has previously been shown that the probable total error in a series of accidental errors tends to vary as the square root of the number of chances for the error to occur. In leveling, therefore, the probable total error varies as the square root of the number of setups of the instrument. For this discussion it is assumed that the number of setups is the same in any mile as in any other mile and thus the total probable error varies as the square root of the distance.

It can be shown that corrections for such errors should be proportional to the square of the probable errors, or, in this case, proportional to $(\sqrt{\text{distance}})^2$ and thus proportional to the distance. It follows that the logical correction to the measured elevation of a particular point in a level circuit should be to the total correction as the distance to that point from the beginning is to the total circuit distance.

For this discussion, the line of levels shown in Fig. 8-3 is considered. The surveyor starts from BM_1 with its known elevation and establishes BM_2, BM_3, and BM_4 and levels back to BM_1.

The total distance leveled from BM_1 around the circuit and back to BM_1 was 15 miles. It is assumed that the total error in that distance was $+0.30$ ft. If the error is assumed to be the same in any 1 mile of leveling as in any other mile, the following calculations can be made:

$$\text{Total correction to be made in 15 miles} = -0.30 \text{ ft}$$

$$\text{Correction per mile} = \frac{-0.30}{15} = -0.02 \text{ ft/mile}$$

Then the correction in the elevation of any one of the bench marks equals the number of miles from the beginning point (BM_1) to the bench mark in question times the correction per mile. For instance,

$$\text{Correction for } BM_2 \text{ elevation} = (3)(-0.02) = -0.06 \text{ ft}$$

$$\text{Correction for } BM_3 \text{ elevation} = (7)(-0.02) = -0.14 \text{ ft}$$

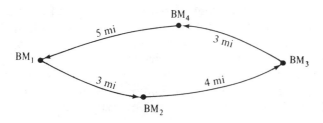

Figure 8-3

TABLE 8-1

Point	Distance from BM_1 (mi)	Observed elevation (ft)	Correction (ft)	Most probable elevation (ft)
BM_1	0	200.00	0.00	200.00
BM_2	3	209.20	−0.06	209.14
BM_3	7	216.44	−0.14	216.30
BM_4	10	211.86	−0.20	211.66
BM_1	15	200.30	−0.30	200.00

The observed elevation for each of the bench marks of Fig. 8-3 is shown in Table 8-1 together with the calculations of their most probable elevations.

Levels over Different Routes

If several lines of levels are run over different routes from a common beginning point to a common ending point where it is desirable to establish a bench mark, it is obvious that different results will be obtained. It is desirable to obtain the most probable elevation of this new point.

As mentioned previously, leveling errors vary approximately as the square of the distance leveled. Thus, in comparing different lines of levels to the same point, the shorter a particular route, the greater will be the importance or the weight given to its results. In other words, the shorter a line, the more accurate its results should be. Thus the weight of an observed elevation varies inversely as the length of its line.

For the example shown in Fig. 8-4, route a is run for 4 miles from BM_1 in order to establish BM_2, where an elevation of 340.21 ft is obtained. A second route, labeled b, is also run from BM_1 to BM_2 over a distance of 2 miles and a measured elevation of 340.27 ft is obtained.

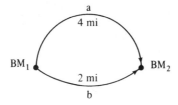

Figure 8-4

Because the second route is half as long as the first one, twice as much weight should be given to its observed value. From this information the most probable elevation of BM_2 can be obtained with the following expression:

$$\frac{(1)(340.21) + (2)(340.27)}{3} = 340.25 \text{ ft}$$

TABLE 8-2

Route	Length (mi)	Measured elevation of BM$_2$ (ft)	Measured elevation − 340.00 (ft)	Weight of route	Weighted difference (ft)
a	4	340.21	0.21	$\frac{1}{4} = 0.25$	$\frac{0.25}{0.75} \times 0.21 = 0.07$
b	2	340.27	0.27	$\frac{1}{2} = 0.50$	$\frac{0.50}{0.75} \times 0.27 = 0.18$
				$\Sigma = 0.75$	$\Sigma = 0.25$

Most probable elevation of BM$_2$ = 340.25 ft

The preceding calculation is perfectly acceptable, but when several routes are involved and the numbers are not as simple, the calculations might be more convenient to handle, as shown in Table 8-2. Here the weight given to each route equals the reciprocal of its length in miles.

Another example, somewhat more complicated, is presented in Table 8-3 and is shown in Fig. 8-5. Again, BM$_1$ has a known elevation and it is desired to set and establish the elevation of BM$_2$. To accomplish this objective, four different routes are run from BM$_1$ to BM$_2$. The lengths of each of these routes and the values of the elevations obtained for BM$_2$ are given in the table and the most probable elevation of BM$_2$ is determined.

For a detailed discussion of level net adjustments the reader is referred to pages 157–167 of the book *Surveying for Civil Engineers*, 2nd ed., by the late Phillip Kissam (New York: McGraw-Hill Book Company, 1981). Adjustments for

TABLE 8-3

Route	Length (mi)	Measured elevation of BM$_2$ (ft)	Measured elevation − 106.00 (ft)	Weight of route	Weighted difference (ft)
a	1	106.50	0.50	$\frac{1}{1} = 1.00$	$\frac{1.00}{1.93} \times 0.50 = 0.26$
b	3	106.44	0.44	$\frac{1}{3} = 0.33$	$\frac{0.33}{1.93} \times 0.44 = 0.08$
c	2	106.52	0.52	$\frac{1}{2} = 0.50$	$\frac{0.50}{1.93} \times 0.52 = 0.13$
d	10	106.36	0.36	$\frac{1}{10} = 0.10$	$\frac{0.10}{1.93} \times 0.36 = 0.02$
				$\Sigma = 1.93$	$\Sigma = 0.49$

Most probable elevation of BM$_2$ = 106.49 ft

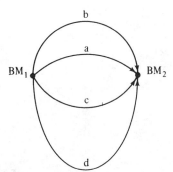

Figure 8-5

complicated nets are usually performed with computers using the method of least squares (see Appendix B).

8-3 PRECISE LEVELING

The term *precise leveling* is usually applied to the leveling practiced by the National Geodetic Survey. The purpose of this section is not to describe in detail how precise leveling is performed but rather to indicate methods by which the average surveyor working with ordinary equipment can substantially raise the precision of his or her work. The surveyor can apply one or more of the procedures presented in the following paragraphs with improvements in his or her leveling for each one.

Miscellaneous

Several practices that improve the precision of leveling were mentioned in Chapter 7. These include the following: setting tripod legs firmly in solid ground, allowing the least time possible between BS and FS readings for each setup, using clearly marked solid TPs, limiting sight distances to 300 ft or less, careful plumbing of level rods, and avoiding leveling during strong winds and during severe heat waves. Another useful practice is to keep lines of sight at least 2 ft above the ground because of atmospheric refraction effects.

Shading the Level

If a level is used on warm days in the direct sunlight, the result may be unequal expansion of different parts of the level with consequent errors. For instance, if one end of a bubble tube becomes warmer than the other end, the bubble will move toward the warmer end. This problem may be greatly minimized by shading the instrument from the direct rays of the sun with an umbrella while observations are being made and while the instrument is being moved from one point to another.

Figure 8-6   Precise leveling observations. (Courtesy of National Geodetic Survey.)

Although the umbrella may make the surveyors more comfortable, that is not its purpose (see Fig. 8-6).

Three-Wire Leveling

A method called three-wire leveling is occasionally used when levels are equipped with stadia hairs in addition to the regular cross hairs. All three horizontal hairs are read and the average of the three readings is taken as the correct reading. It is helpful to balance fairly well the BS and FS distances, and this is easily done by reading the interval between the stadia hairs.

For many years three-wire leveling was used only for very precise leveling and not for ordinary work because it is such an agonizingly slow process when a target is used. Today, however, it is much more frequently used even for projects which require ordinary precision. A target is probably not used. The three readings are taken with the telescope and are either estimated by eye to about the nearest 0.002 ft or better they are taken with an optical micrometer (previously discussed in Section 6-7). To prevent mistakes in taking each set of 3 readings it is quite important to check the difference in readings between the top and middle hair and the middle and bottom hair. The differences should be almost identical.

Precise Leveling Rods

For ordinary leveling, the usual level rod is satisfactory, but precision can be improved by using a rod treated in some manner against contraction and expansion in order to minimize the effects of changes of temperature and humidity. The preferable solution involves the use of a graduated Invar tape which is independent of the main part of the rod so that it is free to slide in grooves on each side of the rod if the rod changes in length. Precise leveling rods are equipped with a thermometer and a bull's-eye level or a pair of ordinary levels placed at 90° to each other for plumbing the rod.

DOUBLE RODDED LINE ALONG GREEN STREET

Oct. 24, 1990
Clear, mild 60°
Buff dumpy level #63210

J.B. Johnson T
R.C. Knight Ø
N.T. Hanson N

Sta.	BS	HI	FS	Elev.	Mean
BM₁	6.442	680.806	X	674.364	
	6.442	680.806			
TP₁H	7.174	684.359	3.621	672.185	
TP₁L	8.626	684.349	5.083	675.723	
TP₂H	4.570	683.042	5.987	678.472	
TP₂L	5.904	683.028	7.225	677.124	
TP₃H	1.041	674.269	9.814	673.228	
TP₃L	2.432	674.255	11.205	671.823	
BM₂			8.642	665.627	665.620
BM₂			8.642	665.613	
	+42.631		-60.119		

Conc. monument N.E. corner of intersection of Green and Oak Streets

Spike in telephone pole 80'N of McDonald's on Green St.

CHECK

$$(\Sigma \, BS + \Sigma \, FS) \div 2 = (+42.631 - 60.119) \div 2 = -8.744$$

674.364 - 665.620   = -8.744 ✓

N.T. Hanson

Figure 8-7

The National Geodetic Survey uses 3-m-long one-piece rods. They are marked in black-and-white checkerboard fashion so that the cross hairs will always fall on a white space, where its position can be precisely estimated. Furthermore, these rods have the characteristic of being clearly visible in both light and dark conditions. The smallest divisions on the rod are usually 5 mm. With optical micrometers and first-order levels, readings can be taken at least to 0.1 mm.

Double-Rodded Lines

The use of two level rods and two sets of TP, BS, and FS readings may improve leveling precision a little. Two sets of notes are kept, and the observed elevations of the points in question are averaged. When this procedure is used, it is desirable to use TPs for the two lines which have elevation differences of at least 1 ft or more so that the possibility of making the same 1-ft mistakes in both lines is reduced. To minimize systematic errors, rods may be swapped between the lines at convenient intervals, for example, on alternate instrument setups. Double-rodded lines are particularly useful when it is necessary to run a quick set of differential levels in order to establish an elevation when there is not sufficient time to check the work by returning to the initial point or to some other point whose elevation has previously been established. An example set of notes for a double-rodded line is shown in Fig. 8-7. In these notes the turning points are listed as H or L (high route or low route) to identify the BS and FS values on the two different lines.

8-4 PROFILE LEVELING

For purposes of location, design, and construction it is necessary to determine elevations along proposed routes for highways, canals, railroads, water lines, and similar projects. The process of determining a series of elevations along a fixed line is referred to as *profile leveling*.

Profile leveling consists of a line of differential levels with a series of intermediate shots taken during the process. The instrument is set up at a convenient point and a BS taken to a point of known elevation in order to determine the HI. If a bench mark of known elevation is not available, it is necessary either to establish one by differential levels from an existing BM or to set one up and give it an assumed elevation. (The latter procedure might not be satisfactory for some projects, for example, those involving water.)

After the HI is established, a series of foresights are taken along the center line of the project. These readings are referred to herein as *intermediate foresight readings* (IFS). Other surveyors may call them either *ground rod readings* or just rod readings. These readings are taken at regular intervals, say 50 to 100 ft, and at points where sudden changes in elevations occur, such as at the tops and bottoms of river banks, edges and center lines of roads and ditches, and so on.

In other words, shots are taken where necessary to give a true picture of the ground surface along the route.

When it is no longer possible to continue with the IFS readings from the instrument position, it is necessary to take a FS to a TP and move the instrument to another position from which another series of readings can be taken. A portion of a typical set of profile level notes is shown in Fig. 8-8. The author has included a partial math check for the differential leveling part of these notes. The check includes the BS and FS values from the elevation of BM_{78} to the HI elevation just beyond TP_2.

It will be noted that when surveys are made along fixed routes such as these, the distances from the starting points are indicated by stationing. The usual practice is to set stakes along the center line of the project at regular intervals, say 50 to 100 ft. Points along the route with even multiples of 100 ft are referred to as *full stations* as 0 + 00, 1 + 00, 2 + 00, and so on. Intermediate stations are referred to as *plus stations*. For instance, a point 234.65 ft from the beginning station would be designated as 2 + 34.65. Some surveyors like to assign a station

		PROFILE LEVELING				Oct 31, 1990	J.B Johnson ⊤
FOR		PROPOSED CONGAREE HIGHWAY				Overcast, mild 55°	R.C. Knight ⦿
						Topcon Auto Level #R26077	N.T. Hanson N
Sta.	BS	HI	FS	IFS	Elev.		
BM₇₈	3.11	103.11	✕	✕	100.00	Brass monument 208'N of sta 0+00	
0+00				8.6	94.5		
+50				7.3	95.8		
1+00				5.9	97.2		
+50				5.8	97.3		
2+00				6.1	97.0		
+50				7.3	95.8		
TP₁	4.73	105.67	2.17		100.94	Stone	
3+00				3.7	102.0		
+22.7				3.2	102.5	Top of ditch	
+24				8.7	97.0	Bottom of ditch	
+35.6				8.6	97.1	Bottom of ditch	
+38				5.5	100.2	Top of ditch	
+50				5.3	100.4		
4+00				7.0	98.7		
+50				8.0	97.7		
TP₂	6.07	110.68	1.06		104.61	Stump	
5+00				11.9	98.8		
+50				10.0	100.7		
	+13.91		-3.23				
	-3.23	Elev.	BM₇₈ = 100.00				
	+10.68	Elev.	HI = 110.68				
		Checks	+10.68				M.T. Hanson

Figure 8-8

of, say, 15 + 00 or 20 + 00 to the starting point of a project so that if the project is expanded in the other direction there probably will not be any negative stations.

It is usually unnecessary to take the intermediate foresight readings with as high a degree of precision as that needed for the regular FS values. If the surface is irregular, as in a field or a forest, readings to the nearest hundredth of a foot are unnecessary and perhaps a little misleading. They are usually taken to the nearest tenth of a foot, as in the example presented in Fig. 8-8. When surfaces are smooth and regular, as along a paved highway, it is not unreasonable to take the IFS readings to the nearest 0.01 ft.

During profile leveling it is usually very wise to set a series of BMs because they can be very useful at a later date, for example, when grades are being established for construction. These control points should be set a sufficient distance from the proposed project center line so that they hopefully will not be obliterated during construction operations. When the profile is completed, it is necessary to check the work by tying into another BM or by running a line of differential levels back to the beginning point.

8-5 PROFILES

The purpose of profile leveling is to provide the information necessary to plot the elevation of the ground along the proposed route. A *profile* is the graphical intersection of a vertical plane, along the route in question, with the earth's surface. It is absolutely necessary for the planning of grades for roads, canals, railroads, sewers, and so on.

Figure 8-9 shows a typical profile of the center line of a proposed highway. It is plotted on profile paper especially made for this purpose, and having horizontal and vertical lines printed on it to represent distances both horizontally and vertically. It is common to use a vertical scale much larger than the horizontal one (usually 10:1) in order to make the elevation differences very clear. In this figure a horizontal scale of 1 in. = 200 ft and a vertical scale of 1 in. = 20 ft are used. The plotted elevations for the profile are connected by freehand because it is felt that the result is a better representation of the actual ground shape than would be the case if the points were connected with straight lines.

The author has sketched a trial grade line in the figure for the proposed highway. The major purpose of plotting the profile is to allow the establishment of grade lines. On an actual job, different grade lines are tried until the cuts and fills balance sufficiently within reasonable haul distances with grades that are not too excessive.

For highway projects, it is normally convenient to plot the plan and profile on the same sheet. The top half of the paper is left plain so that the plan can be plotted above the profile. The plan is not shown in Fig. 8-9.

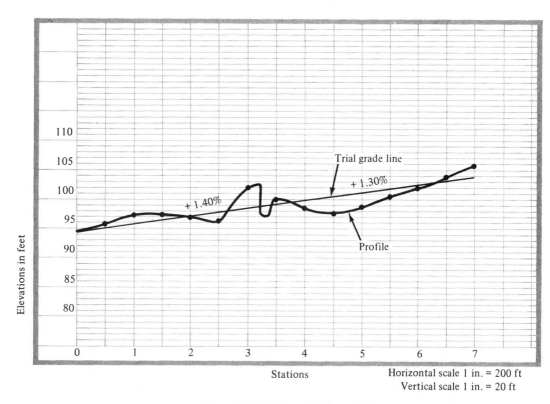

Figure 8-9   Profile and trial grade lines.

8-6 CROSS SECTIONS

Cross sections are lines of levels or short profiles made perpendicular to the center line of a project. They provide the information necessary for estimating quantities of earthwork. There are two general types of cross sections: the ones required for route projects such as roads and the ones required for borrow pits. This section is devoted to the first of these two types; the latter is discussed in Section 18-3.

For route surveys, cross sections are taken at regular intervals such as the 50- or 100-ft stations and at sudden changes in the center-line profile. There is a tendency among surveyors to take too few sections, particularly in rough country. To serve their purpose, the sections must extend a sufficient distance on each side of the center line so that the complete area to be affected by the project is included. Where large cuts or fills seem probable, greater distances from the center line should be sectioned. To attempt to give the reader a feeling as to where sights should be taken, Fig. 8-10 is presented. In this figure suggested locations for rod readings are shown for these different locations along a proposed highway.

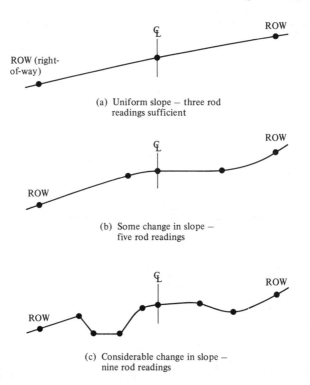

(a) Uniform slope – three rod
readings sufficient

(b) Some change in slope –
five rod readings

(c) Considerable change in slope –
nine rod readings

Figure 8-10   Suggested locations for rod readings for cross-section levels.

The necessary elevations can be determined with a regular level, with a hand level, or with a combination of both. Sometimes a hand level is held or fastened on top of a stick or board about 5 ft tall (called a Jacob staff) and is used to determine the needed elevations. Elevations are usually taken at regular intervals as say 25, 50, 75 ft, and so on, on each side of the route center line as well as at points where significant changes in slopes or features (as streams, rocks, etc.) occur. *The height of the telescope line of sight above the ground is referred to as the h.i. throughout this book (the term HI is used to give the elevation of the line of sight with respect to a reference datum, normally sea level).*

In Fig. 8-11 the level or Jacob staff is set up at the center line of the project. The height of the telescope of the instrument above the ground (h.i. in the figure) is measured and found to be 5.1 ft. The rodman is sent out 15 ft to the left and a FS reading of 11.2 is taken. The difference in elevation from the center line to the point in question is thus −6.1 ft. Then the rodman moves out 18 more ft to a point 33 ft from the center line. The FS reading is 11.5 ft and the difference in elevation is −6.4 ft. If the difference in elevation becomes too large for the level rod height or if in going uphill the elevation differences are larger than the h.i. of the instrument, it will then be necessary to take one or more turning points to obtain all the necessary readings.

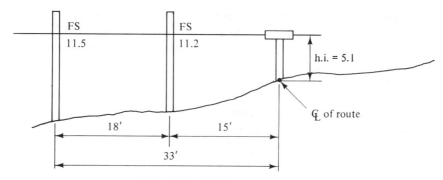

Figure 8-11 Taking cross sections.

A convenient form of recording cross-section notes is shown in Fig. 8-12. This figure includes profile notes on the left-hand page and cross-section notes on the right-hand page. On the cross-section page, the numerators are the differences in elevations from the center-line stations to the points in question; the denominators are the distances from the center line. The surveyor must record

STA	BS	HI	FS	IFS	ELEV	Left		₵	Right	
+87				6.2	276.4	+$\frac{0.7}{33}$ +$\frac{0.4}{15}$		−$\frac{0.2}{18}$	+$\frac{0.1}{33}$	
+50				6.3	276.3	+$\frac{0.4}{33}$ +$\frac{0.3}{15}$		−$\frac{0.4}{18}$	−$\frac{0.6}{33}$	
13+00				6.9	275.7	+$\frac{0.5}{33}$ +$\frac{0.2}{16}$		+$\frac{0.2}{15}$	+$\frac{0.2}{33}$	
TP	7.32	282.58	4.30		275.26					
+50				4.9	274.7	−$\frac{0.3}{33}$ −$\frac{1.4}{15}$	−$\frac{0.3}{12}$	+$\frac{0.2}{16}$	+$\frac{0.5}{33}$	
12+00				5.1	274.5	−$\frac{8.2}{33}$ −$\frac{3.1}{33}$	−$\frac{2.3}{15}$	+$\frac{0.2}{17}$	+$\frac{0.8}{33}$	
+50				5.2	274.4	−$\frac{9.0}{39}$ −$\frac{4.6}{33}$	−$\frac{3.0}{19}$	$\frac{0.0}{20}$	$\frac{0.0}{33}$	
+23				5.2	274.4	−$\frac{7.3}{37}$ −$\frac{6.0}{33}$	−$\frac{4.0}{15}$	$\frac{0.0}{15}$	+$\frac{0.8}{33}$	
11+00				5.3	274.3	−$\frac{6.3}{40}$ −$\frac{6.4}{33}$	−$\frac{6.1}{15}$	+$\frac{0.3}{15}$	+$\frac{0.1}{33}$	
BM₅	5.87	279.56	3.31		273.69					

CROSS SECTIONS FOR ROAD 10A

Nov 7, 1990
Overcast, Mild 50°
Topcon Auto level #R20017

J.B.Johnson,π,N
R.C.Knight,∅

Notes - N.T. Hanson

Nail in root 12" pecan 90' left sta 10+50

J.B. Johnson

Figure 8-12 Profile and cross-section notes.

carefully the signs for these numbers. A plus sign is given to the points that are higher than the center line and a minus sign to those below.

Many surveyors plot their level notes running up the page from the bottom. Such a form is quite logical in that, as the surveyor faces forward along the center line, the area to the right of the center line is shown on the right-hand side of the page and the area to the left is shown on the left-hand side. Thus the notes show the configuration of the ground as the surveyor looks at it. Close stations are at the bottom of the page and far stations are at the top of the page. The notes of Fig. 8-12 are shown in this fashion.

8-7 MISTAKES IN NONCLOSED LEVELING ROUTES

Mistakes in leveling for closed level loops are an annoying problem, but they can be discovered and eliminated. Mistakes in leveling that is not part of a closed loop are a more serious problem. Examples of such cases include profile leveling, cross sections, construction grades, and so on. Just imagine the time and cost involved if we have to tear out a reinforced concrete footing for a building and rebuild it at a different elevation because of a mistaken FS reading. Similar remarks can be made about other construction work, such as bridge abutments, culverts, and other structures. The reader is reminded to be exceptionally careful in this type of work and to repeat many measurements.

PROBLEMS

8-1. In reciprocal leveling across a large ravine the following sets of rod readings were taken. Point A on one side has a known elevation of 882.48 ft; B is a point on the other side whose elevation is desired. Determine the elevation of point B.

Instrument near A: BS on A = 4.32, average FS on B = 6.28

Instrument near B: average BS on A = 5.12, FS on B = 6.37

(*Ans.:* 880.88 ft)

8-2. Reciprocal leveling across a river between points A and B gives the values shown. If the elevation of point A is 3050.67 ft, what is the elevation of point B?

Instrument near A: BS = 5:12; FS readings = 6.66, 6.68, 6.67

Instrument near B: BS readings = 3.23, 3.22, 3.24; FS = 5.19

8-3. Adjust the following unbalanced level circuit.

Point	Distance from BM$_1$ (miles)	Observed elevation (ft)
BM$_1$	0	864.20
BM$_2$	3	868.34
BM$_3$	9	859.42
BM$_4$	15	863.80
BM$_1$	18	863.66

(*Ans.*: BM$_2$ = 868.43 ft)
(*Ans.*: BM$_3$ = 859.69 ft)
(*Ans.*: BM$_4$ = 864.25 ft)

8-4.   Adjust the unbalanced level circuit shown in the accompanying table.

Point	Distance from BM$_1$ (miles)	Observed elevation (ft)
BM$_1$	0	714.77
BM$_2$	3	718.96
BM$_3$	7	729.64
BM$_4$	9	732.44
BM$_5$	14	712.53
BM$_1$	22	715.21

8-5.   A line of levels was run to set the elevation of several bench marks. The following values were obtained:

Point	Distance from BM$_1$ (miles)	Observed elevation (ft)
BM$_1$	0	834.30
BM$_2$	2	819.62
BM$_3$	5	801.38
BM$_4$	7	817.70
BM$_1$	11	833.97

Is the closure satisfactory for a leveling requirement $= \pm 0.12\sqrt{M}$? Adjust the elevations for all of the bench marks.

(*Ans.*: BM$_2$ = 819.68 ft)
(*Ans.*: BM$_3$ = 801.53 ft)
(*Ans.*: BM$_4$ = 817.91 ft)

8-6. A closed loop of differential levels was run to establish the elevations of several bench marks with the following results:

Distance (ft)		Observed elevation (ft)	
A to B	6000	BM_A	1348.22
B to C	5000	BM_B	1364.72
C to D	3000	BM_C	1352.66
D to E	4800	BM_D	1312.88
E to A	2700	BM_E	1329.77
		BM_A	1348.65

8-7. BM_1 has a known elevation and it is desired to establish the elevation of BM_2 by running differential levels over three different routes from BM_1, as shown in the accompanying table. What is the most probable elevation of BM_2?

Route	Length (miles)	Measured Elevation of BM_2 (ft)
a	2	969.94
b	4	969.88
c	8	969.82

(*Ans.*: 969.90 ft)

8-8. Several lines of levels are run over different routes from BM_1 in order to set BM_2 and establish its elevation. The lengths of these routes and the value of the elevations determined are shown in the accompanying table. Determine the most probable elevation of BM_2.

Route	Length (miles)	Measured Elevation of BM_2 (ft)
a	2	314.66
b	5	314.58
c	13	314.50
d	19	314.61

8-9. Set up and complete the level notes for a double rodded line from BM_{11} (elevation 1642.324) to BM_{12}. In the following rod readings H refers to high route and L to low route: BS on BM_{11} = 7.342; FS on TP_1H = 2.306; FS on TP_1L = 4.107; BS on TP_1H = 9.368; BS on TP_1L = 11.162; FS on TP_2H = 3.847; FS on TP_2L = 5.111; BS on TP_2H = 8.339; BS on TP_2L = 9.619; FS on TP_3H = 4.396; FS on TP_3L = 5.448; BS on TP_3H = 7.841; BS on TP_3L = 8.896; FS on BM_{12} = 3.329.

(*Ans.*: 1661.342)

8-10. The centerline for a proposed highway has been staked. With an HI of 854.32 ft the following intermediate foresights are taken at full stations beginning at Sta. 0 + 00: 3.1, 3.6, 4.1, 6.0, 6.9, 7.7, 8.6, 9.3, 9.4, 9.8, 9.2, 8.7, 8.2, 7.9, 8.0, 8.3, 8.8, and 9.4. Plot the profile for these stations.

8-11. Complete the following set of profile level notes.

Station	BS	HI	FS	IFS	Elevation
BM$_1$	4.86				892.11
TP$_1$	5.04		8.32		
0 + 00				6.8	
1 + 00				6.6	
2 + 00				7.4	
2 + 70.5				8.7	
2 + 76.2				8.5	
3 + 00				7.3	
TP$_2$	3.12		6.22		
4 + 00				6.2	
5 + 00				5.8	
5 + 55.0				5.2	
5 + 65.2				4.9	
6 + 00				4.3	
BM$_2$			5.16		

(*Ans.:* BM$_2$ = 885.43)

CHAPTER NINE

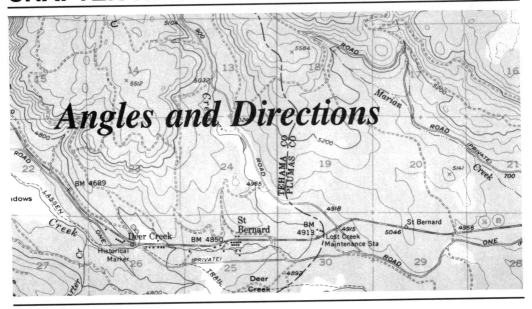

Angles and Directions

9-1 MERIDIANS

In surveying, the direction of a line is described by the horizontal angle that it makes with a reference line or direction. Usually this is done by referring to a fixed line of reference called a *meridian*. There are three types of meridians: astronomic, magnetic, and assumed. An astronomic meridian is the direction of a line passing through the astronomic north and south poles and the observer's position as shown in Fig. 9-1.

Astronomic north is based on the direction of gravity and the axis of rotation of the earth. It is determined from observations of the sun or other stars whose astronomical positions are known (the sun and the north star, Polaris, being the most common). Sometimes the term *geodetic north* is used. It is a direction determined from a mathematical approximation of the earth's shape. It is slightly different from astronomic north and that difference can be as much as 20 arc-seconds in some parts of the western United States. The subject of astronomic directions is discussed in Chapter 16. A *magnetic meridian* is the direction taken by the magnetized needle of a compass at the observer's position; an *assumed meridian* is an arbitrary direction taken for convenience.

Astronomic meridians should be used for all surveys of large extent and, in fact, are desirable for all surveys of land boundaries. They do not change with time and can be reestablished decades later. Magnetic meridians have the disadvantage of being affected by many factors some of which vary with time. Furthermore, there is no precise method available for establishing what magnetic north was years ago in a given locality. An assumed meridian may be used sat-

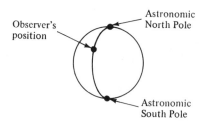

Figure 9-1   Astronomic meridian.

isfactorily for many surveys which are of limited extent. The direction of an assumed meridian is usually taken roughly in the direction of an astronomic meridian. Assumed meridians have a severe disadvantage—the problem of reestablishing their direction if the points of the survey on the ground are lost.

Sometimes for surveys of limited extent another type of meridian is used. A line through one point of a particular area is selected as a reference meridian (usually astronomic) and all other meridians in the area are assumed to be parallel to this, the so-called *grid meridian*. The use of a grid meridian eliminates the need for considering the convergence of meridians at different points in the area.

9-2 UNITS FOR MEASURING ANGLES

Among the methods used for expressing the magnitude of plane angles are the sexagesimal system, the centesimal system, and the methods using radians, and mils. These systems are described briefly in the following paragraphs.

Sexagesimal System.   In the United States as in many other countries the sexagesimal system is used in which the circumference of circles is divided into 360 parts or degrees. The degrees are further divided into minutes and seconds. Thus an angle may be written as 36°27′32″. The National Geodetic Survey as it uses the metric system continues to use the sexagesimal system for angles and directions.

Centesimal System.   In some countries, particularly in Europe, the centesimal system is used in which the circle is divided into 400 parts called *gon*. (These were until recently referred to as *grads*.) It will be noted that 100 gon = 90°. An angle may be expressed as 122.3968 gon (which when multiplied by 0.9 will give us 110.15712° or 110°09′25.6″).

The Radian.   Another measure of angles frequently used particularly for calculation purposes is the radian. One radian is defined as the angle subtended at the center of a circle by an arc length exactly equal to the radius of the circle. This definition is illustrated in Fig. 9-2. The circumference of a circle equals 2π

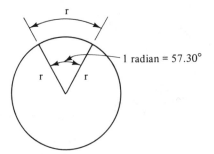

Figure 9-2 The radian.

times its radius r and thus there are 2π radians in a circle. Therefore, 1 radian equals

$$\frac{360°}{2\pi} = 57.30°$$

(One gon is equal to 0.01571 radian.)

The Mil.   Another system of angular units divides the circumference of the circle into 6400 parts or mils. This particular system of angle measurement is used primarily in military science.

9-3 AZIMUTHS

A common term used for designating the direction of a line is the *azimuth*. The azimuth of a line is defined as the clockwise angle from the north end or the south end of the reference meridian to the line in question. For ordinary plane surveys, azimuths are generally measured from the north end of the meridian. This will

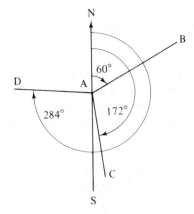

Figure 9-3

be the case for the problems presented in this book. Azimuths are occasionally measured clockwise from the south end of the meridian as for some geodetic and astronomic projects. When azimuths are measured from the south end of the meridian, that fact must be clearly indicated.

The magnitude of an azimuth can fall anywhere from 0 to 360°. The north azimuths of several lines AB, AC, AD are shown in Fig. 9-3. The values are respectively 60°, 172°, and 284°.

Every line has two azimuths (forward and back). Their values differ by 180° from each other, depending on which end of the line is being considered. For instance, the *forward azimuth* of line AB is 60° and its *back azimuth* or the azimuth of line BA, which may be obtained by adding or subtracting 180°, equals 240°. Similarly, the back azimuths of lines AC and AD are 352° and 104°, respectively.

Azimuths are referred to as astronomic, magnetic, or assumed, depending on the meridian used. The type of meridian being used should be clearly indicated.

9-4 BEARINGS

Another method of describing the direction of a line is to give its *bearing*. The bearing of a line is defined as the smallest angle which that line makes with the reference meridian. It cannot be greater than 90°. In this manner, bearings are measured in relation to the north or south ends of the meridian and are placed in one of the quadrants so that they have values of NE, NW, SE, or SW.

In Fig. 9-3 the bearing of line AB is N60°E, that of AC is S8°E, and that of AD is N76°W. As with azimuths, it will be noticed that every line has two bearings, depending on which end of the line is being considered. For instance, the bearing of line BA in Fig. 9-3 is S60°W.

Depending on the reference meridians being used, bearings may be astronomic, magnetic, or assumed. Thus it is important, as it is for azimuths, to indicate clearly the type of reference meridian being used.

It is correct to say that a bearing is N90°E, but it is more common to say that it is due east. Similarly, the other three cardinal directions are usually referred to as due south, due west, and due north.

We have seen that directions may be given by either bearings or azimuths and we have learned that they are easily interchangeable. Until the last few decades American surveyors generally favored the use of bearings over azimuths and most legal documents reflected this favoritism. Today, however, programmable pocket calculators and electronic computers are used almost everyday by surveyors. To make calculations with these marvelous devices it is usually simpler to use the straight numerical values of azimuths rather than getting involved with bearings, which require knowledge of quadrants and the signs of the trigonometric functions. As a result, surveyors generally use azimuths today instead of bearings.

9-5 THE COMPASS

Man has been blessed on earth with a wonderful direction finder, the magnetic poles. The earth's magnetic field and the use of the compass have been known to navigators and surveyors for many centuries. In fact, before the sextant and transit were developed, the compass was the only means by which the surveyor could measure angles and directions.

For many centuries there has been a legend which says that the compass was originally developed by a Chinese emperor who had to fight a battle in a heavy fog. The story goes that he invented a chariot that always pointed toward the south and was thereby able to locate his enemy. Actually, no one knows who first developed the compass, and the honor is claimed by the Greeks, the Italians, the Finns, the Arabs, and a few others. Regardless of its inventor, it is known that the compass was available to sailors during the Middle Ages.

The magnetic poles are not points but oval areas that are located not at the geographic poles but approximately 1000 miles away. Today, the magnetic north pole is located approximately 1000 miles south of the astronomic north pole in the Canadian arctic near Bathurst Island. The compass needle lines up with magnetic north; in most places this means that the needle points slightly east or west of astronomic north, depending on the locality. The angle between astronomic north and magnetic north is referred to as the *magnetic declination*. (Navigators call it the *variation* of the compass.) The magnetic lines of force in the northern hemisphere are also inclined downward from the horizontal toward the magnetic north pole. The magnetized needles of compasses are counterbalanced with a little coil of brass wire on their south ends against the *dip* of the needles at the north end.

Not only are the magnetic fields not located at the geographic poles of the earth, but magnetic directions are also subject to several variations: long-term variations, annual variations, daily variations, as well as variations caused by magnetic storms, local attractions, and so on.

Modern transits and theodolites are usually manufactured without compasses because of all the inaccuracies involved in using them. Nevertheless, a discussion of compasses is included in this chapter because a knowledge of their use can be very helpful to today's surveyor, particularly in the area of land surveying.

9-6 VARIATIONS IN MAGNETIC DECLINATION

The angle of declination at a particular location is not constant but varies with time. For periods of approximately 150 years there is a gradual unexplainable shift in the earth's magnetic fields in one direction after which a gradual drift occurs in the other direction to complete the cycle in the next 150-year period. This variation, called the *secular variation*, can be very large, and it is quite

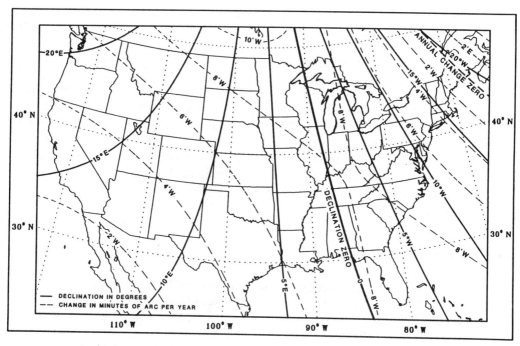

Figure 9-4 Magnetic declination in the United States, 1985. (Courtesy of U.S. Geological Survey.)

important in checking old surveys whose directions were established with a compass. There is no known method of predicting the secular change and all that can be done is to make observations of its magnitude at various places around the world. Records kept in London for several centuries show a range of magnetic declination from 11°E in 1580 to 24°W in 1820.[1] The time period between extreme eastern and western declinations varies with the locality. It can be as short as 50 years or even less and as long as 180 years or more.

   Maps are available that provide the values of magnetic declination throughout the United States. The chart shown in Fig. 9-4 was prepared from observations at more than 8500 stations throughout the country. On this chart points of equal declination are connected with lines called *isogonic lines*. For some parts of the country magnetic declinations are zero and the lines connecting them are called *agonic lines*. Also given on the chart are dashed lines called *isopors*, which show places of equal changes in magnetic declination. It is thought that this chart provides magnetic declinations within about 15′ of their correct values for more than half of the country. For some extreme cases, however, they may be in error by as much as 1°. The chart shows that the declination varies from over 20°W in

---

[1] R. C. Brinker and P. R. Wolf., *Elementary Surveying*, 7th Ed. (New York: Harper & Row Publishers, Inc., 1984), p. 159.

parts of Maine to over 20°E in parts of Washington State—a range of over 40°. This figure is a reduced version of MAP GP-986-D "The Magnetic Field in the United States, 1985, Declination Chart," which is produced by the U.S. Geological Survey. Updated versions, which are published approximately every five years, may be obtained from the U. S. Department of the Interior, Geological Survey, Box 25046, M.S. 968, Denver Federal Center, Denver, CO 80225-0046.

In addition to secular variations in the magnetic declinations, there are also annual and daily variations of lesser importance. The *annual variations* usually amount to less than a 1′ variation in the earth's magnetic field. Daily there is a swing of the compass needle through a cycle, causing a variation of as much as approximately one-tenth of 1°. The magnitude of this *daily variation* is still so small in comparison to the inaccuracies with which the magnetic compass can be read that it can be neglected.

9-7 LOCAL ATTRACTIONS

The direction taken by a compass needle is affected by magnetic attractions other than that of the earth's magnetic field. Fences, underground pipes, reinforcing bars, passing cars, nearby buildings, iron ore deposits under the earth's surface, and other steel or iron objects may have a considerable effect on compass readings. In addition, the effect of power lines, particularly because of variations in voltage, may be so great that compasses are useless in some areas. Even the surveyor's wristwatch, pen, belt buckle, taping pins, or steel tape may have a distorting effect on compass readings.

On many occasions the surveyor may not realize that the magnetic bearings he or she reads with the compass have been affected by local attractions. To detect local attraction, he or she must read both forward and back bearings for each line to see if they correspond reasonably well. To read a forward bearing, the surveyor sights along a line in the direction of the next point. To read a back bearing for the same line, he or she moves to the next point and sights back to the preceding point. If the two readings vary significantly from each other, local attraction has probably been the cause. This topic is continued in Section 9-9.

9-8 READING BEARINGS WITH A COMPASS

Compass readings may be used for surveys in which speed is important and when only limited precision is required. In addition, they may be used for checking more precise surveys or for rerunning old property lines that were originally run with compasses. They may also be used for preliminary surveys, rough mapping, timber cruising, and checking angle measurements.

For many years the surveyor's compass (see Fig. 9-5) was used for determining directions. This instrument was originally set up on a single leg called a

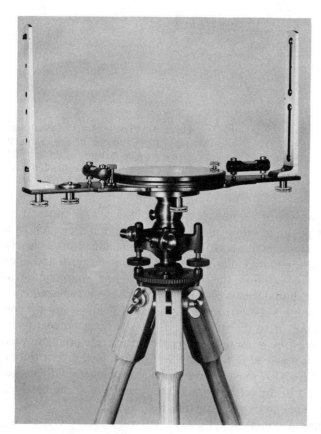

Figure 9-5 Surveyor's compass. (Courtesy of Teledyne Gurley.)

Jacob staff. Later, tripods were used. The reader should particularly notice the folding upright sight vanes or peep sights that were used for alignment. These sights were fine slits running nearly the whole length of the vanes. This obsolete instrument (now a collector's item on the antique circuit) was formerly used for laying off old land boundaries. Today, it is used occasionally when attempting to reestablish old property lines that were originally run with the same type of instrument.

To read a bearing with a surveyor's compass or with the compass on a transit, the surveyor sets up the instrument at a point at one end of the line whose bearing is desired, releases the needle clamp, and sights down the line to a point at the other end of the line. He or she then reads the bearing on the circle at the north end of the needle and reclamps the compass needle. When the compass is unclamped, a jeweled bearing at the center of the needle rests on a sharp pivot point, allowing the needle to swing freely. It is important to take care of the point and keep it from becoming dull and causing a sluggish needle. This is avoided by reclamping the needle, that is, by raising the needle off the pivot when the compass is not in use to minimize wear.

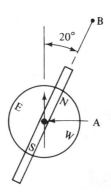

Figure 9-6 The telescope is over point A and sighted toward B. The bearing is N20°E. Notice that the scale turns with the telescope, but the needle continues to point toward magnetic north.

The condition or sensitivity of a compass needle can easily be checked by drawing the needle out of position with a piece of magnetic material such as a knife blade. If when the magnetic material is removed, the needle returns closely to its original position, it is in good condition. If the needle is sluggish and does not return closely to its original position, the pivot point can be resharpened with a stone—a rather difficult task and it is probably better to have it replaced.

In Fig. 9-6 the instrument is assumed to be located at point A and the telescope is sighted toward point B so that bearing AB can be determined. It will be noticed that whichever way the telescope is directed, the needle points toward magnetic north and bearings are read at the north end of the needle. For this to be possible, the positions of E and W must be reversed on the compass. Otherwise, when the telescope is turned to the NE quadrant, the needle would be between N and W on the compass. This fact can be seen in Fig. 9-6.

Sometimes in handling a compass, its glass face will become electrically charged because of the friction created by rubbing it with the hands or a piece of clothing. When this happens, the needle may be attracted to the undersurface of the glass by the electric charge. This charge may be removed from the glass either by breathing on it or by touching the glass at several points with wet fingers.

Many compasses (e.g., the old ones on transits) are equipped so that they may be adjusted for magnetic declinations; that is, the scale or circle under the compass which is graduated in degrees may be rotated in relation to the telescope by the amount of the declination. In this way the compass may be used to read directly approximate astronomic bearings even though declinations may be quite large.

In rerunning old surveys that were originally done with compasses, it is frequently necessary to try to reestablish some obliterated lines with the compass. In such cases it is absolutely necessary (if many years have elapsed) to estimate changes that have occurred in magnetic declinations. Such information may be obtained from old U.S.G.S. magnetic declination charts.

For new surveys it is usually desirable to work with astronomic bearings. This practice provides a permanence to surveys and greatly simplifies the work of future surveyors. An astronomic bearing may be determined for one line (probably from observations of the sun or Polaris), and the astronomic bearings for the other lines may be calculated from the angles measured with the transit or theodolite to the nearest minute or closer. Of course, if there are a great many lines involved, it is well to recheck astronomic directions every so often during the survey. (Magnetic bearings may well be used for present-day surveys by reading the magnetic bearing for one line and by computing the bearings for the other sides from the measured angles, but astronomic bearings are much to be preferred.)

Despite the increasing use of astronomic directions, it is not a bad practice to read magnetic bearings for each line if the instrument has a compass. Most new instruments do not. If mistakes are made in angle measurements, the magnetic bearings may frequently be used to determine where the mistakes occurred.

Although compass readings are estimated to the nearest 15 to 30′, it is doubtful if our accuracy at best is better than to the nearest degree. Transits and theodolites may be read to the nearest minute or closer. In addition, compass readings may be decidedly affected by local magnetic attractions. As a result, *the precision obtained is quite limited, and compasses are becoming more nearly obsolete for surveying purposes each year.*

9-9 DETECTING LOCAL ATTRACTION

When magnetic bearings are read with a compass, local attraction can frequently be a problem; thus all readings should be carefully checked. This is normally accomplished by reading the bearing of each line from both its ends. If the forward and back bearings differ by 180°, there probably is no appreciable local attraction. If the bearings do not differ by 180°, local attraction is present and the problem is to discover which bearing is correct.

For this discussion, consider the traverse of Fig. 9-7. It is assumed that the forward and back bearings were read for line AB as being S81°30′E and N83°15′W, respectively. Obviously, local attraction is present at one or both ends of the line.

One method of determining the correct value is to read bearings from points A and B to a third point as perhaps C in the figure and then to move to that third point and read back bearings to A and B. If forward bearing AC agrees with back bearing CA, this shows that there is no appreciable local attraction at either A or C. Therefore, the local attraction is at B and the correct magnetic bearing from A to B is S81°30′E.

It should be noted that at a particular point, local attraction will draw the needle a certain amount from the magnetic meridian. Thus all readings taken from

Figure 9-7

that point should have the same error, due to local attraction (assuming that the attraction is not a changing value as, say, voltage variations in a power line), and the angles computed from the bearings at that point should not be affected by the local attraction.[2]

9-10 TRAVERSE AND ANGLE DEFINITIONS

Before proceeding with a discussion of angle measurement, several important definitions concerning position and direction need to be introduced. In surveying, the relative positions of the various points (or stations) of the survey must be determined. Historically this was done by measuring the straight-line distances between the points and the angles between those lines. We will learn in Section 11-6, however, that another procedure (called radiation) is becoming more and more commonly used today.)

A *traverse* may be defined as a series of successive straight lines that are connected together. They may be *closed*, as are the boundary lines of a piece of land, or they may be *open*, as for highway, railroad, or other route surveys. Several types of angles are used in traversing, and these are defined in the following paragraphs.

An *interior angle* is one that is enclosed by the sides of a closed traverse (see Fig. 9-8).

An *exterior angle* is one that is not enclosed by the sides of a closed traverse (see Fig. 9-8).

An *angle to the right* is the clockwise angle between the preceding line and the next line of a traverse. In Fig. 9-9 it is assumed that the surveying party is proceeding along the traverse from A to B to C, and so on. At C the angle to the right is obtained by sighting back to B and measuring the clockwise angle to D.

A *deflection angle* is the angle between the extension of the preceding line and the present one. Two deflection angles are shown in Fig. 9-10. There the

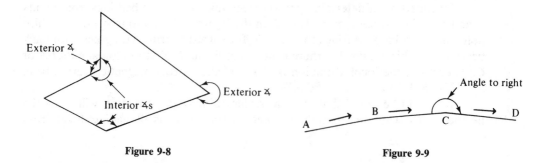

Figure 9-8   Figure 9-9

[2] C. B. Breed and G. L. Hosmer, *Elementary Surveying*, 11th ed. (New York: John Wiley & Sons, Inc., 1977), pp. 27–28.

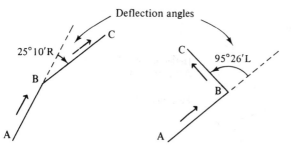

Figure 9-10

necessity for designating them as being either right (R) or left (L) should be noted. In each of these cases it is assumed that traversing is proceeding from *A* to *B* to *C*, and so on. The use of deflection angles permits easy visualization of traverses and facilitates their representation on paper. In addition, the calculations of successive bearings or azimuths is very simple (azimuths being easier to obtain).

9-11 TRAVERSE COMPUTATIONS

If the bearing of one side of a traverse has been determined and the angles between the sides have been measured, the bearings of the other sides can easily be calculated. Actually, several possible methods of solving such a problem can be used, but regardless of which procedure is chosen, preparation of a careful sketch of the known data is required. Once the sketch is made, the required calculations are obvious.

One way to solve most of these problems is by making use of deflection angles. Example 9-1 illustrates the situation in which the bearing of one line and the angle to the next line are given and it is desired to find the azimuth and bearing of that second line.

Example 9-1

For the traverse shown in Fig. 9-11, the bearing of side *AB* is given as well as the interior angle at *B*. Compute the north azimuth and the bearing of side *BC*.

Solution   A sketch is made of point *B* showing the north–south and east–west directions; the location of *AB* is extended past *B* (as shown by the dashed line), and the deflection angle at *B* is determined (see Fig. 9-12). With the value of this angle known, the azimuth and the bearing of side *BC* are obvious.

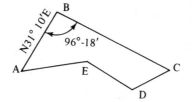

Figure 9-11

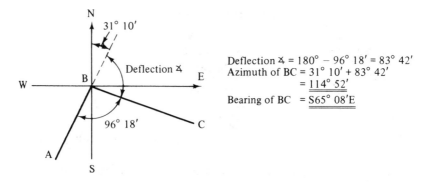

Deflection $\angle = 180° - 96° 18' = 83° 42'$
Azimuth of BC $= 31° 10' + 83° 42'$
$= \underline{114° 52'}$
Bearing of BC $= \underline{\underline{S65° 08'E}}$

Figure 9-12

If the bearings of two successive lines are known, it is easy to compute the angles between them. A sketch is made, the deflection angle is calculated, and the value of any angle desired becomes evident. Example 9-2 presents the solution of this type of problem.

Example 9-2

For the traverse shown in Fig. 9-13, the bearings of sides *AB* and *BC* are given. Compute the interior angle at *B*.

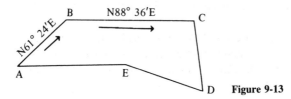

Figure 9-13

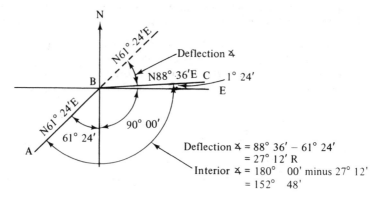

Deflection $\angle = 88° 36' - 61° 24'$
$= 27° 12'$ R
Interior $\angle = 180°$ 00' minus $27° 12'$
$= 152°$ 48'

Figure 9-14

Solution A sketch is made (Fig. 9-14), the deflection angle from the line *AB* extended to line *BC* is calculated, and any needed angle is immediately obvious.

At the end of this chapter are given several exercise problems dealing with angles, bearings, and azimuths. Before proceeding with the chapters that follow, the reader should be sure that he or she can handle all of these problems. He or she should be able to compute azimuths of lines from their bearings, and vice versa, calculate deflection angles from bearings, work from angles to bearings and bearings to angles, and be able to solve the magnetic declination problems discussed in Section 9-12.

9-12 MAGNETIC DECLINATION PROBLEMS

The reader may become confused when first attempting to calculate the astronomic bearing of a line from its magnetic bearing for which the magnetic declination is known. A problem presenting similar difficulties is the calculation of the magnetic bearing of a line today when its magnetic bearing at some time many years ago is known, as are the entirely different magnetic declinations at the two times. These problems, however, may easily be handled if the student remembers one simple rule: *Make a careful sketch of the data given.* Example 9-3 presents the solution of a magnetic declination problem.

As noted in Figs. 9-15 and 9-16, many persons illustrate magnetic north on drawings with a half arrowhead—the half which is shown being located on the same side of the line that magnetic north is from astronomic north. Such a procedure is not sufficient and meridian arrows on drawings should be clearly designated as astronomic, magnetic, or assumed.

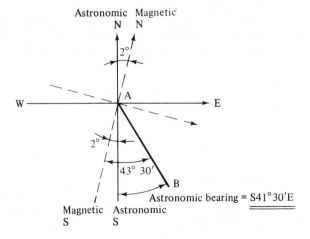

Figure 9-15

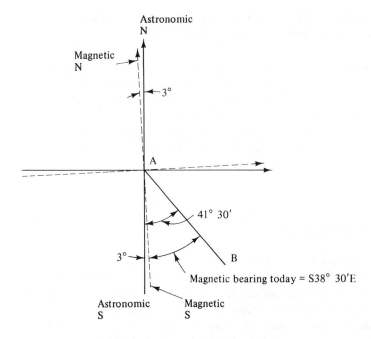

Figure 9-16

Example 9-3

The magnetic bearing of line *AB* was recorded as S43°30'E in 1888. If the declination was 2°00'E, what is the astronomic bearing of the line? If the declination is now 3°00'W, what is the magnetic bearing of the line today?

Solution   A sketch is made (Fig. 9-15) in which the astronomic directions (N, S, E, W) are shown as solid lines and the magnetic directions are shown as dashed lines. Magnetic north is shown 2°00'E or clockwise of astronomic north and line *AB* is shown in its proper positions 43°30'E of magnetic south. Once the sketch is completed, the astronomic bearing of the line is obvious.

To determine the magnetic bearing of the line today another sketch is made (Fig. 9-16), showing the astronomic bearing of the line and the present declination. From this completed sketch the magnetic bearing of the line today is obvious.

PROBLEMS

9-1. **(a)** Convert 36.4285ᵍ to sexagesimal units

(*Ans.*: 32°47'08")

(b) Convert 61°17'15" to centesimal units.

(*Ans.*: 68.0972ᵍ)

9-2. A circular arc has a radius of 420.00 ft and a central angle of 41°16'30". Determine the central angle in radians and the arc length.

9-3. Three lines have the following north azimuths: 141°33′, 254°19′, and 325°44′. What are their bearings?

(Ans.: S38°27′E, S74°19′W, N34°16′W)

9-4. Determine the north azimuths for sides *AB*, *BC*, and *CD* in the accompanying sketch for which the bearings are given.

9-5. Calculate the north azimuth for sides *OA*, *OB*, *OC*, and *OD* in the accompanying figure.

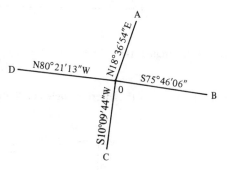

(Ans.: 18°36′54″, 104°13′54″, 190°09′44″, 279°38′47″)

9-6. Find the bearings of sides *BC* and *CD* in the accompanying figure.

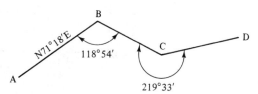

9-7. Compute the bearings of sides *BC* and *CD* in the accompanying figure.

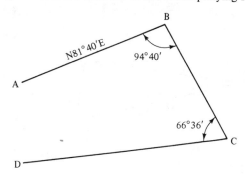

(Ans.: S13°00′E, N79°36′W)

9-8. What are the bearings of sides *CD*, *DE*, *EA*, and *AB* in the accompanying figure?

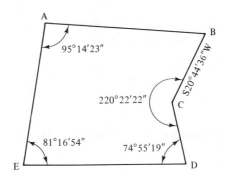

9-9. Determine the angles *AOB*, *BOC*, and *DOA* for the figure of Problem 9-5.

(*Ans.*: 85°37′00″, 85°55′50″, 98°58′07″)

9-10. Compute the value of the interior angles at *B* and *C* for the figure shown.

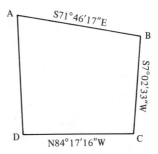

In Problems 9-11 to 9-14, compute all of the interior angles for each of the figures shown.

9-11.

N79°33′E

N73°14′W

S9°52′E

A

B

C

(*Ans.*: A = 27°13′, B = 89°25′, C = 63°22′)

9-12.

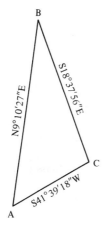

9-13.

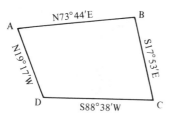

(*Ans.*: B = 91°37′, C = 73°29′)

9-14.

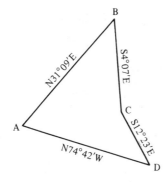

9-15. Compute the deflection angles for the traverse of Problem 9-12.
(*Ans.*: A = 147°31′09″R, B = 152°11′37″R, C = 60°17′14″R)

9-16. Compute the deflection angles for the traverse of Problem 9-14.

9-17. From the data given, compute the missing bearings.

1 − 2 = _____
2 − 3 = _____
3 − 4 = N4°10′W
4 − 1 = _____
Interior ∢ at 1 = 52°31′
Interior ∢ at 2 = 36°18′
Interior ∢ at 4 = 224°13′

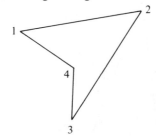

(*Ans.*: 1 − 2 = N79°06′E, 4 − 1 = N48°23′W)

9-18. From the data given, compute the missing bearings.

1 − 2 = N28°12′E
2 − 3 = _____
3 − 4 = S36°14′W
4 − 5 = _____
5 − 6 = N82°44′W
6 − 1 = _____
Interior ∢ at 1 = 121°36′
Interior ∢ at 2 = 81°17′
Interior ∢ at 4 = 238°54′

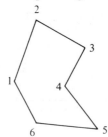

9-19. For the accompanying figure, compute the following:
(a) Deflection angle at *B*.

(*Ans.*: 70°28′R)

(b) Interior angle at *B*.

(*Ans.*: 109°32′)

(c) Bearing of line *CD*.

(*Ans.*: S24°18′E)

(d) North azimuth of *DA*.

(*Ans.*: 287°08′)

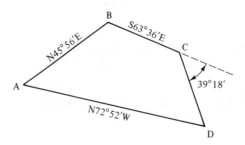

9-20. For the figure shown, compute the following:
  (a) Bearing of line *AB*.
  (b) Interior angle at *C*.
  (c) North azimuth of line *DE*.
  (d) Deflection angle at *B*.

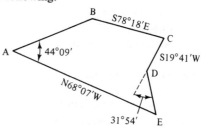

9-21. For the figure shown, compute the following:
  (a) Deflection angle at *B*.

  (b) Bearing of *CD*.

  (c) North azimuth of *DE*.

  (d) Interior angle at *E*.

  (e) Exterior angle at *F*.

(*Ans.*: 90°09′R)

(*Ans.*: S55°45′W)

(*Ans.*: 160°44′)

(*Ans.*: 61°00′)

(*Ans.*: 249°11′)

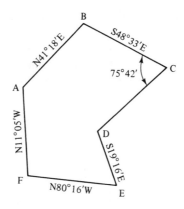

9-22. The following values are deflection angles for a closed traverse: 110°33′R, 84°27′R, 128°51′R, 81°37′L, and 132°54′R. If the bearing of side *CD* is S31°17′W, compute the bearings of the other sides.

9-23. The magnetic north azimuth of a line is 128°27′ while the magnetic declination is 3°30′E. What is the astronomic azimuth of the line?

(*Ans.*: 131°57′)

9-24. At a given place the magnetic bearings of two lines are N44°57′E and S59°08′E. If the magnetic declination is 4°20′W, what are the astronomic bearings of the lines?

9-25. The magnetic bearings of two lines are N14°30′E and S85°30′E. If the magnetic declination is 6°15′W, what are the astronomic bearings of the lines?

(*Ans.*: N8°15′E, N88°15′E)

9-26. The astronomic bearings of two lines are N86°33′E and S45°17′W. Compute their magnetic bearings if the magnetic declination is 5°30′E.

9-27. Change the following astronomic bearings to magnetic bearings for a 4°15′W magnetic declination: N5°17′E, N18°18′W, and S88°33′E.

                 (*Ans.*: N9°32′E, N14°03′W, S84°18′E)

9-28. The magnetic north azimuth of a line was 134°30′ in 1890 when the magnetic declination was 7°15′W. If the magnetic declination is now 2°30′E, determine the astronomic azimuth of the line and its magnetic azimuth today.

9-29. In 1860 the magnetic bearing of a line was S83°30′E and the magnetic declination was 3°15′W. Compute the magnetic bearing of this line today if the magnetic declination is now 4°30′E. What is the astronomic bearing of this line?

          (*Ans.*: Astronomic = S86°45′E Magnetic = N88°45′E)

From the information given in Problems 9-30 to 9-33, determine the astronomic bearing of each line and its magnetic bearing today.

	Magnetic Bearing in 1905	Magnetic Declination in 1905	Magnetic Declination Today
9-30.	S59°30′W	3°15′W	4°30′E
9-31.	S87°15′E	5°15′E	4°45′W
9-32.	N3°30′W	4°45′E	9°30′E
9-33.	N4°30′E	8°30′W	4°15′W

       (*Ans. 9-31*: Astronomic = S82°00′E, Magnetic today = S77°15′E)

       (*Ans. 9-33*: Astronomic = N4°00′W, Magnetic today = N0°15′E)

CHAPTER TEN

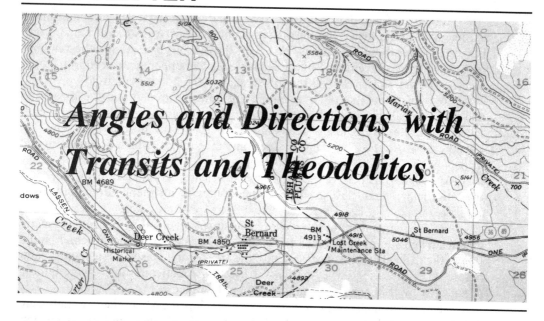

Angles and Directions with Transits and Theodolites

10-1 INTRODUCTION

The instruments used for measuring horizontal and vertical angles are divided into two groups—transits and theodolites—but the distinction between the two is not clear. Originally, both these instruments were called theodolites. The source of the term *theodolite* is not known with certainty. In any case, the theodolite is an instrument used for measuring horizontal and vertical angles. The first theodolites were manufactured with long telescopes that could not be inverted, but as time passed some were made with shorter telescopes that could be inverted end for end or "transited," and they were called transits.

Today, most instruments have telescopes that can be inverted; thus the original distinction between the two no longer applies and to a large extent what they are called is a matter of local usage. It seems to be the convention that the more precise and optically read instruments are called theodolites. Several European and Japanese companies produce superb, optically read instruments with which American firms have been unable to compete.

The earlier theodolites used verniers and micrometer microscopes for reading angles, but modern theodolites have optical systems with which the user may obtain both horizontal and vertical angles through an eyepiece located near the telescope. Optical theodolites have several features quite distinct from normal transits: The scales are lighted for night use; there are three leveling screws; and there are no telescope levels and no compasses. Furthermore, they have not only enclosed optical systems for reading angles, but they also have optical systems for centering the instruments over points.

Actually, theodolites have horizontal and vertical circles for angle measurements, as do transits, but the circles are made of glass instead of metal. Light passes through the glass circles, and with the aid of glass prisms the readings from the circles are reflected into the eyepiece. The upper scale seen in the eyepiece is the vertical circle and the lower scale is the horizontal circle.

Some instruments combine features of both the transit and the theodolite and they are called *optically read transits*. They have glass circles as do theodolites, but angles are read by means of glass verniers and magnifying eyepieces. Finally, there are the very popular electronic theodolites, which provide a digital display of the horizontal and vertical angles.

Although transits have been used by American surveyors for many decades, they are being steadily replaced with theodolites. With theodolites, better precision is obtained in less time at less cost with fewer mistakes.

In 1571 the concept of the modern transit was evolved by the English mathematician and surveyor Thomas Digges.[1] The Danish astronomer Roemer built a theodolite in 1690 for the observation of the stars, but it was about a century later before it was used for surveying work.[2] The first American transits were built in about 1831 by William J. Young and Edmund Draper.[3]

The first optically read theodolite was developed early in the twentieth century by Heinrich Wild of Switzerland. In the 1970s advances in electronics permitted the development of theodolites with digital readout of both horizontal and vertical angles on liquid-crystal displays. The values so obtained can be recorded in field books or stored in electronic data collectors.[4]

In this book, to avoid confusion, the author refers to transits as being the American-style instruments with four leveling screws and silvered horizontal and vertical scales which are equipped with plumb bobs for centering over points. The term *theodolite* is used to refer to the European and Japanese-style instruments with three leveling screws and horizontal and vertical glass circles that may be read directly or with an optical micrometer and those instruments providing digital displays of angle readings. Theodolites are equipped with optical plummets.

The surveyor's transit (Fig. 10-1) is a versatile instrument that may be used to measure vertical and horizontal angles, prolong straight lines, perform leveling of a precision almost equivalent to that obtained with levels, determine magnetic bearings, and measure distances by stadia. This universal instrument is used for land surveying, mapping, construction surveys, and astronomical observations.

---

[1] M. O. Schmidt and K. W. Wong, *Fundamentals of Surveying*, 3rd ed. (Boston: PWS Engineering, 1985), p. 163.

[2] A. R. Legault, H. M. McMaster, and R. R. Marlette, *Surveying* (Englewood Cliffs, N.J.: Prentice-Hall, Inc., 1956), p. 83.

[3] R. C. Brinker and R. Minnick, editors, *The Surveying Handbook* (New York: Van Nostrand Reinhold Company, Inc., 1987), p. 136.

[4] Ibid., p. 136.

Figure 10-1 Transit. (Courtesy of Teledyne Gurley.)

For decades it was recognized as the American surveyor's most important instrument.

The transit is typically used to measure angles to the nearest 1' of arc, but better instruments are available (at somewhat higher costs) with which angles may be read to the nearest 30, 20, 15, or even 10". Theodolites with which angles may be read to the nearest 1" are common, but some theodolites are available with which angles may be read to the nearest 0.2". Sections 10-2 to 10-11 are devoted to transits and Sections 10-12 to 10-17 are concerned with theodolites.

10-2 PARTS OF THE TRANSIT

The various parts of a transit and their use may best be learned by handling and working with them. For this reason, only a brief introduction to these parts is presented here. Detailed discussions of their operations and manipulations are given in the following sections of this chapter. The transit consists of three fundamental parts: the alidade, the horizontal circle, and the leveling head assembly. See Fig. 10-2 for a picture of a transit and these parts.

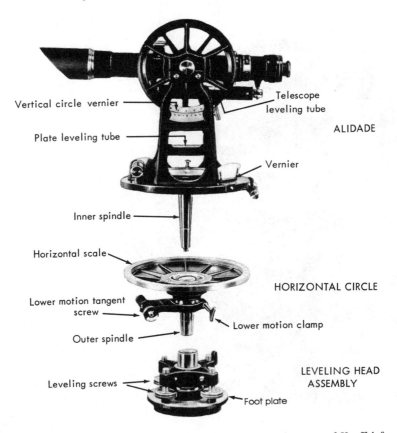

Vertical circle vernier

Plate leveling tube

Telescope leveling tube

ALIDADE

Vernier

Inner spindle

Horizontal scale

HORIZONTAL CIRCLE

Lower motion tangent screw

Lower motion clamp

Outer spindle

LEVELING HEAD ASSEMBLY

Leveling screws

Foot plate

Figure 10-2 Transit showing three fundamental parts. (Courtesy of Keuffel & Esser—a Kratos Company.)

The *alidade*, or upper plate, is the top part of the instrument and includes a circular cover plate that is rigidly connected to a solid conical shaft called the *inner spindle*. It also includes the telescope (with a magnification power of approximately 18 to 28 diameters), the telescope level tube, the vertical circle (which is mounted on the telescope for measuring vertical angles), two plate leveling tubes (one parallel to the telescope and one perpendicular to it), two pairs of verniers for reading the horizontal circle, and probably a compass. The telescope rotates up or down on a pair of side shafts called *trunnions*. This word was originally used to refer to the same-side shafts on cannons.

The *horizontal circle* or lower plate is the scale with which horizontal angles are measured. It is usually graduated in degrees and halves of degrees and numbered every 10° clockwise and counterclockwise. The underside of the horizontal circle is attached to a hollow, vertical, tapered spindle called the *outer spindle* into which the inner spindle fits. In addition, there is a clamp (called the *lower motion clamp*) that is used to permit or prevent rotation of the outer spindle.

The *leveling head assembly* consists of the four leveling or foot screws and the footplate. The leveling screws are set into cups to prevent them from scoring the footplate. Included in the assembly is a device that permits the transit to be moved laterally from $\frac{1}{4}$ to $\frac{3}{8}$ in. without movement of the tripod.

The tripods used with transits are generally about $4\frac{1}{2}$ or 5 ft in height, but there are available types of extension legs with which this height can be increased another 4 or 5 ft. These might be useful on occasions where sights need to be taken over obstacles such as bushes, walls, and mounds.

10-3 CARE OF THE TRANSIT

Although the general rules given here for taking care of the transit are closely related to the ones previously given for the level, the subject is of sufficient importance to warrant some repetition. The major rule, as for the level, is "don't drop the instrument" because very serious damage will almost surely result. The following list presents other important items to remember in caring for the transit:

1. When the transit is being transported in a vehicle, the instrument should be held in the lap or kept in its box.
2. The transit is removed from its box and held either by its standards or its leveling head and placed on the tripod. Do not pick it up by the telescope.
3. Place the tripod with its legs well apart and sunk firmly in the ground.
4. Do not place the instrument on a smooth, hard surface such as a concrete slab unless some provision (such as a triangular frame) is made to keep the tripod legs from slipping.
5. Do not turn its screws tightly. If more than fingertip force is needed to turn them, the instrument either needs cleaning or repairs. Overtightening may lead to appreciable damage to the instrument.
6. Never leave an instrument unattended because it may be upset by wind, vehicles, children, or farm animals, or it may be stolen.
7. Hold the transit in the arms in a horizontal position with the transit ahead when carrying it inside a building. This enables one better to avoid obstacles.
8. If precipitation occurs, put the dustcap over the objective lens. In addition, it is a good idea to carry a waterproof silk bag with which to cover the transit.
9. If the instrument gets wet, let it dry because wiping may damage the finish as well as the scale graduations, and the lens may even be scratched.
10. Optical glass is not very hard and is easily scratched. If the lenses get dirty, they should be brushed carefully with a clean camel's hair brush. Fingers should not be placed on the lenses because our skin oil holds dust. If dusting with the brush is not sufficient, special lens cleaning tissue and perhaps a

lens-cleaning fluid obtained from an optician may be used. Ether and alcohol may be used on lens for cleaning, but any other solvents will be too harsh.

11. The working parts of a transit are lubricated with graphite and thus only a *very very* slight amount of oil or none at all should be used on the instrument. Oil tends to collect dust and it also becomes sticky at low temperatures.

12. Before the transit is returned to its case or box we should first check to be sure that it is dry and clean. Then center the leveling head, adjust the leveling screws to approximately the same lengths, and be sure that none of them is too tight.

10-4 SETTING UP THE TRANSIT

To measure angles, the transit is set up over a definite point such as an iron pin or a tack in a stake. For centering purposes, a plumb bob is suspended from a hook or chain beneath the instrument. Many modern transits are equipped with optical plumbing devices that permit very accurate centering. With an optical plummet a line of sight is passed through a prism along the transit's vertical axis. When the transit is leveled, this line of sight will be vertical and the instrument can be accurately centered without a plumb bob. Nevertheless, it is convenient to use plumb bobs to center transits roughly before the optical plummets are used.

Figure 10-3 Transit set on a steep slope. Notice the manner in which the tripod is set with one leg uphill and two legs downhill.

The plumb bob is then removed and the fine centering is done with the plummet. After it is centered the instrument is releveled, the plummet recentered, and the procedure is continued until the instrument is concurrently centered and leveled.

The transit is leveled in very much the same way that the level is, but there is one exception. Because the transit has two plate bubble tubes at right angles to each other, the instrumentman has only to turn the telescope until the axis of each of these tubes is parallel to a line through opposite leveling screws. He or she may then proceed with the leveling operation without having to turn the telescope as he or she works from one pair of leveling screws to the other.

When both bubble tubes are properly centered, the telescope may be turned through 180° to see if the bubbles remain centered. If the bubbles move, the instrument is out of adjustment, but if the movement is slight, for example, less than one division, the instrumentman is wasting time in trying to recenter the bubble every time he or she moves the telescope.

To center the transit over a particular point with a plumb bob, the instrumentman places the tripod so that the plumb bob hangs approximately over the point. In doing this he or she tries to keep the leveling head approximately level. The instrumentman roughly levels the instrument to see where the plumb bob will be. If it is an appreciable distance off center, he or she picks up the tripod without moving the relative position of the legs and moves the instrument in the desired direction. The instrumentman repeats the rough leveling process, after which the instrument may have to be moved again. When the instrumentman gets the plumb bob fairly close to center, he or she may move it closer and closer by pressing one or two tripod legs more firmly into the ground. (Some surveyors use transits that have adjustable leg tripods that permit the lengthening or shortening of the legs as required to center the plumb bob.)

Finally, when the plumb bob is within approximately $\frac{1}{4}$ to $\frac{3}{8}$ in. of its proper position, two adjacent leveling screws are loosened. This loosens all four of the leveling screws and the head of the instrument may be pushed or slid over the required distance. When the plumb bob is centered, the instrumentman retightens the screws, relevels the instrument, and checks to see if the plumb bob remains centered. If it does not, he or she may have to repeat this last part of the procedure.

10-5 READING TRANSIT VERNIERS

In Section 7-5 the reading of level rod verniers was discussed. In this section the subject is continued but in relation to angle measurements. The verniers used for reading horizontal and vertical angles are identical in principle with target rod verniers. In other words, a person who can use the level rod vernier can use a transit vernier just as well.

The graduated scale on the lower plate is provided with two sets of numbers, one increasing in a clockwise direction from 0 to 360° and the other increasing from 0 to 360° in a counterclockwise direction. For reading these scales, two pairs of verniers (*A* and *B*) are provided. (Verniers that can be read in either direction are called *double-direct verniers.*) These are located on opposite sides of the upper plate, the *A* vernier being located below the eyepiece for convenience. There are two *A* verniers and two *B* verniers because one vernier is needed at each location for the clockwise numbers and one for the counterclockwise numbers. These verniers are located in windows in the cover plate covered with glass.

STYLES OF GRADUATIONS

The circles and verniers of transits are graduated in various ways. The usual styles of graduation and the method of numbering the horizontal circle and vernier are shown below.

GRADUATED 30 MINUTES READING TO ONE MINUTE
DOUBLE DIRECT VERNIER

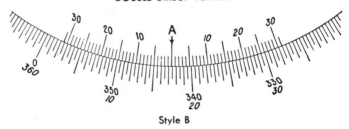

Style B

Style B represents the usual graduation of the horizontal circle of a transit with its vernier, as furnished on K & E Transits Nos. P5081C and P5136. This is an ordinary double direct vernier, reading from the center to either extreme division (30). The circle is graduated to half degrees, and the vernier (from 0 to 30) comprises 30 divisions; therefore, the value of one division on the vernier is 30 minutes ÷ 30 = 1 minute.

The figure reads $17° + 25' = 17° 25'$ from left to right and $342° 30' + 05' = 342° 35'$ from right to left.

GRADUATED 20 MINUTES READING TO 30 SECONDS
DOUBLE DIRECT VERNIER

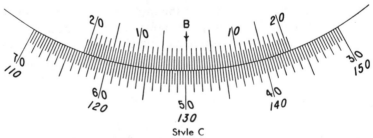

Style C

Style C represents the graduation and vernier of a 6¼ in. transit such as No. P5085C. This is also a double direct vernier, reading from the center to either extreme division (20). The circle is graduated to 20 minutes or 1200 seconds and there are 40 divisions in the vernier; consequently, the value of one division on the vernier is 1200 seconds ÷ 40 = 30 seconds.

The figure reads $130° 00' + 9' 30'' = 130° 9' 30''$ from left to right, and $49° 40' + 10' 30'' = 49° 50' 30''$ from right to left.

Figure 10-4   Styles of graduations. (Courtesy of Keuffel & Esser, a Kratos Company.)

For most angle measurements, the *A* vernier is the only one used, but for more precise work, both *A* and *B* verniers are used and the mean of the readings is used. The purpose of taking the two sets of readings is to reduce instrumental errors caused by imperfections in the scale and vernier.

For the average transit scale, the smallest divisions are $\frac{1}{2}°$ and the verniers are so constructed that 30 divisions on the vernier cover 29 divisions on the main scale. As a result, the smallest subdivision that can be read is

$$D = \frac{s}{n} = \frac{\frac{1}{2}°}{30} = 1'$$

Pictures of two transit verniers, together with example readings, are illustrated in Fig. 10-4. When the telescope has been turned in a clockwise direction, the numbers increasing in that direction will be read on the horizontal scale. Similarly, the verniers whose numbers are increasing in the same direction (clockwise) will be used. The letter *A* at the 0° mark signifies which of the verniers is being used.

To reduce the possibility of mistakes, it is wise to make a rough estimate by eye of the fractional part of a circle division involved to check against the vernier reading. This practice will greatly reduce the number of mistakes occurring in angle measurement. Some transits are equipped with attached magnifying glasses to aid the instrumentman in reading the verniers. If this is not the case, a hand magnifying glass or reading glass should be used. A surveyor using a magnifying glass can estimate readings to the nearest $\frac{1}{2}$ of the vernier's least division.

If not permanently attached to the transit or theodolite, a magnifying glass is conveniently kept on a string around the user's neck. If the user keeps it in a pocket, he or she will often have to fumble around to find it, or may take it home and fail to bring it back the next day, or could leave it lying on the instrument and might drop it.

Other transits are available in which the horizontal scale can be read to the nearest 30″ or less. Different instruments have different arrangements of verniers and the instrumentman should carefully study them before using the instrument. Vertical circles normally have verniers with which vertical angles can be read to the nearest minute.

10-6 POINTING THE INSTRUMENT

The surveyor should point at targets and not aim at them. Aiming, which is excessive pointing at a target, causes eyestrain and makes the target look as though it is moving. The most reliable sighting is usually the first trial when the target first appears to be aligned with the cross hairs. When aligning the cross hairs on the target with the tangent screw or slow motion screw, it is desirable that the

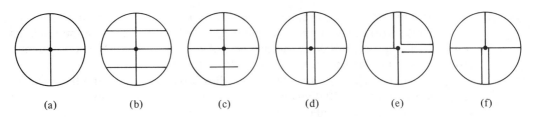

Figure 10-5 Some cross-hair arrangements.

last movement be in a clockwise direction. Such a procedure prevents the occurrence of small errors due to slack in the slow-motion screw.

All pointings should be made with the target close to the intersection of the vertical and horizontal cross hairs. Figure 10-5 shows some of the common cross-hair arrangements for transits and theodolites. The black dot in each case indicates the recommended position of the target. The first three arrangements (a), (b), and (c) are usually found on less precise transits. In (b) the top and bottom horizontal hairs are the stadia hairs. Sometimes to avoid confusion with the middle hair they are shortened as shown in (c). Parts (d), (e), and (f) show double hairs. These permit the centering of distant targets instead of covering them with the hairs. These are usually used on more precise transits.

To measure an angle, the telescope is pointed at one target and then turned to another. If, as is often true, the targets in question are represented by tacks in wooden stakes, and if they are reasonably close to the instrument, it is often possible to leave the tacks sticking up where they may be seen with the telescope. If this is not feasible, a nail or taping pin may be held on top of the tack. When the stakes, iron pins, or whatever are too low for sighting, a plumb bob may be held over the point and the telescope lined up with the string. Sometimes a card or paper is attached to the string so that it will be more visible.

For longer sights, it is necessary to use a larger item for sighting, usually a range pole. The alternate feet of these wooden, metal, or fiberglass poles are painted red and white to make them more visible. The rodman should stand directly behind the rod, and when the bottom of the rod is not visible to the instrumentman, the rodman must be sure that the rod is vertical. If the rodman is needed at some other point, he or she may stick the rod with its pointed end into the ground behind the point. Where the sight distances are not too large, the rodman may set the pole so that it leans across the point and tie the string of a plumb bob around the pole so that the plumb bob hangs over the point.

10-7 MEASUREMENT OF HORIZONTAL ANGLES

For this discussion it is assumed that a transit is located at point A, as shown in Fig. 10-6, and it is desired to measure the angle between the lines AB and AC. Both the upper motion and the lower motion clamps are released and the hori-

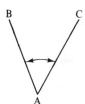

Figure 10-6

zontal scales are adjusted by turning the instrument on its spindle with the hands until the *A* vernier is near zero. The clamps are then tightened and the upper motion tangent screw is turned until the vernier reading is zero. The lower motion is released and the telescope is sighted close to a range pole or other object at point *B*, after which the lower motion is clamped and the line of sight is pointed precisely on point *B* with the lower motion tangent screw. The reading at the *A* vernier should still be zero.

To turn the angle, the upper motion is released and the telescope pointed toward *C*. The fine adjustment of the line of sight to *C* is made by clamping the upper motion and turning the upper motion tangent screw, after which the angle is read at the *A* vernier, as described previously.

10-8 CLOSING THE HORIZON

Before the student attempts to use the transit to measure horizontal angles for an actual traverse, he or she probably needs a good practice session to be sure of fully understanding the operation of his or her instrument. One excellent exercise is referred to as "closing the horizon" or "measuring the angles about a point."

The transit is set up at a convenient point and several taping pins are stuck in the ground at convenient distances from the transit. The angles between the taping pins are measured, the scale being set back to zero for each measurement. When the work is completed, the angles are added together to see if their sum equals 360°. Once the student is able to do this correctly, he or she will have learned how to use properly the upper and lower motions of the transit.

Sample notes for this field exercise are shown in Fig. 10-7. For practice in reading the compass and as a rough check on the horizontal angles, the bearings of the lines from the transit to the chaining pins are recorded and the angles between them are computed. It should be realized that this part of the problem is not feasible for most new transits and theodolites, which do not have compasses, for the reasons described in Section 9-5. In the next section of this chapter, measurement of the interior angles of a closed traverse is described and bearings read with an old transit are shown as part of the problem.

Closing the horizon is useful for checking the measurement of any angle. If the surveyor has measured one or more angles for some position of the instrument, he or she can set the scale to zero and measure the angle required to complete

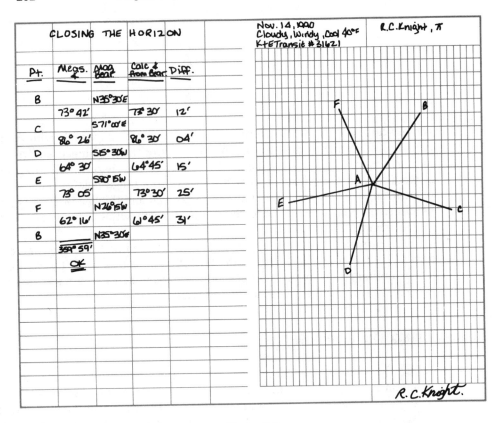

Pt.	Meas. ∠	Mag. Bear	Calc. from Bear	Diff.
B		N35°30'E		
	73°42'		73°30'	12'
C		S71°00'E		
	86°26'		86°30'	04'
D		S15°30W		
	64°30'		64°45'	15'
E		S80°15W		
	73°05'		73°30'	25'
F		N26°15W		
	62°16'		60°45'	31'
B		N35°30E		
	359°59'			
	OK			

CLOSING THE HORIZON

Nov. 14, 1990
Cloudy, Windy, Cool 40°F
K+E Transit # 31621   R.C.Knight, T

R. C. Knight.

Figure 10-7

the circle. If the surveyor adds the angles together to see if the total is 360°, he or she has a very good check on the work.

10-9 MEASURING ANGLES BY REPETITION

All surveyors on more occasions than they would care to admit will make mistakes in measuring angles. After they have measured an angle, they would like to be as sure as possible that no mistakes were made so they will not have to return and repeat the measurement. Usually, it is much easier to prevent mistakes than it is to figure out where they occurred at a later time. A method that nearly always eliminates mistakes in angle measurement is that of *measuring angles by repetition*. After an angle is measured, the lower motion clamp and the lower motion tangent screw are used to sight the telescope on the initial point again. This means that the reading at the vernier should be the same as when the angle was initially measured.

With the aid of the upper motion clamp and tangent screw, the telescope is sighted on the second point and the reading is taken with the vernier. Obviously, the measured value should equal twice the value obtained initially for the angle. If not, a mistake has been made. The student will notice that the sample notes in Fig. 10-8 for angle measurement provide a column for doubling angles. The instrumentman should do his or her best to forget the value recorded for the single angle and not compare the values until the double angle has been read and recorded. Otherwise, he or she will have lost much of the value of this check.

For more precise surveying, horizontal angles may be repeated six or eight times. The value of the angle is added each time to the previous value on the vernier. The resulting value is divided by the number of measurements in order to obtain the angle. Actually, an angle can be measured with a 1″ transit to approximately the nearest ±30″. If an angle is turned six times and if the final total is read, it should contain roughly the same error. Dividing it by 6 should then give a reading within ±05″. Actually, repeating an angle more than six or eight times does not appreciably improve the precision of its measurement because of accidental errors in centering and pointing and instrumental errors of the transit.

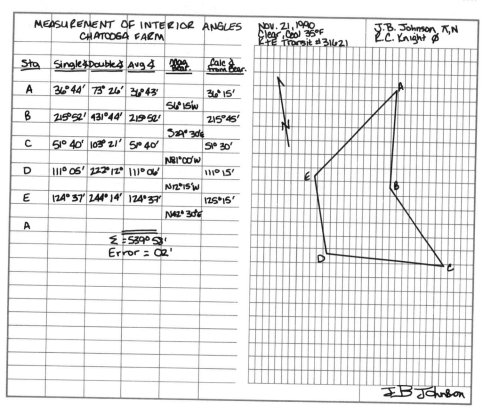

Figure 10-8

When sights are relatively short, 300 ft or less, there is actually little advantage in repetition because the errors made in pointing the telescope and in setting up over the points take away all or almost all of the increased precision supposedly obtained by repetition. Similarly, the precision of a 1″ theodolite is probably wasted if the sight distances are short.

10-10 LAYING OFF AN ANGLE BY REPETITION

In the preceding section the measurement of an angle by repetition to a precision greater than the least reading obtainable with the vernier was described. In this section a related procedure is described for laying off an angle to an equivalent precision.

For this discussion it is assumed that with a 1′ transit it is desired to lay off an angle equal to 42°16′10″ from an existing line *AB*. The transit is used to turn an angle α of 42°16′ with the existing line *AB*. Point *C* is set several hundred feet (say, 500 ft here) from the transit in the new direction, as shown in Fig. 10-9.

The newly established angle is measured by repetition six times. It is assumed here that the result is 42°15′50″. Thus the established angle is in error by 20″. A new point *C′* (see Fig. 10-8) is set a distance from *C* equal to (500) (tan 0°0′20″) = 0.048 ft. To check the work, the new angle is measured again by repetition.

10-11 MEASURING VERTICAL ANGLES

Vertical angles may be measured with the transit in much the same way that horizontal angles are measured. The horizontal plate levels are carefully leveled, the horizontal cross hair is sighted on the point to which the vertical angle is to be measured, and the vertical scale and its vernier are used to read the angle.

As described in Chapter 14, the usual procedure involves (1) measuring the h.i. of the transit telescope by holding a level rod on the ground by the instrument, (2) sighting the telescope on the level rod at the other point with the center hor-

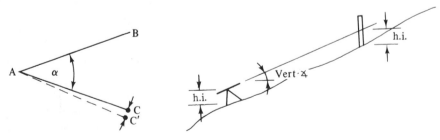

Figure 10-9   Figure 10-10

izontal cross hair adjusted on the rod to the h.i. of the instrument, and (3) reading the vertical angle. This procedure is illustrated in Fig. 10-10.

The transit is not constructed to permit the measurement of vertical angles by repetition. Vertical angles are referred to as either plus or minus (the sign must be carefully recorded), depending on whether the telescope is sighted above or below the horizontal.

A transit has a bubble tube attached to its telescope. The bubble tube is as sensitive as the plate bubble tube. When the telescope bubble is centered, the vernier on the vertical scale should read 0°00′. Actually, it is easy for these scales to get out of adjustment. The discrepancy is called the *index error* of the vertical circle.

For an instrument that has a full vertical circle, the index error problem is easily handled by the so-called *double-centering* procedure. The angle is measured once with the telescope in its normal position and once with the telescope inverted. The average of these two readings is taken. For transits that have vertical arcs instead of full vertical scales, it is necessary to correct the readings by an amount equal to the index error. Great care must be taken to use the correct sign for such corrections.

10-12 THEODOLITES

The vernier transit, which has been discussed in the preceding sections of this chapter and which has been used in the United States for many decades is steadily being replaced with theodolites. Theodolites are manufactured to accomplish the same purposes as transits, that is, to determine horizontal and vertical angles and to prolong straight lines. Theodolites, however, enable the user to make single observations of angles as precisely or more precisely and in less time than those which can be made by several repetitions with a transit.

Theodolites have horizontal and vertical circles for angle measurements, as do transits, but the circles are made of glass instead of metal. Light passes through the glass circles, and with the aid of glass prisms the readings from the circles are reflected into the eyepiece. The values are greatly magnified, enabling the user to make readings without eyestrain. All the readings are taken from the eyepiece end of the telescope so that the user does not have to keep moving around the instrument. The tribrach has three leveling screws and a bull's-eye level. In addition, there is a level on the alidade.

Theodolites usually have short telescopes that can be transited. These telescopes are internally focusing and enable the user to have clear sharp views for both long and short sight distances.

In this section brief discussions are presented concerning scale-reading theodolites, repeating optical theodolites, directional optical theodolites, and electronic theodolites.

Scale-Reading Theodolites

With a scale-reading theodolite (Fig. 10-11) the vertical and horizontal circles are read directly with an optical microscope to the nearest minute and with inter- polation can be read to $\pm 6''$. A scale-reading theodolite is very well suited for layout work. As do transits, it has a horizontal circle that is graduated clockwise and counterclockwise and also a vertical circle. To reduce mistakes caused by reading the wrong scale, they are usually colored differently, such as yellow for the horizontal scale and white for the vertical scale.

The vertical circle for the scale-reading theodolite reads 0° when the tele- scope is pointed directly overhead and therefore zenith angles are read. (As de- scribed in detail in Section 10-17 a vertical angle is the angle up or down from a horizontal line to a particular point. A zenith angle is the angle from a vertical line down to a particular point.) These instruments have a compensator which automatically indexes the vertical circle so that if the telescope is horizontal the vertical circle reading will be 90°00′. Should the instrument be out of level, say more than $\pm 6'$, a red warning screen will become visible on the vertical circle image.

Figure 10-11 Engineer's scale reading theod- olite K1-S. (Courtesy of Kern Instruments, Inc.)

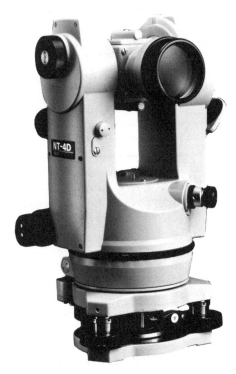

Figure 10-12 NT-4D repeating optical theod- olite. (Courtesy of Nikon, Inc.)

Repeating Optical Theodolites

Angle-measuring instruments are often referred to as repeating instruments or as direction instruments. The transits discussed in the last few sections of this chapter are repeating instruments. They have upper and lower motions and can be used to add or accumulate angles on the horizontal circles, as described with the repetition procedure in Section 10-9.

Theodolites may be either repeating or directional. As with transits the user of a repeating theodolite can measure angles as many times as he or she likes by adding them successively on the instrument circle (see Fig. 10-12). Repeating theodolites have a horizontal circle graduated from 0 to 360° clockwise, and some of them are also graduated counterclockwise. The vertical circle is graduated from 0 to 360°, with the 0° mark usually corresponding to the instrument's *zenith* or the condition when the instrument is sighted vertically upward. Therefore, the vertical scale reads 90° when the telescope is in the normal or horizontal position, or 270° when it's inverted. These theodolites have automatic compensators so that the circle will correctly read 90° when the telescope is horizontal. The horizontal and vertical circles are usually graduated in 20″ or 1′ divisions. They are read directly with an optical microscope to 30″, 10″, or even 6″. They can be estimated to even smaller values.

Directional Optical Theodolites

The directional theodolite (Fig. 10-13) is a nonrepeating type of instrument that does not have a lower motion. The horizontal circle remains fixed during a series of readings. The telescope is sighted on each of the points in question and directions to those points are read. Horizontal angles are determined by calculating the differences between the directions.

Repeating theodolites have a double vertical axis, but directional theodolites have only one axis and they cannot be used to measure angles by repetition. The surveyor sights on a point and reads the direction and then sights on the next point and reads that direction. The difference between the two readings is the angle between the lines. These types of instruments permit the user to read the circle from diametrically opposite positions. The values can be averaged, thus minimizing errors due to imperfect circles. Directional theodolites have micrometer scales that enable the surveyor to make readings directly to 1″ and with estimation as close as 0.1″ or 0.2″.

Electronic Theodolites

During the past few years advances in electronics have led to the advanced electronic theodolites (Fig. 10-14). As these instruments provide a visual display of horizontal and vertical angles, there is no need for microscopes. They are extremely popular with their users. With electronic theodolites the measurements

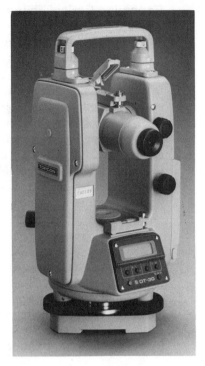

Figure 10-13 DKM2-A 1″ directional optical theodolite with digital circle readout. (Courtesy of Kern Instruments, Inc.)

Figure 10-14 Topcon DT-30 electronic digital transit with liquid-crystal display. (Courtesy of Topcon Instrument Corporation of America.)

are displayed digitally and may be recorded in field books, and they can be used with electronic data recorders to store the information for later use.

The circle readings can be set to zero by touching a button. For some instruments the horizontal circle can be set to any desired value (0° or otherwise). Then the angle is measured by sighting to the next point and its value is automatically displayed by the instrument. The angles can be repeated any desired number of times with the instrument in the normal or reversed positions and the averages taken. Some of these instruments can be read to about ±0.2″, but when atmospheric and pointing errors are considered the standard deviations are ±0.4″ or 0.5″.

10-13 SETTING UP THE THEODOLITE

The purpose of this section is to provide a general discussion applicable to the setting up of all theodolites. More detailed information for a specific type of theodolite can be found in its operation manual. Before the theodolite is unpacked, the tripod should be carefully set over the point to be occupied with the plate approximately level. The theodolite is carefully removed from its carrying case by picking it up using its standards or the handles provided for this purpose on some theodolites. The instrument is set up on the tripod, and the centering screw, which is located under the tripod head, is screwed up into the base of the instrument—otherwise, the instrument will fall off the tripod when picked up. The instrument can be shifted laterally when the centering screw is loose. It is centered over the tripod and the screw is tightened.

The bases of theodolites, or *tribrachs*, are usually designed so as to permit the interchange of various instruments and accessories such as theodolites, EDMIs, targets, subtense bars, and so on without disturbing the centering of the instrument.

The tripod is located over the point to be occupied as closely as possible, the same as described earlier for the transit. Once the instrument is approximately centered over the point with the tripod legs it can be centered more exactly with a plumb bob or preferably with an optical plummet. If the plumb bob is not over the point or if the cross hairs of the optical plummet are not centered on the point, the centering screw can be loosened and the instrument shifted over the point.

Next the theodolite is approximately leveled with the rather insensitive bull's eye or circular level and then it is very carefully leveled with the plate level. The alidade is turned so that the level vial is parallel to two of the three leveling screws and the bubble is brought to the center with these two screws. The instrument is rotated 90° and the bubble is centered using a third screw. Then the instrument is rotated back 90° to its position parallel to the two leveling screws to see if the bubble is still centered. After a few trials the bubble is centered and the alidade is rotated through 180°. If the bubble remains centered, the instrument is level. If not, it is out of adjustment.

Unless the bubble moves a large distance, the instrument probably should not be adjusted because it takes considerable time and repeated adjustments tend to cause significant wear on the adjusting screws. Instead, a reversing procedure is recommended. After the normal leveling procedure is completed, the alidade is rotated so that the bubble axis is in line with one of the foot or leveling screws. The position of the lower end of the bubble (i.e., the end of the bubble toward the vertical circle) is noted. The alidade is rotated through 180° and the position of the lower end of the bubble is again noted. The mean of the two positions is the *reversing position* of the bubble.

The alidade is now turned 90° so that the axis of the bubble tube is parallel to the other foot screws. By turning the two screws in opposite directions, the bubble is brought to its reversing position. To level the instrument once the reversing position is known, the user needs only to bring the instrument to its reversing position for each of the two directions of the alidade 90° apart.

Once the reversing position is determined and the theodolite leveled by this method, the vertical axis will be truly vertical and the bubble will remain in that same position whichever way the telescope is pointed. The reversing position may gradually change and thus its position should be checked before and after each series of observations.[5]

10-14 FORCED CENTERING

Theodolites have a convenient feature that enables us to detach them from the tribrach and interchange them with EDMIs or targets for EDMIs and so on. Whichever instrument is on the tribrach can be removed and replaced with any one of the other instruments. The new instrument will be automatically centered, and level or nearly so. This interchangeable system is called *forced centering*.

Some surveying crews use several tripods and tribrachs. They can set up the theodolite at one point, sight on a target that is set on a tribrach at the next point, make their readings, remove the theodolite from the tribrach, move it to the next point, attach it to the tribrach, make their readings, and so on. Forced centering speeds up operations and reduces the number of setups and minimizes the accidental errors at setups.

10-15 READING THEODOLITE SCALES

In this section sample readings are shown for a repeating optical theodolite, a directional optical theodolite, and an electronic theodolite. Initially readings are shown in Fig. 10-15 for a repeating optical theodolite.

Readings for a directional theodolite are shown in Fig. 10-16. With this type of instrument the observer actually views both sides of the circle simultaneously by means of internal instrument optics. Each of these readings therefore represents the mean of two opposed sides of the circle. This is in effect equivalent to averaging the readings of the *A* and *B* verniers of the transit.

The upper scale seen in the eyepiece is the vertical circle, while the lower scale is the horizontal circle. With some instruments, however, only one circle

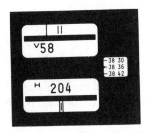

Example of horizontal angle reading 204° 38′ 36″

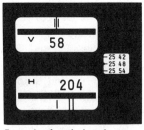

Example of vertical angle reading 58°25′48″

Figure 10-15 Sample readings for horizontal and vertical angles for a repeating-type TH-06D Pentax theodolite. (Courtesy of Pentax Corporation.)

[5] J. F. Dracup et al., *Surveying Instrumentation and Coordinate Computation Workshop Lecture Notes* (Falls Church, Va.: American Congress on Surveying and Mapping, 1973), pp. 1–5.

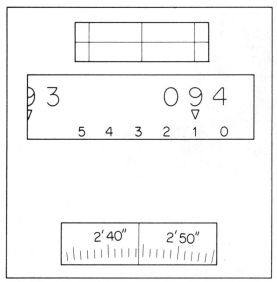

Vertical circle reading (360°) 94° 12′ 44.3″

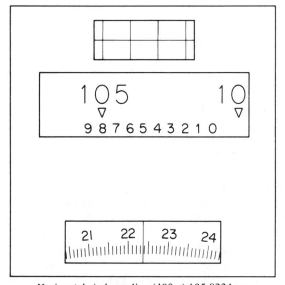

Horizontal circle reading (400 g) 105.8224 gon

Figure 10-16 Sample angle readings for a Wild T-2 directional theodolite. (Courtesy of Wild Heerbrugg Instruments, Inc.)

can be seen at a time and an inverter knob is available to select one circle or the other.

There are quite a few optically read direction theodolites on the market today, but they all have approximately the same basic characteristics. The readings shown in Fig. 10-16 are for a Wild T-2 theodolite. The student can easily verify the readings given in the figure. For the several other types of optical

Figure 10-17 Sample readings for vertical angle (VA) and horizontal angle (HA) for an NE-20S Nikon electronic digital theodolite. (Courtesy of Nikon, Inc.)

theodolite scales in use, the surveyor can study the manufacturer's service manual before making readings.

For the sample readings shown, the vertical scale is based on the 360° or sexagesimal system, which is used almost exclusively in the United States. The horizontal circle shown in the figure is based on the 400 grad or centesimal system, which is widely used in Europe. These systems were discussed in Section 9-2.

10-16 MEASURING WITH A DIRECTION THEODOLITE

An example set of readings taken with a direction theodolite is presented in Fig. 10-18. The instrument was located at station A and directions were read from that point to stations B, C, and D. These notes include directions from station A to each of the other three stations with four different positions. Two readings were taken on each position (one with the telescope normal or direct and one with the instrument reversed or plunged). Thus a total of eight readings were made to each station.

Although directional theodolites do not have a lower motion, they can be approximately set to certain desired values. For position 1 the direction to station B was set near 0°. (It is difficult to set the circle very close to zero but it is not really necessary to do so anyway.) The observer made a pointing on station B and read the direction and then proceeded to make pointings in a clockwise order to stations C and D. The person then reversed the telescope and sighted again on station D and then in a counterclockwise order sighted on stations C and B. These observations make up one position.

Additional positions are set so they are spread around the circle at intervals approximately equal to $180°/n$ where n is the number of positions taken. For this illustration four positions were planned to make up the "set"; therefore, the circle settings were varied by approximately $180°/4 = 45°$. The purpose of spreading out the positions around the circle is to minimize possible errors in the circle graduations and any eccentricity of the plate. The number of positions to be used

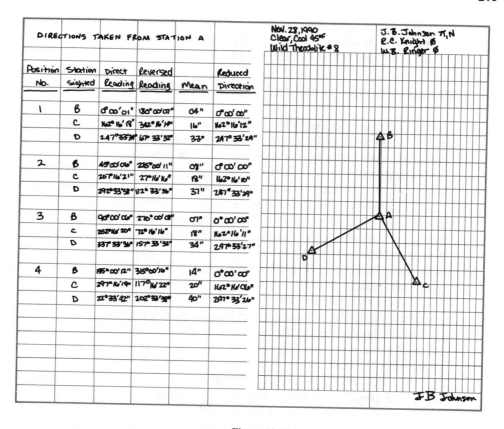

Figure 10-18

for a particular survey depends on the required accuracy for the work. The number is usually specified for high-order government surveys, as described in Chapter 20.

For position 1 with the telescope normal the instrument was pointed at station *B* and the direction was read at 0°00′01″. Then it was pointed at stations *C* and *D* and the readings were 162°16′18″ and 247°33′34″, respectively.

The telescope was reversed and again sighted on station *D*. The circle reading should have changed by 180° but as shown in the notes the difference was not quite 180° due to sighting and instrumental errors. In this case the difference was 6″. The other pointings were made in a counterclockwise order.

The average or mean value (04″ for position 1) is shown in the "Mean" column. The mean direction reading to station *B* is actually 0°00′04″, but this was reduced to 0°00′00″ in the "Reduced Direction" column of the notes. Similarly, the mean directions to stations *C* and *D* for position 1 were reduced by 04″ in the last column. A similar procedure was followed for the reversed readings.

For position 2 the mean direct reading to station *B* was 45°00′06″. This was reduced to 0°00′00″, and the directions to stations *C* and *D* were similarly adjusted by the same amount to 162°16′10″ and 247°33′29″, respectively.[6]

10-17 MEASURING ZENITH ANGLES WITH A THEODOLITE

As described previously, a vertical angle is the plus or minus angle from a horizontal plane to the point in question. Sometimes a plus angle is referred to as an *elevation angle* and a negative angle is called a *depression angle*. A *zenith angle* is the angle from a vertical line to the point in question. These angles are illustrated in Fig. 10-19. Engineer's transits are usually constructed to measure vertical angles conveniently, while theodolites are usually constructed to read zenith angles. As Section 10-11 was devoted to the measurement of vertical angles with a transit, this section is devoted to the measurement of zenith angles. It should be noted that theodolites have a 90° reading when the telescope is horizontal. Some of them have the 0° reading mark at the zenith (high point), whereas others have it at the nadir (low point).

Newer theodolites have an *automatic compensator* (controlled by gravity) which will cause the vertical scale to be in its proper position when the instrument is correctly leveled. Some theodolites have a spirit level attached to their vertical circles for the purpose of getting the circle in its proper position when the instrument is leveled.

The instrument is leveled and sighted on the point in question using the horizontal cross hair with the fine setting made with the vertical tangent screw. If the theodolite does not have an automatic compensator, the spirit level is centered and the zenith angle is read. The telescope is rotated 180° about its vertical axis, inverted, and the procedure repeated. The average of the two angles is computed and the index error is theoretically removed. For greater accuracy several direct readings (telescope in normal position) and several reverse readings (telescope inverted) may be taken and averaged.

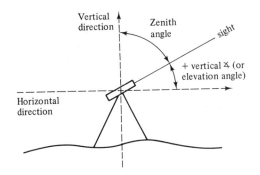

Figure 10-19

[6] R. C. Brinker and P. R. Wolf, *Elementary Surveying*, 8th ed. (New York: Harper & Row Publishers, Inc., 1989), p. 217.

PROBLEMS

10-1. The least divisions on a transit scale are one fifth of a degree. If 24 divisions on the vernier cover 23 divisions on the main scale, what is the least count of the vernier?

(*Ans.*: 0.5′)

10-2. The smallest divisions on a transit scale are one third of a degree. How many divisions should a vernier have in order to enable its user to make a reading to the nearest 30 seconds?

10-3. A horizontal angle was measured by repetition six times with a transit. If the reading on the scale was 21°33′ after the angle was read once and if the final reading was 129°17′, determine the value of the angle measured.

(*Ans.*: 21°32′50″)

10-4. A horizontal angle was measured by repetition eight times with a transit. The initial reading was 76°17′ and the final value was 250°14′. What is the value of the angle measured?

10-5. A 30-second transit was used to measure an angle by repetition ten times. If the first reading was 71°36′30″ and the final value was 356°7′30″, what is the value of the angle?

(*Ans.*: 71°36′45″)

CHAPTER ELEVEN

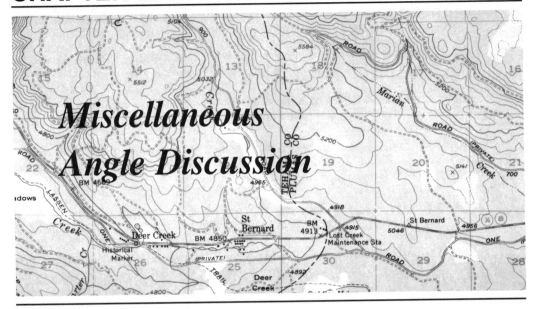

Miscellaneous Angle Discussion

11-1 COMMON ERRORS IN ANGLE MEASUREMENT

Most of the errors commonly made in angle measurement are probably obvious enough, but they are nevertheless listed here together with comments on their magnitudes and methods of reducing them. These are divided into the usual categories: personal, instrumental, and natural.

Personal Errors

Most of the inaccuracy in angle measurement is caused by these errors. Personal errors are accidental in nature and cannot be eliminated. They can, however, be reduced substantially by following the suggestions made herein. Perhaps the largest personal errors occur in pointing, in setting up the instruments, and in reading scales.

1. *Instrument not centered over point.* If the instrument is not centered exactly over a point, an error will be introduced into an angle measured from that position. Here it is necessary for a person to use his or her sense of proportion. If the points to be sighted are distant, errors caused by imperfect centering will be small. If, however, sight distances are very short, centering errors may be very serious. Should a sight be taken on a point 300 ft away and should the instrument be 1 in. off the theoretical line of sight, the angle will be in error by approximately 1'.

216

2. *Pointing errors.* If the vertical cross hair in the telescope is not perfectly centered on observed points, errors similar to those described for imperfect centering of the instrument will occur. The most important method of reducing errors in pointing is to keep the sight distances as long as possible. In fact, this is a basic principle of good surveying—*avoid short sight distances if at all possible.*

   If the points sighted are close to the instrument, the width of a range pole is an appreciable factor. Either a plumb bob may be held over the point with observations being made on the plumb bob string or it may actually be possible to sight on tacks in the stakes, and so on, as described in Section 10-6. A good rule to follow in this regard is to sight only on vertical targets (range poles, taping pins, plumb bob strings, etc.) which appear to be only a little wider than the cross-hair thickness when looking through the telescope.

   In using the telescope, the longer a person stares at the point the more difficult it will be to obtain a good reading because after a while the point will seem to move. As described previously, the surveyor should use the first clear sighting of the target, since that will in all probability be the most accurate.

3. *Unequal setting of tripod.* Tripod legs should be pushed firmly into the ground to provide solid support for the transit. The instrumentman must be careful not to brush against the instrument and not to step too closely to the tripod legs if the ground is soft. A good practice, as in leveling, is to check the bubble tubes before and after readings to make sure that they are still centered. In very soft or swampy ground it may actually be necessary to provide special supports for the tripod legs, for example, stakes driven into the ground.

4. *Improper focusing of telescope (parallax).* To minimize errors caused by improper focusing, the instrumentman should carefully focus the eyepiece until parallax disappears.

5. *Imperfect vernier or micrometer readings.* Accidental errors are made when vernier readings are taken because it is impossible to read them perfectly. For 1′ transits, these errors range in magnitude up to ±30″, with average values in the range of ±20″. Vernier errors may be reduced significantly if angles are measured by repetition. In addition, it is essential to take readings with magnifying glasses.

Instrumental Errors

Since no intrument is perfect, there are instrumental errors. If the instruments are out of adjustment, the magnitudes of these errors will be increased, but they are greatly reduced by the process of double sighting in which the readings are taken once with the telescope in its normal position and once in its inverted position. Then the results are averaged. The operation of rotating the telescope about its horizontal axis is called *plunging* or *inverting* the telescope.

It is impossible to manufacture perfect instruments, and these imperfect instruments get out of adjustment. The geometrical relations that should be present between the various components of transits and theodolites and the adjustments to be made if these relationships do not exist are presented in Sections 11-12 and 11-13.

Natural Errors

In general, natural errors are not sufficiently large to affect work of ordinary precision. For more precise work some things can be done to reduce natural errors. These are included in the list that follows this paragraph. Should weather conditions become unusually severe, work should be discontinued.

1. *Temperature changes.* Use an umbrella over the instrument or do the work at night.
2. *Horizontal refraction.* Try to keep sights away from items that radiate considerable heat such as pipes, tanks, buildings, and so on.
3. *Vertical refraction.* Read vertical angles from both ends of a line and average the readings—the uphill angle will be too large by the amount of the refraction error: the downhill angle will be too small by the amount of the refraction error.
4. *Wind.* Shield instruments as much as possible and use an optical plummet if available for centering the instrument over the point.

11-2 COMMON MISTAKES IN MEASURING ANGLES

Listed in this section are the most common mistakes made in angle measurement with transits. If angles are measured two or more times by the repetition method, mistakes caused by any of the first four items will be discovered and can be eliminated by repeating the measurement.

Reading the Wrong Vernier. The correct vernier to read is the one whose numbers are increasing in the same direction as the numbers are increasing on the scale being used.

Misreading the Vernier. The average transit has a horizontal scale divided with $\frac{1}{2}°$ divisions. To read an angle the vernier reading should be added to the last main scale division. If the vernier reading is 12 and the last division on the main scale is 36°30′, the total reading should be 36°42′, but occasionally the instrumentman will read this as 36°12′. A good habit to form is to estimate the value of an angle from the scale by eye before the vernier is read. This practice will help to eliminate these $\frac{1}{2}°$ mistakes. In addition, the measurement of angles by repetition will reveal mistakes that have been made so that the work can be repeated.

Reading the Wrong Circle.   Sometimes the surveyor may read the angle on the wrong circle, particularly if the angle is near 180°. For angles in this range, the instrumentman should carefully check the adjacent numbers to be sure that he or she is reading the scale whose numbers are increasing in the proper direction.

Using Wrong Tangent Screw.   This is probably the most common mistake made by the beginner. He or she learns by experience the uses of the clamp and tangent screws. The measurement of angles by repetition will reveal if mistakes have been made because of this or one of the preceding three items. The angles may be redone until the repetition process checks.

Recording Wrong Values.   If the recorder calls out the angles aloud while writing them down, he or she should be able to eliminate this kind of mistake. Of course, writing down the initial value of the angle and the value after repetition will also reveal these mistakes.

Vertical Angles Not Recorded as Plus or Minus.   Obviously, surveyors must record this information so that anyone using the notes will know whether the sights were taken uphill or downhill. Oral checks by the instrumentman with the rodman should prevent this mistake.

11-3 ANGLE–DISTANCE RELATIONSHIPS

For a particular survey it is logical for the angles and distances to be measured with comparable degrees of precision. It is not sensible to go to a great deal of effort to obtain a high degree of precision in distance measurements and not do the same with the angle measurements, or vice versa. If the distances are measured with a high degree of precision, the time and money spent have been partially wasted unless the angles are measured with a corresponding precision.

If an angle is in error by 1′, it will cause the line of sight to be out of position by 1 ft at a distance approximately 3440 ft (i.e., 3440 times the tangent of 1′ = 1 ft). Therefore, an angle that is in error by 1′ is said to correspond to a precision in taping of 1/3440. It should be noted that angles measured with the 1′ transit are usually measured a little closer than 1 ′, so that reading them to the nearest 1′ (without repetition) probably corresponds to a precision of approximately 1/5000 in taping.

A similar discussion may be made for the relative precision obtained for angles measured to the nearest 30″, 20″, and so on, or for angles measured with 1′ transits by repetition. Table 11-1 presents the angular errors that correspond to the degrees of precision described in this section. It will be remembered that

TABLE 11-1

Angular error	Angular precision
5′	$\dfrac{1}{688}$
1′	$\dfrac{1}{3440}$
30″	$\dfrac{1}{6880}$
10″	$\dfrac{1}{20,600}$
1″	$\dfrac{1}{206,000}$

the precision obtained with stadia measurement of distance varies between 1/250 and 1/1000. The table shows that in order to obtain comparable precision in the angle measurements, the angles should be measured to approximately the nearest 5′.

It should be realized that this is not the whole story on angle precisions. For instance, the trigonometric functions do not vary directly with angle sizes. In other words, the tangent of an angle of 1°11′ that is 1′ in error does not miss its correct value by the same amount as the tangent of an angle of 43°46′ that is 1′ in error. Nevertheless, the approximate relations given in Table 11-1 provide a satisfactory guide for almost all surveying work.

11-4 TRAVERSING

As described previously, a traverse consists of a series of successive straight lines that are connected together. The process of measuring the lengths and directions of the sides of a traverse is referred to as *traversing*. Its purposes are to find the positions of certain points.

Open traverses, which are normally used for exploratory purposes, have the disadvantage that arithmetic checks are not available. For this reason, extra care should be used in making their measurements. The angles should be measured by repetition if a repeating instrument is being used. For all instruments they should be measured several times, half with the telescope in its normal position and half with it inverted. A rough check can also be made on the angles by reading magnetic bearings for each line with a compass (if the instrument has a compass) and computing the angles between the bearings. The use of astronomic directions with occasional checks made by observations of the sun is a much better procedure. Distances should be measured forward and back whether tapes or EDMIs are being used.

In Chapter 15 another method is presented for checking long open traverses. The positions of the end points and perhaps some intermediate points are established using the Global Positioning System (GPS). As described in Chapter 15 closure can be checked using these points rather than by traversing back to the starting point along a different set of lines. A procedure similar to the GPS one can be used if there are nearby NGS monuments whose positions are known.

A *closed traverse* is one that begins and ends at the same point. A closed traverse could also be one which starts at a known point and ends at another known point, provided both points are on the same coordinate system. Whenever feasible a closed traverse is much to be preferred over an open one because it offers simple checks for both angles and distances, as will be seen in Chapter 12.

11-5 OLDER METHODS OF TRAVERSING

For many decades traversing was performed for both open and closed figures by taping the distances and determining directions by measuring one of the following: deflection angles, angles to the right, interior angles, or azimuths. These methods of traversing, although perfectly satisfactory, are nevertheless somewhat obsolete. They are each briefly described in the following paragraphs and then the more modern procedures are discussed in the next section. Whichever method is used, it is wise to measure each angle two or more times.

Deflection Angle Traverse

A deflection angle, defined in Section 9-10, is the angle between the extension of the preceding line and the present one. To measure a deflection angle, the telescope is inverted and sighted on the preceding point, and the telescope is then reinverted and turned to the left or to the right as required to sight on the next point and the angle is read. This method permits easy visualization of traverses, facilitates their representation on paper, and in addition, makes the calculations of successive bearings or azimuths very simple. Deflection angles are sometimes used for route surveys: for example, for highways, railroads, or transmission lines. Overall, their use has greatly decreased because of the frequency of mistakes in reading and recording angles as being right or left, especially when small angles are involved. The algebraic sum of all the deflection angles for a closed traverse (with no lines crossing) equals 360°. Note that right and left deflection angles must be given opposite signs in the summation.

Angle to Right Traverse

Perhaps the most common method of measuring the angles for a traverse in the past was the angle to the right method. With this method, which was partly supplanted by the deflection angle method as the years went by, the telescope is first sighted on the preceding traverse corner. Then it is turned in a clockwise direction until the next point is sighted and the angle is read.

Interior Angle Traverse

Obviously, this type of angle measurement applies only to closed traverses. The telescope is sighted on the preceding corner and then is turned either in a clockwise or counterclockwise direction so that the interior angle may be measured. It will be noted that if the surveyor proceeds in a counterclockwise direction while traversing, he or she will always turn clockwise angles.

The sum of the interior angles of a closed polygon is given by the following expression, in which n is the number of sides of the figure:

$$\Sigma = (n - 2)(180°)$$

Thus if all the interior angles of a closed traverse are measured and their total is very close to $(n - 2)(180°)$, we are fairly sure that the angles were measured accurately. Of course, due to the errors that have been described (personal, instrumental, and natural), it will be normal for some difference to exist between the total and the correct value. The difference is called the *angle misclosure*. This value is eliminated by distributing it around the traverse by one of several methods described in Section 12-6.

Azimuth Traverse

An azimuth traverse may be used conveniently for a survey in which a large number of details are to be located: such as, topographic mapping. If the instrument is set up so that azimuths may be read directly to each point, the work of plotting them on a map will be appreciably simplified. To accomplish this purpose the telescope is sighted to the preceding corner of the traverse with the instrument reading to the back azimuth of that line. Then the telescope is sighted to as many points as desired from that instrument position and the forward azimuth to each of the points is read.

11-6 MODERN TRAVERSING

Today with theodolites, EDMIs, and total stations, most traverses are run by a more efficient procedure than the ones described in the preceding section. With this procedure, sometimes called *radiation*, traversing is very seldom conducted directly along the boundary lines of the traverse. Points are selected at convenient positions from which several corners or stations of the traverse are readily visible. Using this procedure most obstacles can be avoided and bush cutting kept to a minimum. From the selected points horizontal and vertical angles and distances

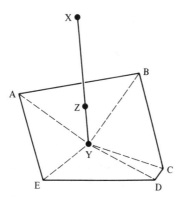

Figure 11-1 Radiation.

are measured and the horizontal and vertical positions of the traverse corners computed.

It is assumed that the position of a nearby point is available, such as point X in Fig. 11-1. This position may be given by coordinates as described in Chapter 12. The azimuth of a line such as XY in Fig. 11-1 may be known from previous work or determined with an astronomic observation or with a compass (the latter choice being a poor one).

If the traverse is fairly small, a convenient point can probably be selected from which all of the corners of the traverse can be sighted. For instance, in Fig. 11-1 point Y is assumed to be such a point. For this particular case it is assumed that the azimuth of line XY is known. The horizontal angles between all of the lines with respect to line YX are measured. In addition, the vertical angles (or the zenith angles for many instruments) and the distances to each corner are measured.

The azimuths of the lines are determined and with the vertical angles horizontal distances are computed. From these azimuths and lengths the positions of all the corners can be computed by a method called latitudes and departures, which is described in Chapter 12. With these values it is but a simple step to the determination of the lengths and azimuths of the sides of the traverse. (In Section 11-4 traversing was said to be the process of measuring the lengths and directions of the sides of a traverse. By that definition radiation is really not traversing because side observations are made to the traverse corners from a convenient point or points and the lengths and directions of the sides of the traverse computed.)

Although the calculations described here can be made with hand-held calculators with reasonable simplicity, there are today a multitude of programs on the market for programmable calculators and computers with which the calculations can be done in minutes or even seconds once the data are input.

If a total station instrument is used (see Chapter 15) the measurements described herein for radiation and the subsequent calculations may be abbreviated

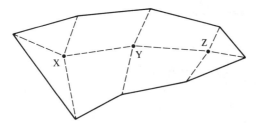

Figure 11-2

even further. Furthermore, with an automatic total station the horizontal and vertical angles are both read electronically for use with the slope distances in internal computers or data collectors. With some total stations it is possible to have considerable data reduction in the field. Other instruments are constructed so that the data reduction and the plotting will be with office computers.

With an automatic total station we can input the coordinates (these may be assumed) of our position along with the reference line (*YX*) in Fig. 11-1. A sight is taken along the reference line and sights are taken to each of the traverse corners. With the total station's microprocessor the distances, elevations, and coordinates are automatically computed and displayed.

We can make a check on our work by moving to a different point and repeating the measurements and the calculations. In Fig. 11-1 point *Z* is shown as the check point. It does not have to be along line *XY* as shown in this figure. If the values determined for coordinates, azimuths, and distances are in good agreement, the final values can be averaged.

If two points are close together in the traverse, such as *C* and *D* in Fig. 11-1, a small error in the length of *YC* or *YD* may have a rather large effect in the direction of side *CD*.[1] For larger traverses it is probable that all of the traverse corners will not be visible from one instrument setup. Nevertheless, the same procedure may be followed except that there will be a need for more setups, as shown in Fig. 11-2.

Auxiliary Closed Traverses

As we have described, it is often quite difficult to work directly on the property lines of a piece of land because of various obstacles. For such cases we may work from one or more points, as shown in Figs. 11-1 and 11-2, or we may have a complete closed auxiliary traverse, as shown in Fig. 11-3. It is often feasible to establish an auxiliary traverse which itself can easily be traversed. The corners of such an auxiliary traverse are located in the vicinity of the property corners and located so that ties can easily be made to them by distances and directions. Such a situation is shown in Fig. 11-3, where the outside traverse represents the land boundaries and the dashed lines represent the auxiliary traverse.

[1] F. H. Moffitt and H. Bouchard, *Surveying*, 8th ed. (New York: Harper & Row Publishers, Inc., 1987), pp. 322–323.

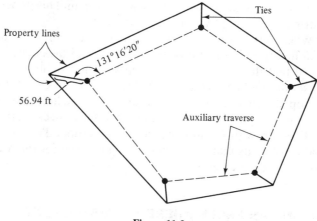

Figure 11-3

    The lengths and directions of the sides of the auxiliary traverse are determined. The lengths and directions of the ties are measured and from this information the location of the property corners are determined. As a last step the lengths and directions of the property lines will be determined. The reader will learn how to make these calculations in the next few chapters.

11-7 INTERSECTION OF TWO LINES

A frequent surveying problem involves the intersection of two lines. For this discussion, reference is made to Fig. 11-4, in which points *A*, *B*, *C*, and *D* are established on the ground. It is desired to find the intersection of lines *AB* and *CD* (shown as point *x*).

    If two instruments are available, the problem can be handled easily. One instrument can be set up at *A* and sighted toward *B*, and the other instrument can be set up at *C* and sighted toward *D*. A rodman is sent to the vicinity of the

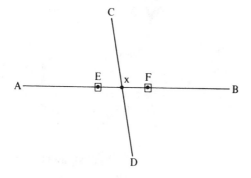

Figure 11-4

intersection point and is waved back and forth until his or her range pole (or plumb bob string) is lined up with both instruments.

When only one instrument is available, the problem can still be handled with little difficulty with the aid of some stakes and tacks and a piece of string. It is assumed that the instrument is set up at *A* and sighted on *B*. Two stakes *E* and *F* are set on the line so that line *EF* straddles line *CD*. A tack is set sticking out of the top of each of these stakes (properly on the line) and a string is tied between them. After the position of the string is carefully checked, the instrument is moved to *C*, lined up with *D*, and sighted on the string. This is the desired point *x*. It should be obvious that the longer the sights used for this work, the more precise will be the measurements.

11-8 MEASURING AN ANGLE WHERE THE INSTRUMENT CANNOT BE SET UP

Another common problem faced by the surveyor is the one of measuring an angle at a point where the instrument cannot be set up. Such a situation occurs at the intersection of fences or between the walls of a building, as illustrated in Fig. 11-5.

To handle this problem, line *AB* is established parallel to the upper fence, and line *CD* is established parallel to the lower fence. The lines are extended and intersected at *E*, as described in Section 11-7, and the desired angle α is measured with the instrument. In establishing line *AB*, point *A* is located at a convenient distance from the fence. The shortest distance to the fence is measured by swinging the tape in an arc with *A* as the center. The desired distance is the perpendicular distance. Similarly, point *B* is located the same distance from the fence. The same process is used to locate points *C* and *D* near the other fence in order to establish line *CD*.

A quicker and more precise method that is frequently used (described in

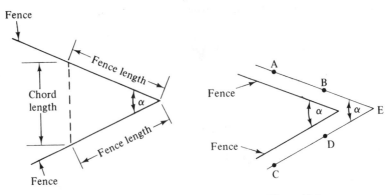

Figure 11-5   Figure 11-6

Section 4-13) involves the measurement of convenient distances up each fence line and the chord distance across from one fence to the other, as shown in Fig. 11-6. The desired angle α may then be computed by trigonometry. If the distances measured up the fence lines are equal, the angle may be obtained from the following expression:

$$\sin \frac{1}{2}\alpha = \frac{\text{chord length}}{2 \text{ fence lengths}}$$

Some type of radiation procedure, such as the ones described in Section 11-6, would be better than the methods described here.

11-9 PROLONGING A STRAIGHT LINE BY DOUBLE CENTERING

An everyday problem of the surveyor is that of prolonging straight lines. Such work is quite common in route surveying, where straight lines may have to be prolonged for considerable distances over rough terrain. In Fig. 11-7 it is assumed that line AB is the line to be extended beyond point B. The instrument is set up at B and backsighted on A. Then the telescope is plunged to set point C'. If the instrument is not in proper adjustment (if the line of sight of the instrument is not perpendicular to its horizontal axis), point C' will not fall on the desired straight line. For this reason *double centering* is the method chosen for this problem.

With the double-centering method the telescope in its normal position (bubble tube down) is sighted on A and plunged to set point C'. The telescope is then rotated horizontally about its vertical axis until point A is sighted. The telescope is now inverted or upside down (bubble tube up). It is once more plunged and point C'' is set. The correct point C is halfway between the two points C' and C''. Of course, if the instrument is properly adjusted, points C' and C'' should coincide if the distances are short. For long distances, however, there will be some displacement between the points, even for well-adjusted instruments. This procedure substantially reduces errors caused by instrument inadjustment and gives the instrumentman a check against the presence of other errors and mistakes. From point C the straight line is continued to point D, and so on.

The technique of double centering is a very important one for the surveyor and one that can be accomplished in a very short time. If one can learn to measure all angles (angles to the right, deflection angles, etc.) by double centering, the precision of the work will be appreciably improved and many blunders will be

Figure 11-7

eliminated. When one is measuring angles by repetition, half of the measurements should be made with the telescope in the normal position and half with the telescope inverted.

11-10 ESTABLISHING POINTS ON A STRAIGHT LINE BETWEEN TWO GIVEN POINTS

Points Intervisible

If the entire line is visible betwen the two points, the surveyor has no problems. He or she can set up the instrument at one end, sight on the other end, and then establish any desired points in between. If a large vertical angle is involved in setting any of the points, the careful surveyor may very well set the points with the telescope in its normal position and then check them with the telescope inverted.

Balancing In

If the end points are not intervisible from each other but there is an area in between from which both the points may be seen (a surprisingly common situation), the process called *balancing in* or *wiggling in* may prove beneficial. In this procedure the surveyor sets up the transit at a point that he or she believes to be on a straight line from *A* to *C* (see Fig. 11-8). One point is sighted and the telescope plunged to see if the other point is in line. If not, the instrument is moved to another position and the steps are repeated. Since it is very difficult to estimate closely the first time, the instrument may have to be moved several times, with the final adjustment probably made by loosening the leveling screws and shifting the head of the instrument. Once the instrument is in line, any desired intermediate points from *A* to *C* may be set. When balancing in is possible, it may very well save the time involved in running trial lines, as described in the following paragraph, and the delay required to make the necessary calculations to finish the problem.

Two Points Not Intervisible

When it is desired to establish intermediate points between two known points that cannot be seen from each other and cannot be seen from any point in between,

Figure 11-8

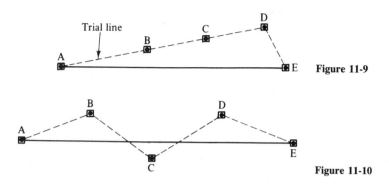

Figure 11-9

Figure 11-10

the surveyor may very well use a random line procedure. This is very common in surveying when hills, trees, and great distances are involved. The surveyor can run a trial straight line (measuring the distances involved) in the general direction from A toward E (Fig. 11-9). He or she sets points B, C, and D (perhaps by double centering), and when near E, measures the distance DE and the interior angle at D. From this information the desired interior angle at A is computed, the straight line from A toward E is run, and points are established in between. If point E is missed slightly, the intermediate points are readjusted proportionately.

Another random line method which may be a little more practical will become obvious after study of Chapter 12 is completed. The surveyor runs a random set of straight lines from A to E, as shown in Fig. 11-10, and then computes the desired length and direction of line AE. Then he or she can return to A and run the straight line AE and then compute the distances required to set points on the desired line by measuring over from the random trial lines.

11-11 CLEANING SURVEYING EQUIPMENT

Equipment will frequently become wet or dirty, and to ensure its precise operation and to lengthen its life, it must be cleaned at regular intervals and at times when it has been subject to unusually severe conditions. The equipment can be sent to a professional repair service for thorough cleaning and lubrication (a good practice every few years), but the surveyor should continually do several things to keep equipment in good condition.

Professional help is usually needed for internal cleaning and lubrication of instruments, but external dust, dirt and water should be removed immediately. Water and dirt can be removed from instruments with cotton balls and pipe cleaners. Dust can be removed from lenses with camel's-hair brushes. If the lenses are streaked, it will be necessary to use an optical glass cleaner and lint-free cloth. Do not use strong cleaners or rub too firmly.

11-12 ADJUSTMENTS TO TRANSITS AND THEODOLITES

The importance of proper instrument adjustments was discussed in Section 7-13. In this section and the two that follow, the common relations that should exist in transits and theodolites are presented together with the adjustments that can normally be made by the surveyor. The relations that should exist are as follows:

1. The vertical and horizontal axes and the line of sight of the instruments should be mutually perpendicular.
2. When the upper plate of the instrument is in a horizontal position, the plate bubble should be centered.
3. If there is a telescope bubble tube, its bubble should be centered when the telescope is horizontal.

11-13 ADJUSTMENTS OF TRANSITS

In this section common field tests and adjustments for transits are presented. Theodolites are discussed in the next section. These tests and adjustments should be conducted under favorable weather conditions and preferably with the transits in the shade or protected with umbrellas.

Plate Bubble Tubes.   The axes of the plate bubble tubes should be perpendicular to the vertical axis of the instrument. This relation and adjustment are identical with those described previously for the dumpy level. The instrument is carefully leveled, with each bubble tube being parallel to a pair of leveling screws. The telescope is rotated about its vertical axis by 180°. If either bubble moves, the bubble is moved halfway back to center with the adjusting screw on the tube in question. Then the test is rerun.

Cross-hair Ring.   The vertical cross hair should lie in a plane perpendicular to the horizontal axis of the instrument. The telescope is sighted so that either the bottom or top of the vertical cross hair is focused on a well-defined point, for example, on a building. The telescope is moved up or down (rotated about its horizontal axis) to see if the point remains on the vertical cross hair. If it does not, the four cross-hair ring capstan screws are loosened and the ring is rotated by light tapping. Then the screws are tightened and the test is rerun.

Line of Sight.   The telescope line of sight should be perpendicular to the horizontal axis. After the instrument is set up and carefully leveled, it is sighted on point *A*, which has been set in the ground (perhaps a chaining pin) at least 250 or 300 ft from the instrument. The telescope is plunged and point *B* is set a similar distance from the instrument. Then the telescope is turned horizontally 180° about its vertical axis and sighted on point *A* again. The telescope is plunged once more. If it is lined up with point *B*, it is properly adjusted. If not, point *C* is set in line with the new sighting and opposite point *B* and another point called *D* is set one-fourth of the distance from *C* to *B*. To adjust the instrument, the cross-hair ring must be moved laterally until the vertical hair is lined up on point *D*. This is done by using the horizontal capstan screws on the cross-hair ring. One is loosened and the other tightened until the necessary adjustment is made.

The Horizontal and Vertical Axes.   The horizontal axis of the instrument should be perpendicular to its vertical axis. The transit is set up near a building or other object where a clearly defined point involving a large vertical angle can be seen. After the instrument is carefully leveled, the vertical hair is sighted on the point. Then the telescope is depressed and point *A* is set on the ground. The telescope is plunged and rotated 180° about the vertical axis and sighted up to the original point. Once again the telescope is depressed. If the vertical hair hits point *A*, no adjustment is required. Otherwise, point *B* is set next to *A* and point *C* is established halfway in between. Point *C* will lie in the same vertical plane with the original point up on the building. The horizontal axis can be adjusted with the capstan screws at one end of the axis. The vertical hair is sighted on point *C* and the telescope raised and sighted on a point opposite the original point. The axis is adjusted until the vertical hair hits the original point. Then the test is rerun to check the correctness of the work.

Telescope Bubble Tube.   The line of sight of the telescope should be parallel to the axis of the bubble tube. Since this test (the "two-peg test") and adjustment are handled exactly as described for the dumpy level in Section 5-13, they are not repeated here.

The Vertical Scale.   When the plate bubble tubes and the telescope bubble tube are centered, the vertical scale should read zero. If this is not the case, the surveyor can skip the adjustment and read the angle as described in Section 10-11, or can adjust the scale to zero with the capstan screws that hold the vernier plate. The screws are loosened and the plate is moved until the zero marks are at the proper position. Then the capstan screws are retightened. When making this adjustment, one wishes one had three hands.

11-14 ADJUSTMENTS OF THEODOLITES

Modern theodolites are manufactured so carefully and their parts fit together so well that they very seldom get out of adjustment unless they are severely abused. Should one of them be damaged or need internal cleaning or reconditioning, it should probably be sent to one of the service departments of the manufacturer for adjustment and repair.

The manuals furnished with theodolites explain a few simple adjustments that can be made by the user. These adjustments, which are similar to those previously described for transits, include adjustments for the plate bubble, telescope bubble, and bull's-eye bubble tubes as well as those for the line of sight, horizontal axis, vertical circle index, and optical plummet. Some comments regarding these checks are made in the remainder of this section.

Plate Bubble Tube. This check is run as it is for transits. If after careful leveling of the instrument, the bubble moves when the telescope is rotated 180° about the vertical axis, it is moved halfway back with the bubble tube adjusting screws.

Bull's-eye Bubble Tube. To check this adjustment the instrument is carefully leveled so that the plate bubble, which was previously checked, is centered. If the bull's-eye bubble is not centered at this time, it is centered with the adjusting screws provided.

Line of Sight, Horizontal and Vertical Axes, Telescope Bubble Tube (if it has one), and Vertical Scale. These checks and adjustments are identical with those for the transit. Most theodolites do not have telescope bubble tubes, but if they do, the adjustments are the same.

Optical Plummet. If the optical plummet is in correct adjustment, it will give the same results as a plumb bob. The optical plummet is built into the tribrach in some cases and is part of the theodolite upper assembly in other cases. If it is in the tribrach, it will remain in position if the theodolite is rotated about its vertical axis; in the other case, it will rotate with the theodolite.

The plummet can be checked by carefully leveling the theodolite, establishing a point beneath an attached plumb bob, removing the plumb bob, and checking to see if the optical plummet mark coincides with the same point. If not, it will have to be adjusted with the adjusting screws, which are located just ahead of the eyepiece. These four screws are used to move the plummet sight or reference mark to the plumb bob mark.

Should the optical plummet be located in the upper assembly, another method of adjustment can be used. The instrument is leveled, a mark is placed on the ground using the optical plummet, and then the telescope is rotated 180° about the vertical axis. If the plummet mark moves from the mark on the ground, it is moved halfway back with the adjusting screws.[2]

---

[2] R. C. Brinker and R. Minnick, editors, *The Surveying Handbook* (New York: Van Nostrand Reinhold Company, Inc., 1987), pp. 228–231.

CHAPTER TWELVE

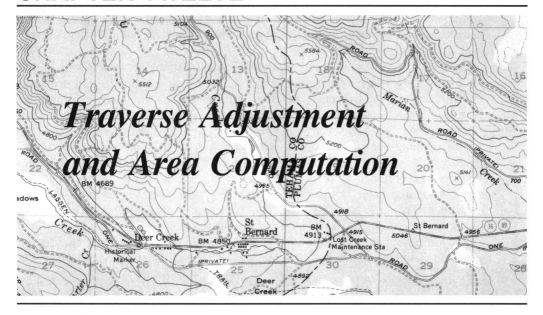

Traverse Adjustment and Area Computation

12-1 COMPUTATIONS

Almost all surveying measurements require some calculations to reduce them to a more useful form for determining distance, earthwork volumes, land areas, and so on. This chapter is devoted to the calculation of land areas. Perhaps the most common need for area calculations arises in connection with the transfer of land titles, but areas are also needed for the planning and design of construction projects. Some obvious examples are the laying out of subdivisions, the construction of dams, and the consideration of watershed areas for designing culverts and bridges.

12-2 METHODS OF CALCULATING AREAS

Land areas may be calculated by several different methods. A very crude approach that should be used only for rough estimating purposes is a graphical method in which the traverse is plotted to scale on a sheet of graph paper and the number of squares inside the traverse are counted. The area of each square can be determined from the scale used in drawing the figure, and thus the area is roughly estimated.

A similar method, which yields appreciably better results but is again satisfactory only for estimating purposes, involves the use of the planimeter (see Section 12-12). The traverse is carefully drawn to scale and a planimeter is used to measure the traverse area on the paper. From this value the land area can be computed from the scale of the drawing. It is probable that with careful work, areas may be estimated by this method within a range of from $\frac{1}{2}$ to 1% of the correct values.

A very useful and accurate method of computing the areas of traverses that have only a few sides is the triangle method. The traverse is divided into triangles and the areas of the triangles are computed separately. The necessary formulas appear in Fig. 12-1, in which several traverses are shown. If the traverse has more than four sides, it is necessary [as shown in part (c) of the figure] to obtain the values of additional angles and distances either by field measurements or by lengthy office computations. For this five-sided figure, the triangle areas ABE and CDE can be obtained as before, but additional information is necessary to determine the area BCE. The problem is amplified for traverses that have more than five sides. For such cases as these, the surveyor is probably better advised to use one of the methods described later in this chapter.

Other methods of computing areas within closed traverses are the double meridian distance, double parallel distance, and coordinate methods discussed in

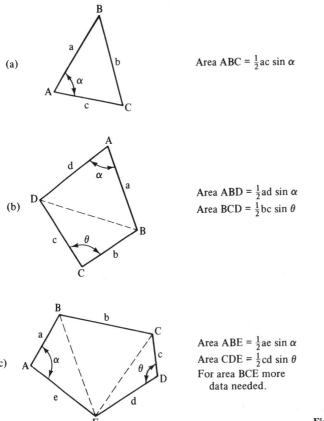

(a)

Area ABC $= \frac{1}{2}ac \sin \alpha$

(b)

Area ABD $= \frac{1}{2}ad \sin \alpha$
Area BCD $= \frac{1}{2}bc \sin \theta$

(c)

Area ABE $= \frac{1}{2}ae \sin \alpha$
Area CDE $= \frac{1}{2}cd \sin \theta$
For area BCE more
data needed.

Figure 12-1

Sections 12-7 through 12-11. In addition, several methods are presented in Section 12-12 for computing the areas of land within irregular boundaries. Computer methods of handling the same problems are discussed in Chapter 13.

12-3 BALANCING ANGLES

Before the area of a piece of land can be computed, it is necessary to have a closed traverse. The first step in obtaining a closed figure is to balance the angles. The interior angles of a closed traverse should total $(n - 2)(180°)$, where n is the number of sides of the traverse. It is unlikely that the angles will add up perfectly to this value, but they should be very close. The usual rule for average work is that the total should not vary from the correct value by more than approximately the square root of the number of angles measured times the least division readable with the vernier (see Section 2-11). For an eight-sided traverse and a 1' transit, the maximum error should not exceed

$$\sqrt{8} \times 1' = 2.83' \qquad \text{say, } 3'$$

(For very good work with the 1' transit, the maximum error should probably not exceed $\pm 30'' \sqrt{n}$.)

It is customary for the instrumentman to check the sum of the angles for the traverse before leaving the field. If the discrepancies are unreasonable, he or she must remeasure the angles one by one until the source of the trouble is found and corrected.

If the angles do not close by a reasonable amount one or more mistakes have been made. If a mistake has been made in only one angle, that angle can often be identified by plotting the lengths and directions of the sides of the traverse to scale. Then if a line is drawn perpendicular to the error of closure, it will often point to the angle where the mistake was made. It can be seen in Fig. 12-2 that if the angle containing the mistake is reduced, it will tend to cause the error of closure to be reduced.

When the angular errors for a traverse have been reduced to reasonable values, they are distributed among the angles so that their sum will be exactly $(n - 2)(180°)$. Each angle may be corrected by the same amount, only certain angles may be corrected because of difficult field conditions, or an arbitrary rule may be used to make the corrections. Example B-2 of Appendix B illustrates the adjustment of the interior angles of a closed traverse using the least squares method.

Should the sum of the angles be in error by 1', the surveyor may very well decide to correct one of the angles by 1'. If there is an angle that is suspect in the surveyor's mind (that is one in which obstructions, short sides, or other problems were involved), that will be the angle corrected.

If the error is 2', two angles may be corrected by 1' each. The two angles selected may be the ones at the end of a short side, or at the ends of a side for

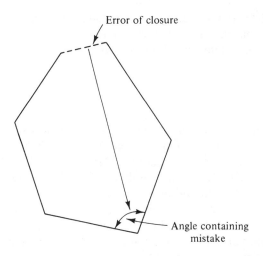

Error of closure

Angle containing
mistake   **Figure 12-2**

which obstructions were a problem, or perhaps angles on opposite sides of the traverse. If a 3' error occurs, three angles may each be corrected by 1'. The angles to be corrected would probably be spread around the traverse so as not to distort one section of the traverse too much.

The very sensible thought may occur to the reader that all the angles in the traverse should be corrected by an equal amount. For a 3' error in 12 angles, each correction would be $3'/12 = 0.25' = 15''$. It is probably not advisable to make the adjustments in units smaller than the least value readable with the instrument. For the example given, if the angles were measured to the nearest 15'' or less, this probably is a reasonable procedure. If the angles were measured to

TABLE 12-1

Point	Measured ∡	Corrected ∡	Calculated bearings from corrected ∡'s
A	36°43'	36°44'	
			S6°15'W
B	215°52'	215°53'	
			S29°38'E
C	51°40'	51°40'	
			N81°18'W
D	111°06'	111°06'	
			N12°24'W
E	124°37'	124°37'	
			N42°59'E
A			
	Σ = 539°58'	Σ = 540°00'	

the nearest 1', however, it does not seem very reasonable to correct them to the nearest 15". After the angles are balanced, the bearings of the sides of the traverse are computed. The initial bearing is preferably an astronomic one, but a magnetic or assumed bearing may be used and the other bearings computed from the balanced traverse angles.

The interior angles measured for the traverse in Fig. 10-8 are added together in Table 12-1 and total 539°58', which is 2' less than the correct value of 540°00' for the interior angles of a closed five-sided figure. These are balanced in the table by merely adding 1' to angles *A* and *B*. If the magnetic bearing of side *AB* is used as a reference, the bearings of the other sides are calculated in the table from the balanced angles.

12-4 LATITUDES AND DEPARTURES

The closure of a traverse is checked by computing the latitudes and departures of each of its sides. The *latitude* of a line is its projection on the north–south meridian and equals its length times the cosine of its bearing. In the same manner, the *departure* of a line is its projection on the east–west line (sometimes called the *reference parallel*) and equals its length times the sine of its bearing. These terms (illustrated in Fig. 12-3) merely describe the *x* and *y* components of the lines.

For the calculations used in this chapter, the latitudes of lines with northerly bearings are designated as being north or plus. Those in a southerly direction are designated as south or negative. Departures are east or positive for lines having easterly bearings and west or negative for lines having westerly bearings. For example, line *AB* in Fig. 12-3(a) has a northeasterly bearing and thus a + latitude and a + departure. Line *CD* in Fig. 12-3(b) has a southeasterly bearing and thus a − latitude and a + departure. Calculations of latitudes and departures are illustrated in Section 12-5.

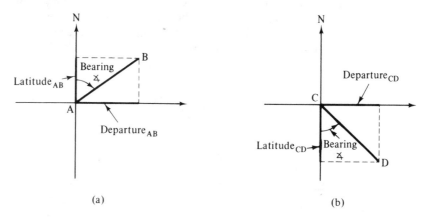

(a)   (b)

Figure 12-3

12-5 ERROR OF CLOSURE

If you start at one corner of a closed traverse and walk along its lines until you return to your starting point, you will have walked as far north as you have walked south and as far east as you have walked west. This is the same thing as saying that for a closed traverse the sum of the latitudes should equal zero and the sum of the departures should equal zero. When the latitudes and departures are calculated and summed, they will never be exactly equal (except by accidental results of canceling errors).

When the latitudes are added together, the resulting error is referred to as the *error in latitude* (E_L); the error that occurs when the departures are added is referred to as the *error in departure* (E_D). If the measured bearings and distances of the traverse of Fig. 12-4 are plotted exactly on a sheet of paper, the figure will not close because of E_L and E_D. The usual magnitude of these errors is greatly exaggerated in this figure.

The error of closure can be easily calculated as follows:

$$E_{closure} = \sqrt{(E_L)^2 + (E_D)^2}$$

and the precision of the measurements can be obtained by the expression

$$precision = \frac{E_{closure}}{perimeter}$$

After the precision is determined the surveyor will make a decision as to whether the work has been done satisfactorily. In most areas of the United States minimum acceptable precisions are provided by law for various kinds of surveys. Typical values are 1/5000 for rural land, 1/7500 for suburban land and 1/10,000 for urban land. If the precision is satisfactory, the surveyor will proceed to balance the errors in latitudes and departures and compute the area of the traverse as described in the next few sections of this chapter.

Should the precision obtained be unsatisfactory for the purposes of the survey, it will be necessary to recheck the work. Of course, as a first step, the math should be checked carefully for mistakes, and if mistakes cannot be found, the field measurements must be checked.

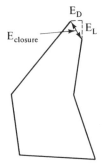

Figure 12-4

TABLE 12-2

Side	Bearing	Length (ft)	Cosine	Sine	Latitude +N	Latitude −S	Departure +E	Departure −W
AB	S6°15′W	189.53	0.9940563	0.1088669	—	188.404	—	20.634
BC	S29°38′E	175.18	0.8692074	0.4944476	—	152.268	86.617	—
CD	N81°18′W	197.78	0.1512608	0.9984939	29.916	—	—	195.504
DE	N12°24′W	142.39	0.9766723	0.2147353	139.068	—	—	30.576
EA	N42°59′E	234.58	0.7315521	0.6817856	171.607	—	159.933	—
		Σ = 939.46			Σ = +340.591	−340.672	+246.550	−246.714

$$E_L = -0.081 \qquad E_D = -0.164$$

$$E_{\text{closure}} = \sqrt{(0.081)^2 + (0.164)^2} = 0.183 \text{ ft}$$

$$\text{Precision} = \frac{0.183}{939.46} = \frac{1}{5138} \quad \text{say, } \frac{1}{5100}$$

Table 12-2 presents the calculations necessary for obtaining the latitudes, departures, E_L, E_D, E_{closure}, and the precision for the traverse used previously. Usually, the degree of precision is rounded to the nearest 100.

If the precision is unsatisfactory the surveyor should study the calculations carefully before rechecking the field measurements. If the angles balance, a mistake in distance is to be suspected. Usually, it will have occurred in a side roughly parallel to the direction of the error of closure line. For instance, if a traverse has a large error in latitude but a small error in departure, one might look for a side running primarily in the north–south direction. If such a side exists, it would be logical to remeasure the distance for that side first. In a similar manner, if the error in departure is twice the error in latitude, the surveyor would want to check the distances for any sides whose calculated departures and latitudes were approximately in that proportion.

Despite these ideas, small mistakes may not be located easily and the surveyor may have to remeasure carefully all or most all of the sides of the traverse. Once the errors of closure are reduced to reasonable values, they will be adjusted so that the traverse will close perfectly (to the number of places being used), as described in Section 12-6.

12-6 BALANCING LATITUDES AND DEPARTURES

The purpose of balancing the latitudes and departures of a traverse is to attempt to obtain more probable values for the locations of the corners of the traverse. If, as described in the preceding section, a reasonable precision is obtained for the type of work being done, the errors are balanced out in order to close the traverse. This usually is accomplished by making slight changes in the latitudes and departures of each side so that their respective algebraic sums total zero.

Theoretically, it is desirable to distribute the errors in a systematic fashion to the various sides, but practically, the surveyor may often use a simpler procedure. He or she may decide to make large corrections to one or two sides where the most difficulties were encountered in making the measurements. This surveyor may look at the magnitude of E_L and E_D and decide, after studying the lengths of the various sides and the sines and cosines of their bearings, that by changing the length of such and such a side or sides the traverse can just about be balanced.

The practical balancing methods just described may not seem to the reader to give very good results. However, they may be as satisfactory as the results obtained with the more theoretical rules described in the following paragraphs, since the latter methods are based on assumptions that are not altogether true and may, in fact, be far from being true.

Very often the surveyor may have no idea which side should get the correction and, in fact, the person performing the calculations may be someone other than the person who made the measurements. For such cases a systematic balancing method such as the compass rule, the transit rule, or one of several others is desirable.

A very popular rule for balancing the errors is the *compass or Bowditch rule,* named after the distinguished American navigator Nathaniel Bowditch (1773–1838), who is given credit for its development. It is based on the assumption that the quality of distance and angular measurements is approximately the same. It is particularly applicable to surveys made with EDMIs and theodolites. It is further assumed that the errors in the work are accidental and thus that the total error in a particular side is directly proportional to its length. The rule states that *the error in latitude (departure) in a particular side is to the total error in latitude (departure) as the length of that side is to the perimeter of the traverse.* For the example traverse, the correction for the latitude of side *AB* is calculated below.

$$\frac{\text{Correction in lat.}_{AB}}{E_L} = \frac{l_{AB}}{\text{perimeter}}$$

$$\text{Correction in lat.}_{AB} = E_L \frac{l_{AB}}{\text{perimeter}} = \frac{(+0.081)(189.53)}{939.46} = +0.017 \text{ ft}$$

If the sign of the error is $+$, the correction will be minus. Actually, the sign of the corrections can be determined by observing what is needed to balance the numbers to zero. In Table 12-3 the latitudes and departures for the traverse are balanced by the compass rule.

A rule that is occasionally used is the *transit rule.* It is based on the assumption that errors are accidental and that the angle measurements are more precise than the length measurements, such as for a stadia traverse. The computations involve corrections to the latitudes and departures in such a manner that the lengths of the sides are changed but not their directions. In this method *the correction in the latitude (departure) for a particular side is to the total correction in latitude (departure) as the latitude (departure) of that side is to the sum of all the latitudes (departures).*

$$\frac{\text{Correction in lat.}_{AB}}{E_L} = \frac{\text{lat.}_{AB}}{\Sigma \text{ latitudes}}$$

Computer programs are readily available for solving all the problems (from balancing to area calculations) described in this chapter. These programs use one of the systematic methods of balancing, such as the compass rule or the transit rule described here, or other methods, such as the Crandall method or the *least squares method.*

The *Crandall method* provides another systematic adjustment procedure, which is rather similar in application to the transit rule. It is particularly applicable to traverses where the angles have been measured more precisely than the distances, stadia surveys again being an example. With this method the angles are corrected equally while a weighted least squares adjustment is made to the lengths.

TABLE 12-3

Side	Latitude correction		Departure correction		Balanced latitudes and departures			
	N	S	E	W	N	S	E	W
AB	—	+0.017	—	+0.033	—	188.387	—	20.601
BC	—	+0.015	+0.030	—	—	152.253	86.647	—
CD	+0.017	—	—	+0.035	29.933	—	—	195.469
DE	+0.012	—	—	+0.035	139.080	—	—	30.551
EA	+0.020	—	+0.041	—	171.627	—	159.974	—
					$\Sigma^2 = 340.640$	340.640	246.621	246.621

The *least squares method* is usually the best method available for adjusting surveying data. It is difficult to apply, however, unless a computer is being used. In each of the other methods (compass, transit, estimation, and Crandall) some type of systematic correction is made to errors that are accidental or random in nature. The least squares method, developed from probability theory, provides the most probable results for the traverses. With this method the angle and distance measurements are adjusted so as to make the sum of the squares of the residuals the least possible.[1] Least squares is presented in Appendix B of this book.

The lengths and bearings of the sides of a traverse will be slightly changed when their latitudes and departures are balanced. The new length of each side can be determined by computing the square root of the sum of the squares of its adjusted latitude and departure ($l_{AB} = \sqrt{(\text{dep.}_{AB})^2 + (\text{lat.}_{AB})^2}$. Its adjusted bearing may be determined by trigonometry. For example,

$$\tan \text{ bearing } \angle = \frac{\text{adjusted departure}}{\text{adjusted latitude}}$$

Space is not taken to show these calculations for the example traverse considered in these sections. The computer programs commonly used today to make precision, balancing, and area computations (which are discussed in Chapter 13) will usually provide the final adjusted lengths and bearings of the sides.

12-7 DOUBLE MERIDIAN DISTANCES (DMDs)

The best-known procedure for calculating land areas with hand-held calculators is the *double meridian distance* method. The *meridian distance* of a line is the distance (parallel to the east–west direction) from the midpoint of the line to the

---

[1] R. C. Brinker and P. R. Wolf, *Elementary Surveying*, 8th ed. (New York: Harper & Row Publishers, Inc., 1989), p. 251.

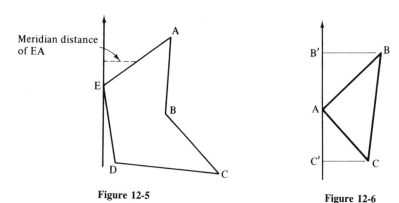

Figure 12-5   **Figure 12-6**

reference meridian. Obviously, the *double meridian distance* (DMD) of a line equals twice its meridian distance. In this section it will be proved that if the DMD of each side of a closed traverse is multiplied by its balanced latitude and if the algebraic sum of these values is determined, the result will equal two times the area enclosed by the traverse.

Meridian distances are considered positive if the midpoint of the line is east of the reference meridian and negative if it is to the west. In Fig. 12-5, the plus meridian distance of side *EA* is shown by the dashed horizontal line. For sign convenience, the reference meridian is usually assumed to pass through the most westerly or the most easterly corner of the traverse. If the surveyor has difficulty in determining the most westerly or the most easterly corners, he or she can solve the problem quickly by making a freehand sketch. He or she starts at any corner of the traverse and plots the departures successively to the east or west for each of the lines until returning to the starting corner. The location of the desired points will be obvious from the sketch.

In Fig. 12-5 it can be seen that the DMD of side *EA* equals twice its meridian distance or equals its departure. The DMD of side *AB* equals two times the departure of *EA* plus two times one-half the departure of *AB*. In this manner the DMD of any side may be determined. By studying this process, however, the reader will develop the following rule for DMDs that will simplify the calculations: *The DMD of any side equals the DMD of the last side plus the departure of the last side plus the departure of the present side.* The signs of the departures must be used and it will be noticed that the DMD of the last side (*DE* in Fig. 12-5) must equal the departure of that side, but it will of necessity be of opposite sign.

To see why the DMD method for area calculation works, we refer to Fig. 12-6. In this discussion, north latitudes are considered plus and south latitudes are considered minus. If the DMD of side *AB*, which equals *B'B*, is multiplied by its latitude, the result will be plus two times the area of the triangle *B'BA*, which is outside the traverse. If the DMD of *BC* is multiplied by its latitude *B'C'*, which is minus, the result will be minus twice the trapezoidal area *B'BCC'*, which is inside and outside the traverse. Finally, the DMD of *CA* times its latitude *C'A*

TABLE 12-4

Side	Departures		DMD	Latitudes		Double areas	
	E	W		N	S	+N	−S
AB	—	20.601	299.347	—	188.387	—	56,393
BC	86.647	—	365.393	—	152.253	—	55,632
CD	—	195.469	256.571	29.933	—	7,680	—
DE	—	30.551	30.551	139.080	—	4,249	—
EA	159.974	—	159.974	171.627	—	27,456	—
						$\Sigma = +39,385$	$-112,025$

$$2A = -72,640$$

$$A = -36,320 \text{ sq ft} = 0.834 \text{ acre}$$

equals plus two times the area ACC', which is outside the traverse. If these three values are added together, the total will equal two times the area inside the traverse because the areas outside the traverse will be canceled.

Table 12-4 shows the area calculations by the DMD method for the example traverse used in this chapter. Because the signs of the latitudes must be used in the multiplications, the table provides one side for positive values under the north column for north latitudes and one for negative values under the south column for south latitudes. The traverse area equals one-half the algebraic sum of the two columns. It does not matter that the final value is either positive or negative. The resulting area may be quickly checked by moving the reference meridian to another corner and repeating the calculations. If it has been assumed at the most westerly corner, it will probably be moved to the most easterly corner. This same problem is solved with a great saving in time and effort in Section 13-7 using a computer and the disk enclosed with the book.

When lengths are in meters the area is expressed in square meters (m²). The SI system does not specify a particular unit for land areas, but the *hectare*, which equals 10 000 m², is used by many countries. One hectare equals 2.47104 acres.

12-8 DOUBLE PARALLEL DISTANCES (DPDs)

The same procedure as that used for DMDs may be used if double parallel distances (DPDs) are multiplied by the balanced departures for each side. The final areas will be the same. The *parallel distance* of a line is the distance (parallel to the north–south direction) from the midpoint of the line to the reference parallel or east–west line. The parallel is probably drawn through the most northerly or the most southerly corner of the traverse.

12-9 RECTANGULAR COORDINATES

Rectangular coordinates are the most convenient method available for describing the horizontal positions of survey points. In the world of computers, just about everyone uses coordinates to define the positions of survey points. In our courthouses coordinate systems are frequently used to describe property corner locations. Dams, highways, industrial plants, and mass-transit systems are located, planned, designed, and constructed on the basis of computerized information, which includes coordinates as well as other information concerning topography, geology, drainage, population, and so on. As a result of these facts, it is absolutely necessary for the surveyor to be familiar with, and able to use, coordinates. The coordinates of a particular point are defined as the distances measured to that point from a pair of mutually perpendicular axes. The axes are usually labeled X and Y, the perpendicular distance from the Y axis to a point is called the x coordinate, and the perpendicular distance from the X axis to the point is called the y coordinate.

In the United States it is common to have the X axis coincide with the east–west direction and the Y axis with the north–south direction. Just the opposite system is used in some countries, particularly in Europe. In Fig. 12-7 the coordinates for two points A and B at the ends of a line are shown. The positive directions are indicated by the arrows on the axes. Point B is below the X axis and thus has a negative Y value.

If the latitudes and departures of a traverse have been computed, a coordinate system can easily be established. For illustration the balanced latitudes and departures of Table 12-4 are used to compute the coordinates of the corners of the traverse considered earlier in the chapter.

A convenient location is selected for the origin. It may be placed at one of the corners of the traverse or at some other convenient point. It is often located at such a point that the entire survey will fall within the first or northeast quadrant. If this is the case, there will be no negative coordinates.

For the balanced latitudes and departures of Table 12-4, the most westerly point, E, is assumed to fall on the Y axis. As a result, the x coordinates of all the points will be positive. They are determined as follows:

$$E = 0$$
$$A = 0 + 159.974 = +159.974$$
$$B = +159.974 - 20.601 = +139.373$$
$$C = +139.373 + 86.647 = +226.020$$
$$D = +226.020 - 195.469 = +30.551$$
$$E = +30.551 - 30.551 = 0$$

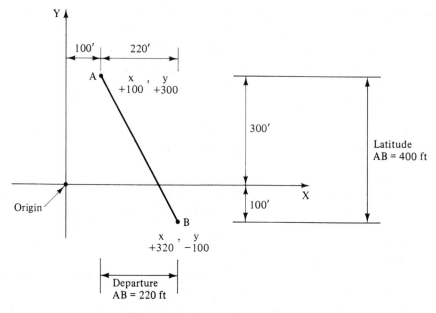

Figure 12-7 Rectangular coordinates.

The most southerly point, C, is assumed to fall on the X axis, so that all y coordinates are positive. They are computed as follows:

$$C = 0$$
$$D = 0 + 29.933 = +29.933$$
$$E = +29.933 + 139.080 = +169.013$$
$$A = +169.013 + 171.627 = +340.640$$
$$B = +340.640 - 188.387 = +152.253$$
$$C = +152.253 - 152.253 = 0$$

The x and y coordinates for each of the corners of the traverse are shown in Fig. 12-9. The use of coordinates has become standard practice for many types of surveys. One of its earliest surveying applications was in mine surveys. Coordinates are also quite useful for plotting maps and for computing land areas.

12-10 AREAS COMPUTED BY COORDINATES

Another useful method for computing land areas is the method of coordinates. Some surveyors like this method better than the DMD method because they feel that there is less chance of making mathematical mistakes. The amount of work

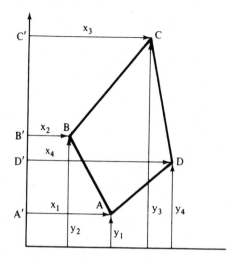

Area ABCD = area C′CDD′ + area D′DAA′ − area C′CBB′ − area B′BAA′

Area ABCD = $(\frac{1}{2})$ $(x_3 + x_4)$ $(y_3 - y_4)$ + $(\frac{1}{2})$ $(x_4 + x_1)$ $(y_4 - y_1)$ − $(\frac{1}{2})$ $(x_3 + x_2)$

$(y_3 - y_2)$ − $(\frac{1}{2})$ $(x_2 + x_1)$ $(y_2 - y_1)$

Multiplying these values and rearranging the results yields

2 area = $y_1(- x_2 + x_4) + y_2(- x_3 + x_1) + y_3(- x_4 + x_2) + y_4(- x_1 + x_3)$

Figure 12-8

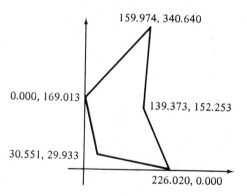

2A = (0.000)(−340.640 + 29.933) + (159.974)(−152.253 + 169.013) +
(139.373)(−0.000 + 340.640) + (226.020)(−29.933 + 152.253) +
(30.551)(−169.013 + 0.000)

2A = 72,640

 A = 36,320 sq ft = 0.834 acres

Figure 12-9

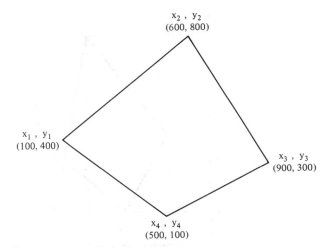

Figure 12-10

is approximately the same for both methods. Instead of computing the DMDs for each side, the coordinates of each traverse corner are computed and then the coordinate rule is applied. This rule is derived in Fig. 12-8.

From this figure it can be seen that to determine the area of a traverse, each y coordinate is multiplied by the difference in the two adjacent x coordinates (using a consistent sign convention such as minus the following plus the preceding). The sum of these values is taken and the result equals two times the area. The math can be checked quickly by taking each x coordinate times the difference in the two adjacent y coordinates. The coordinates of the corners of the example traverse are given in Fig. 12-9 as determined from the previously balanced latitudes and departures, and the area is computed.

12-11 ALTERNATE COORDINATE METHOD

There is available a very simple variation of the coordinate method for area computations which is a little easier to remember and apply. For this discussion reference is made to Fig. 12-10, where the x and y coordinates of the corners of a traverse are shown.

The formula that was presented in the last line of Fig. 12-8 may be rewritten in the following form:

$$2A = x_1y_2 + x_2y_3 + x_3y_4 + x_4y_1 - y_1x_2 - y_2x_3 - y_3x_4 - y_4x_1$$

With this expression it is possible very quickly to compute the area within a traverse by following the steps listed below.

1. For each of the corners of the figure a fraction is written with x as the numerator and y as the denominator. These are listed on a horizontal line *and the fraction for the first or starting corner is repeated at the end of the line.* Next, a solid diagonal line is drawn from x_1 to y_2, from x_2 to y_3, and so on. Then a dashed diagonal line is drawn from y_1 to x_2, from y_2 to x_3, and so on.

$$\frac{x_1}{y_1} \quad \frac{x_2}{y_2} \quad \frac{x_3}{y_3} \quad \frac{x_4}{y_4} \cdots \quad \frac{x_1}{y_1}$$

2. The summation of the products of the coordinates joined by the solid lines minus the summation of the products of the coordinates joined by the dashed lines equals twice the area within the traverse. This is exactly a statement of the formula given earlier in this section.

$2A$ = summation of solid-line products minus the summation

of the dashed-line products

The area within the traverse of Fig. 12-10 is determined by this alternate coordinate method as follows:

$$\frac{100}{400} \quad \frac{600}{800} \quad \frac{900}{300} \quad \frac{500}{100} \quad \frac{100}{400}$$

$2A$ = $(100)(800) + (600)(300) + (900)(100) + (500)(400) - (400)(600)$

$- (800)(900) - (300)(500) - (100)(100) = -570{,}000$

$A = 285{,}000$ sq ft

12-12 AREAS WITHIN IRREGULAR BOUNDARIES

Very often property boundaries are represented by irregular lines, for example, the center line of a creek or the edge or center line of a curving road. For such cases as these, it is often not feasible to run the traverse along the exact boundary line. Instead, it may be practical to run it a convenient distance from the boundary and locate the position of the boundary by measuring offset distances from the traverse line, as shown in Fig. 12-11. The offsets may be taken at regular intervals if the boundary does not change suddenly, but when it does change suddenly, offsets are taken at irregular intervals, as shown by *ab* and *cd* in the figure.

The area inside the closed traverse may be computed by one of the methods described previously, and the area between the traverse line and the irregular boundary may be determined separately and added to the other value. If the land in question between the traverse line and the irregular boundary is carefully plotted to scale, the area may be determined satisfactorily with a planimeter. Other methods commonly used are the trapezoidal rule, Simpson's one-third rule, and

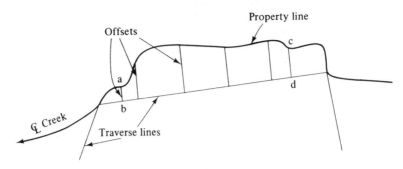

Figure 12-11

another coordinate method involving offsets. All of these methods are described in this section, but it should be recognized that they probably do not yield results that are any more satisfactory than those obtained with a planimeter, because of the irregular nature of the border between the measured offsets.

Planimeter

A polar planimeter (see Fig. 12-12) is a device that can be used to measure the area of a figure on paper by tracing the boundary of the figure with a tracing point. As the point is moved over the figure the area within is mechanically integrated and recorded on a drum and disk. An excellent mathematical proof of the workings of the planimeter is presented in the book by Davis, Foote, and Kelly.[2] When a planimeter is used to determine an area, it is not necessary to compute latitudes, departures, or coordinates. It is not even necessary to have figures consisting of straight lines as is required with the DMD and coordinate methods. It is, however, necessary to plot the figures carefully to scale before the planimeter is used.

    The planimeter is particularly useful for measuring the areas of irregular pieces of land as well as the areas of cross sections. If the operator is careful, he or she can obtain results within 1% or better, depending on the accuracy of the plotted figures, the types of paper used and the carefulness with which the figures are traced.

    The major parts of a planimeter are the tracing point, the anchor arm with its weight and post, the scale bar, and the graduated drum and disk with its vernier. When an area is to be traced, the drawing is stretched out flat so there are no wrinkles and the anchor point is pushed down into the paper at a convenient location so that the operator can trace all or a large part of the area desired at that one location. The tracing point is set at a distinct or marked point on the drawing and the graduated drum is read (or set to zero). Then the perimeter is

   [2] R. E. Davis, F. S. Foote, and J. W. Kelly, *Surveying Theory and Practice,* 5th ed. (New York: McGraw-Hill Book Company, 1966), pp. 67–69.

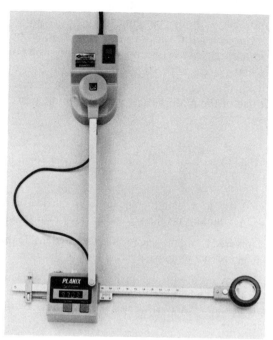

Figure 12-12 Tamaya digital planimeter. (Courtesy of the Leitz Company.)

carefully traced until the tracing point is returned to the starting point, at which time a final drum reading is taken. In traversing around the figure the operator must be very careful to note the number of times the drum reading passes zero. Counters are available with some planimeters with which the number of revolutions is automatically recorded. The initial reading is subtracted from the final reading and represents to a certain scale the area within the figure. If the perimeter is traced clockwise, the final reading will be larger than the initial reading, but it will be smaller if traversed counterclockwise.

If the anchor point is placed outside the area to be traversed, the area of the figure will equal

$$A = Cn$$

where C is a constant and n is the difference from the initial to the final readings on the drum. The constant C is usually given on the top of the tracing arm or on the instrument box. It is equal to 10.00 sq in. for many planimeters. If the user is not sure of C, he or she can easily construct a figure of known area (e.g., a 5 in. by 5 in. square), run the tracing point around the area, and determine what C would have to be to give the correct area when multiplied by the net reading on the drum. For some instruments the value of C is given in SI units.

If the anchor point is placed inside the area to be traversed (as it will often be for large areas), it will be necessary to make a correction to the area computed. It is possible to hold the tracing arm in such a position that the tracing point can be moved completely around 360° without changing the drum reading. The area of this circle, called the *zero circle* or the *circle of correction,* must be added to *Cn* if a figure is traversed in a clockwise direction with the anchor point on the inside. It will be noted that if the area of the figure is less than that of the zero circle, the change in the drum reading will be minus for a clockwise traverse. Example 12-1 illustrates the use of the planimeter for determining areas of figures in a drawing.

Example 12-1

Upon calibration it was found that a given planimeter traversed 10 sq in. for each revolution of its drum.

(a) If a given map area is traversed with the anchor point outside the area and a net reading of 16.242 revolutions is obtained, what is the map area traversed?

(b) If the same area is traversed with the anchor point located inside the area and the net reading is 8.346 revolutions, what is the area of the zero circle?

(c) Another map area is traversed with the anchor point placed inside the area and a net reading of 23.628 revolutions is obtained. If this area was plotted on the map with a scale of 1 in. = 20 ft, what was the actual area on the ground?

Solution

(a) Area = (10)(16.242) = 162.42 sq in. on the map

(b) Area of zero circle = (10)(16.242 − 8.346) = 78.96 sq in.

(c) Area on map = (10)(23.628) + 78.96 = 315.24 sq in.

Area on ground = (20 × 20)(315.24) = 126,096 sq ft

Electronic polar planimeters are available today which provide digital read-outs in large, bright numbers. These devices are simple to read and may easily be set to zero. They may also be used to handle cumulative adding and subtracting of areas. When using a planimeter it is well to measure each area several times and average the results. Modern planimeters will usually compute the average for you.

Trapezoidal Rule

When the offsets are fairly close together, an assumption that the boundary is straight between those offsets is satisfactory and the trapezoidal rule may be applied. With reference to Fig. 12-13, the offsets are assumed to be located at regular intervals and the area inside the figure equals the areas of the enclosed trapezoids, or

$$A = d\left(\frac{h_1 + h_2}{2}\right) + d\left(\frac{h_2 + h_3}{2}\right) + \cdots + d\left(\frac{h_{n-1} + h_n}{2}\right)$$

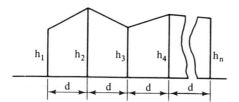

Figure 12-13

from which

$$A = d \left(\frac{h_1 + h_n}{2} + h_2 + h_3 + \cdots + h_{n-1} \right)$$

Simpson's One-Third Rule

If the boundaries are found to be curved, Simpson's one-third rule is considered better to use than the trapezoidal rule. Again, it is assumed that the offsets are evenly spaced. The rule is applicable to areas that have an odd number of offsets. If there is an even number of offsets, the area of all but the part between the last two offsets (or the first two) may be determined with the rule. That remaining area is determined separately, ordinarily assuming it to be a trapezoid.

With reference to Fig. 12-14, Simpson's rule is written as follows:

$$A = (2d) \left(\frac{h_1 + h_3}{2} \right) + \left(\frac{2}{3} \right) (2d) \left(h_2 - \frac{h_1 + h_3}{2} \right) + (2d) \left(\frac{h_3 + h_5}{2} \right)$$

$$+ \left(\frac{2}{3} \right) (2d) \left(h_4 - \frac{h_3 + h_5}{2} \right), \text{ etc.}$$

This reduces to

$$A = \frac{d}{3} [(h_1 + h_n) + 2(h_3 + h_5 + \cdots + h_{n-2}) + 4(h_2 + h_4 + \cdots + h_{n-1})]$$

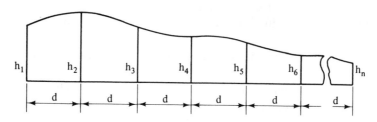

Figure 12-14

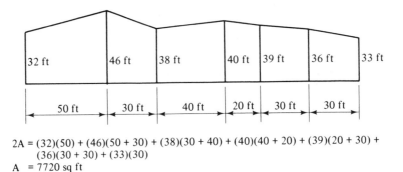

$2A = (32)(50) + (46)(50 + 30) + (38)(30 + 40) + (40)(40 + 20) + (39)(20 + 30) +$
      $(36)(30 + 30) + (33)(30)$
$A = 7720$ sq ft

Figure 12-15

Coordinate Rule for Irregular Areas

When offsets are taken at irregular intervals, the area of each figure between pairs of adjacent offsets may be computed and the values totaled. In addition, the planimeter method is particularly satisfactory here. There are other methods of handling the problem, for example, the coordinate rule for irregular spacing of offsets, which says that twice the area is obtained if each offset is multiplied by the distance to the preceding offset plus the distance to the following offset. It will be noted that for the end offsets the same rule is followed, but there will be only one distance between offsets because the outside one does not exist. An application of this coordinate rule is shown in Fig. 12-15.

Area of Segment of Circle

If we have a tract of land which has a circular horizontal curve for one of its boundaries (as is often the case where the tract is adjacent to a highway or railroad) we can compute the area of the land in question with little trouble as described herein.

For this discussion the traverse ABCDEFA shown in Fig. 12-16 is considered. One method of determining the total area within the figure is to separate the traverse into two parts ABCDEA and the segment of the circle AEFA. In the figure X is the center of the circle, R is the radius whose arc is EFA and I is the angle between the two radii shown.

The area ABCDEA can be determined by one of the methods previously described and added to the area of the segment of the curve which follows

$$\text{Area of segment} = \text{area A}\overline{\text{X}}\text{EFA} - \text{area A}\overline{\text{X}}\text{EA} = R^2 \left(\frac{\pi I^\circ}{360} - \frac{\sin I}{2} \right)$$

An alternate solution is to calculate the area ABCDE$\overline{\text{X}}$A and add to it the area A$\overline{\text{X}}$EFA calculated as follows:

$$\text{Area A}\overline{\text{X}}\text{EFA} = \left(\frac{I^\circ}{360} \right) (\pi R^2)$$

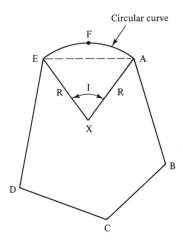

Figure 12-16

PROBLEMS

In Problems 12-1 to 12-4, compute the latitudes and departures for the sides of the traverses shown in the accompanying figures. Determine the error of closure and precision for each of the traverses.

12-1.

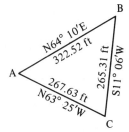

(*Ans.*: E_C = 0.133 ft, precision = 1/6432)

12-2.

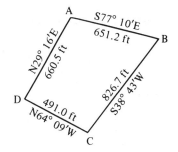

12-3.

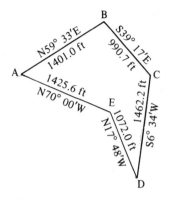

(Ans.: E_C = 1.259 ft, precision = 1/5065)

12-4.

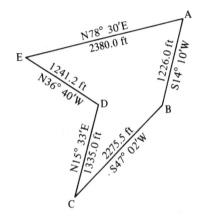

In Problems 12-5 to 12-7, balance each of the sets of latitudes and departures given by the compass rule and give the results to the nearest 0.01 ft.

12-5.

		Latitudes		Departures	
Side	Length	N	S	E	W
AB	400.00	320.00		245.00	
BC	300.00		180.00	235.36	
CA	500.00		140.24		480.00

(Ans.: 320.08; 179.94; 140.14; 244.88; 235.27; 480.15)

12-6.

		Latitudes		Departures	
Side	Length	N	S	E	W
AB	600.00	450.00		339.00	
BC	450.00		285.00	259.50	
CA	750.00		164.46		599.22

12-7.

Side	Length	Latitudes		Departures	
		N	S	E	W
AB	220.40	185.99		118.26	
BC	287.10		234.94	165.02	
CD	277.20		181.25		209.73
DE	200.10	187.99		68.55	
EA	147.90	42.01			141.81

(*Ans.: AB* 186.03, 118.20; *EA* 42.04, 141.85)

In Problems 12-8 to 12-10, balance by the compass rule the latitudes and departures computed for each of the traverses of Problems 12-1 to 12-3.

(*Ans.:* Problem 12-9 for *AB* 144.79, 635.20; for *CD* 213.97, 441.66)

In Problems 12-11 to 12-13, from the given sets of balanced latitudes and departures calculate the areas of the traverses in acres using DMDs with the meridians through the most westerly points.

12-11.

Side	Balanced latitudes		Balanced departures	
	N	S	E	W
AB	600		200	
BC	100		400	
CD	0	0	100	
DE		400		300
EA		300		400

(*Ans.:* 4.13 acres)

12-12.

Side	Balanced latitudes		Balanced departures	
	N	S	E	W
AB	100			200
BC	100		200	
CD	200		100	
DE		700	100	
EA	300			200

12-13.

Side	Balanced latitudes		Balanced departures	
	N	S	E	W
AB	200		100	
BC	100		200	
CD		150	150	
DE		50		200
EA		100		250

(*Ans.:* 1.29 acres)

For Problems 12-14 to 12-17, with the latitudes and departures balanced with the compass rule, compute the areas in acres using DMDs with the meridians through the most westerly points.

12-14. Problem 12-1.

12-15. Problem 12-2.

(*Ans.:* 4.277 acres)

12-16. Problem 12-3.

12-17. Problem 12-4.

(*Ans.:* 65.50 acres)

12-18. Repeat Problem 12-11 with the meridian passing through the most easterly point.

12-19. Repeat Problem 12-12 using double parallel distances (DPDs) with the reference parallel passing through the most northerly point.

(*Ans.:* 1.95 acres)

12.20. Repeat Problem 12-13 using DPDs with the reference parallel passing through the most southerly point.

In Problems 12-21 and 12-22, for the given coordinates compute the length and bearing of each side.

12-21.

Point	x (ft)	y (ft)
A	0	0
B	+200	+150
C	+500	−100
D	+300	−200

(*Ans.:* *AB* = N53°08′E, 250.00 ft; *BC* = S50°12′E, 390.51 ft)

12-22.

Point	x (m)	y (m)
A	+300	+200
B	+400	−300
C	+100	0

In Problems 12-23 to 12-25, compute the area in acres by the method of coordinates for each of the traverses whose corners have the coordinates given.

12-23.

Point	x (ft)	y (ft)
A	+200	+400
B	+300	−100
C	+100	−300
D	−150	−250
E	−100	+150

(*Ans.:* 4.68 acres)

12-24.

Point	x (ft)	y (ft)
A	0	0
B	+400	+150
C	+500	−200
D	+200	−150
E	+100	−300

12-25.

Point	x (ft)	y (ft)
A	+100	+250
B	+350	+300
C	+400	−250
D	−100	−300
E	−300	−200
F	+100	−100

(*Ans.:* 4.53 acres)

12-26. Repeat Problem 12-11 using the coordinate method.

12-27. Repeat Problem 12-12 using the coordinate method.

(*Ans.:* 1.95 acres)

12-28. Repeat Problem 12-13 using the coordinate method.

12-29. Determine the area of the traverse of Problem 12-21 in acres using the coordinate method.

(*Ans.:* 1.89 acres)

12.30. Using the coordinate method compute the area in hectares (1 ha = 10 000 m²) for the traverse of Problem 12-22.

12-31. A given planimeter has a constant of 2; that is, one revolution of the drum equals 2 sq in.

   (a) If a map area is traversed with the anchor point outside the area and a net reading of 16.422 revolutions is obtained, what is the map area?

(*Ans.:* 32.844 sq in.)

(b) If the same area is traversed with the anchor point inside the area and the net reading is 8.324 revolutions, what is the area of the zero circle?

(*Ans.:* 16.196 sq in.)

12-32. Upon calibration, it was found that a given planimeter traversed 10 sq in. for every revolution of its drum and the area of the zero circle was 72.00 sq in. With the anchor point inside the area the reading before traversing an unknown area was 4.832 revolutions and after traversing the area was 8.986 revolutions.

(a) What is the area in square inches?

(b) If the map scale is 1 in. = 50 ft, what is the area on the ground?

12.33. A given planimeter has a constant of 6. A map area is traversed with the anchor point outside the area and a net reading of 16.420 revolutions is obtained. The same area is traversed with the anchor point inside the area and a net reading of 8.320 revolutions is obtained. The scale of the map on which the area is being measured is 1 in. = 50 ft. Determine the following:

(a) Map area traversed.

(*Ans.:* 98.52 sq in.)

(b) Area of the zero circle.

(*Ans.:* 48.60 sq in.)

(c) Actual ground area.

(*Ans.:* 5.65 acres)

In Problems 12-34 and 12-35, compute the area (in square feet) of the irregular tracts of land shown using the trapezoidal rule.

12-34.

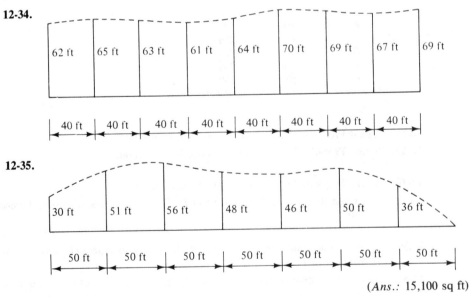

12-35.

(*Ans.:* 15,100 sq ft)

12-36. Repeat Problem 12-34 using Simpson's one-third rule.

12-37. Repeat 12-31 using Simpson's one-third rule.

(*Ans.:* 15,333 sq ft)

12-38. To determine the area between a base line *AB* and the edge of a lake the following offset distances were measured at 50-ft intervals. Compute the area (in square feet) of the tract by using Simpson's one-third rule. Offset distances: 16 ft, 35 ft, 52 ft, 63 ft, 71 ft, 68 ft, 60 ft, 55 ft, 41 ft, 37 ft.

For Problems 12-39 to 12-41, plot to scale the referenced information and determine the areas (in square feet) using a planimeter.

12-39. Problem 12-34.

(Ans.: 21,056 sq ft)

12-40. Problem 12-35.

12-41. Problem 12-38.

(Ans.: 23,775 sq ft)

For Problems 12-42 to 12-44, the offsets are taken at irregular intervals. Determine the area (in square feet) between the traverse line and the boundary for each case. Use the coordinate method. The distances given are measured from the origin in each case.

Problem 12-42		Problem 12-43		Problem 12-44	
Distance	Offset	Distance	Offset	Distance	Offset
0	30	0	18	0	17
30	50	20	22	10	30
50	35	40	28	20	45
80	45	60	34	30	56
90	58	70	46	50	50
100	50	75	58	100	62
120	68	80	64	150	71
150	60	85	66	200	78
		100	60	210	63
		120	55	230	52
				250	50

(Ans.: 4905 sq ft)

CHAPTER THIRTEEN

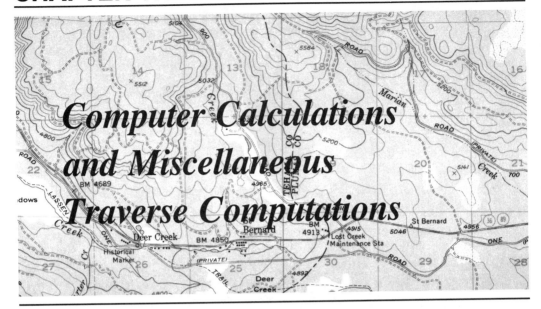

Computer Calculations and Miscellaneous Traverse Computations

13-1 ELECTRONIC COMPUTERS

The types of calculations presented in the preceding chapter and this one are almost "everyday" problems for the practicing surveyor. Until several decades ago these calculations were made laboriously using trigonometric tables and logarithms. As the years passed, large mechanical calculators and then smaller and smaller calculators were developed with which the numbers were handled. Since the 1960s, however, hand-held calculators and digital computers have made the other equipment obsolete. The latter devices, which have built-in trigonometric tables, have appreciably simplified the computational work of the surveyor, both in the office and in the field.

Programmable hand-held calculators are available for only a few hundred dollars. They have both keyboard programming and card reader programs in which cards can be punched and read into the calculator in seconds. Furthermore, libraries of card programs for different problems are available.

In field work the surveyor often makes some measurements and then has to stop and make what are sometimes rather involved calculations (frequently of the types discussed in Chapter 12 and later in this chapter) before he or she can proceed with more field work. A great advantage of small programmable calculators (and small portable computers) is that they may conveniently be used in the field. Thus the necessary calculations can be made quickly without the delays and inconvenience involved in returning to the office, making calculations and returning to the field.

One slight disadvantage of programmable calculators is their lack of capacity. It may be necessary to run some programs in steps. Some data are put into the calculator and certain steps performed; those results are replugged into the calculator and some additional instructions are provided and more steps are performed. After a little practice this stepped procedure is not quite as bad as the author seems to indicate.

13-2 PROGRAMS

To apply the computer to a particular problem, it is necessary to have a program. This means that someone must translate the language of the problem into the language of the computer. Preparing such a program even for the simplest problem is a formidable task which the average practicing surveyor has neither the time nor the background to perform.

Fortunately, surveying problems for which the computer is commonly used fall into a few general classes and many organizations have successfully prepared programs for them and have made them available throughout the country. The result is that today the average surveyor uses "canned programs" that have been prepared by someone else.

13-3 AVAILABILITY OF PROGRAMS

There are today an increasing number of programs that the surveyor may use. To prepare a list of all of the companies having such programs available is an almost impossible task and the list would be obsolete before this chapter is printed. It is sufficient to say that a surveyor in any part of the United States can find suitable programs available from one organization or another for almost any problem that he or she faces. Many of the programs are supplied by the manufacturers of the various computers; they can often be obtained through user groups (organizations of users of a particular computer model); the libraries of service bureaus are quite extensive; and they can sometimes be obtained from other surveying firms.

Regardless of where a program is obtained, almost any surveyor would want to take a short problem or two that he or she has previously solved and check out the correctness of the program. There are few surveyors who would be willing to stake their professional reputations blindly on the claims made by the representative of a software company or other organization without making a preliminary check.

Of particular interest to the surveyor is the fact that some computer programs are available that will not only make the computations necessary for precision, balancing, areas, and so on, but will also print out a drawing or plat of the traverse and show the bearings and distances of its sides. Other programs are available that subdivide property into lots according to given limitations and print out a drawing of the results.

13-4 MICROCOMPUTERS

Microcomputers, also called desktop, home or personal computers, seem to be just what the surveyor needs, whether working in a small or a large company. They can handle quite easily almost any problem that will be encountered. Should an unusually large or unusually difficult problem be faced that is beyond the capacity of an individual microcomputer, that computer can be used as a terminal with a time-sharing company.

Once a surveyor obtains a microcomputer, he or she will learn that it can be used not only for surveying calculations but also for other items of importance in the business, such as payrolls, time and cost records associated with various jobs, accounts receivable, and other administrative tasks. Programs are easily obtainable to handle almost all common surveying computations on microcomputers and the machines can be directly programmed to handle other problems. The languages used for most microcomputers are BASIC (Beginner's All-purpose Symbolic Instruction Code) and FORTRAN (FORmula TRANslation), which are easy to use and to program.

There are available several microcomputers that can be used in vehicles with battery packs. They, together with their programs, can be taken to the field and calculations made in minutes for almost any situation that is encountered. It does not take much imagination to see all sorts of things that can be accomplished with this equipment.

13-5 BUYING A MICROCOMPUTER

Probably the most important thing to look for before buying a computer is the software or programs available that can be useful to the buyer. Once this information is obtained, the next step is to find a microcomputer with which that software is compatible. The average surveyor needs to buy a computer together with as many of the programs needed as possible. Perhaps he or she will some day be able to write programs or modify the available programs, and that is good, but in the meantime a computer is needed that can be used and benefited from immediately.

Before buying a microcomputer the surveyor should try to think in terms of what future uses he or she may make of the device. Perhaps the most common remark heard in relation to computer purchase is: "I started too small." A person contemplating the purchase of a computer should try out a system for a day or two before signing on the dotted line. Arrangements of this type can be made with almost all companies.

One final suggestion concerns the learning process for using computers. For most persons microcomputers that have extensive video prompts are much easier to use than are those for which it is necessary to struggle through detailed manuals step by step.

There are several million microcomputers in use in the United States today, and the number is increasing rapidly. A small personal computer including a monitor or screen, disk drive, printer, and some software requires an investment on the order of $1000 to $3000. This amazingly low price range for significant computing capability makes microcomputers economically feasible for all surveyors. Furthermore, the amount of software available to solve surveying and other types of problems on microcomputers has been increasing at an astronomical rate for the past several years.

13-6 APPLICATION OF THE COMPUTER PROGRAM SURVEY

In this section the reader is introduced to a set of computer programs called SURVEY, which were prepared for use on IBM and IBM-compatible microcomputers. These programs, which are written in BASIC, are furnished on the diskette enclosed with this book.

Although an interested user may obtain a listing of the various steps involved in the programs, SURVEY is designed to operate as a "black box"; that is, the user need only supply the specified input data and the computer will perform the calculations and supply the appropriate results automatically. *Even though the reader may have had little or even no previous computer experience he or she can learn to apply these programs in a very short time (minutes, not hours).*

In this section and the next, the very simple bits of information needed to apply SURVEY to a problem of the type handled in Chapter 12 (precision, balancing, and area calculations) are presented. Later in this chapter the same diskette is used for an omitted-line problem, in Chapter 14 it is used to make stadia calculations and in Chapter 16 calculations are made for astronomical observations. Although SURVEY is very easy to use and the screen prompts are quite clear as to the steps to be taken, the procedure is described in some detail in the next few paragraphs.

For the example to follow, the information and instructions to be inputted in response to the computer prompts are circled. (Inputting of data consists of typing the required information and then pressing the RETURN key.) The DOS diskette is initially placed in drive A of the computer, and then the computer, screen, and printer are switched on. After DOS is booted up, the diskette is removed from drive A and is replaced with the SURVEY diskette.

> Then we type in > BASIC (or BASICA) and hit return
> Followed by > LOAD"SURVEY and hit run key.

The computer screen shows a list of the programs available. The number of the program desired is selected from the list and is entered into the computer.

For this first example the number 1 (for *traverse computations*) is typed, the RETURN key is pressed, and the directions given on the screen are followed.

13-7 COMPUTER EXAMPLE

In this section the example traverse problem of Chapter 12 is reworked using SURVEY. It is assumed that any error in the angles has been distributed as described in Section 12-3 and the azimuth or bearing of each side determined.

In answer to a screen prompt, the number of sides of the traverse is furnished. The reader is next asked to select coordinates for the starting point. Any set of values may be used. Unless coordinates have been determined from previous work you may like to input large positive numbers so that all points will have positive values. Then the user is asked to specify the length units in feet or meters by inputting the letter f or m. If feet are used, the resulting area will be given in square feet and in acres (1 acre = 43,560 sq ft). If meters are specified, the computed area will be in square meters and in hectares (1 hectare = 10 000 m²). After the errors in latitudes and departures are computed, they are balanced with the compass or Bowditch rule. The computer will then calculate and print out the area of the figure, the balanced lengths of the figure sides and the coordinates of the figure corners.

To make the program work, the reader needs to follow very carefully the instructions for inputting lengths and directions, particularly as to separating the entries with commas. If this is not done, the user will be prompted to repeat that line of data.

It may be that the bearings or azimuths were determined only in degrees and minutes. Should this be the case, it is still necessary to provide the number of seconds (in this case 00) for each side.

Example 13-1

Using the program SURVEY, determine the errors of closure and the precision for the traverse of Fig. 13-1. Then balance the closure errors and determine the area of

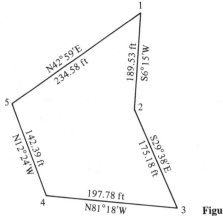

Figure 13-1

the figure. In addition it is desired to determine the coordinates of the corners of the traverse from the balanced latitudes and departures as well as the balanced lengths and directions of the sides. This is the same traverse that we considered in Chapter 12.

Solution

PROJECT TITLE: CHATOOGA FARM

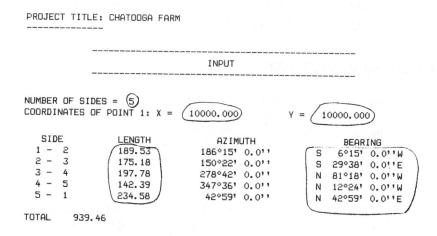

```
                    --------------------------------------------------
                                          INPUT
                    --------------------------------------------------


NUMBER OF SIDES =  ⑤
COORDINATES OF POINT 1: X =   10000.000        Y =   10000.000

     SIDE          LENGTH           AZIMUTH                BEARING
     1 - 2         189.53        186°15' 0.0''        S   6°15' 0.0''W
     2 - 3         175.18        150°22' 0.0''        S  29°38' 0.0''E
     3 - 4         197.78        278°42' 0.0''        N  81°18' 0.0''W
     4 - 5         142.39        347°36' 0.0''        N  12°24' 0.0''W
     5 - 1         234.58         42°59' 0.0''        N  42°59' 0.0''E

TOTAL     939.46
```

```
                    --------------------------------------------------
                                         OUTPUT
                    --------------------------------------------------
              UNBALANCED                            BALANCED
        --------------------------            --------------------------
      LATITUDE        DEPARTURE           LATITUDE        DEPARTURE
     -188.4040        -20.6330           -188.3900        -20.6000
     -152.2680         86.6180           -152.2500         86.6500
       29.9160       -195.5040             29.9300       -195.4700
      139.0680        -30.5770            139.0800        -30.5500
      171.6080        159.9330            171.6300        159.9700


ERROR IN LATITUDE   =  -.08

ERROR IN DEPARTURE  =  -.16


ERROR OF CLOSURE =  .181   Feet

PRECISION = 1 IN 5166

AREA =   .832   Acres
AREA =   36241.92  Square Feet
```

```
Corner coordinates                     Adjusted lengths and directions
--------------------------             --------------------------------------
Point      X            Y              Course      Length     Direction
   1   10000.000    10000.000          1 - 2       189.513   S  6°14'25.3''W
   2    9979.401     9811.611          2 - 3       175.181   S 30°21'16.1''E
   3   10066.050     9659.361          3 - 4       197.748   N 82°42'19.2''W
   4    9870.580     9689.290          4 - 5       142.396   N 13°36'40.6''W
   5    9840.030     9828.370          5 - 1       234.622   N 42°59'10.3''E
```

Concerning the use of the programs on the enclosed diskette, the author would like to make the usual disclaimer: *The reader may use the SURVEY programs in any way that he or she sees fit, but neither the author nor the publisher take responsibility for any difficulties arising as a result.*

It is almost impossible for a programmer to predict all the ways that users will attempt to apply their programs. As a result, the author would be grateful to receive any comments, criticisms, or suggestions from the users of SURVEY.

13-8 OMITTED MEASUREMENTS

Occasionally, one or more angles and/or lengths are not measured in the field and their values are computed later in the office. There are several reasons for not completing the field measurements such as difficult terrain, obstacles, hostile landowners, lack of time, sudden severe weather conditions and so on.

Should all of the measurements for a closed traverse be completed and an acceptable precision obtained, it is perfectly permissible to compute bearings and distances within that traverse. For instance, in Fig. 13-2 it is assumed that a group of lots (1 to 4) are being laid out in the field. The perimeter of the lots, represented by the solid lines, has been run with an acceptable precision and the errors of closure balanced. For such a case it is perfectly permissible to compute the lengths and bearings of the missing interior lot lines, which are shown dashed in the figure. Furthermore, a great deal of time may be saved where there is an appreciable amount of underbrush along the lines and/or where the ends of the lines are not visible from each other due to hills, trees, and so on.

The omission of some of the measurements for one or more sides of a closed traverse is a very undesirable situation. Even though it is possible to calculate the values of up to two missing lengths or two missing bearings (that is the same as three angles missing) or one length and one bearing (two angles), the situation is much to be avoided. The trouble with such calculations is that there is no way to calculate the precision of the field measurements that were made, since the figure was not closed. The measurements which were taken are assumed to be "perfect" in order that the missing quantities may be calculated. As a result, severe blunders can be made in the field, causing the computed values to be meaningless.

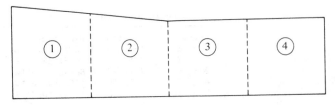

Figure 13-2

From this discussion it is evident that if the measurements of any lengths or angles that are part of a closed traverse are omitted in the field, it is advisable and almost essential to use some kinds of approximate checks on the computed values. Such things as stadia distance readings, angle measurements or compass bearings, and even eyeball estimates can be critical to the success of a survey.

In the next several sections of this chapter the calculations for the following types of omitted measurement problems are described: length and bearing of one side missing, length of one side and bearing of another missing, lengths of two sides missing, and bearings of two sides missing.

13-9 LENGTH AND BEARING OF ONE SIDE MISSING

For the traverse of Fig. 13-3, the lengths and bearings of sides *AB, BC,* and *CD* are known but the length and bearing of side *DA* are unknown. This is the same thing as saying that the angles at *D* and *A* are missing as well as the length of *DA*. From the assumption that the measurements for the three known sides are perfect, it is easily possible to compute the missing information for side *DA*. The problem is exactly the same as the one handled in Table 12-2, where the error of closure of a traverse was determined.

The latitudes and departures of the three known sides are computed, summed, and the "error of closure," represented by the dashed line *DA* in the figure, is determined. This is the length of the missing side; the tangent of its bearing angle equals the horizontal component of its distance divided by its vertical component, and may be written as follows:

$$\text{tan of bearing } \angle = \frac{\text{error of departure}}{\text{error of latitude}}$$

From this expression the bearing can be obtained as illustrated in Example 13-2. A simple way to determine the direction of the missing side is to examine the latitude and departure calculations. To close the figure of this example, the sum of the south latitudes needs to be increased, as does the sum of the east departures. Thus the bearing of the line is southeast.

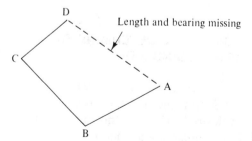

Figure 13-3

Since everyone is subject to making mathematical errors, it is well to compute the latitude and departure for this newly calculated line to determine if the traverse closes. Such a mathematical check is desirable for each of the omitted measurement problems described in this chapter.

Example 13-2

Determine the length and bearing of side *BC* of the traverse shown in Fig. 13-4. The lengths and bearings of the other sides are known.

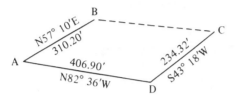

Figure 13-4

Solution   Compute the latitudes and departures of the known sides.

					\multicolumn{2}{c	}{Latitudes}	\multicolumn{2}{c}{Departures}
Side	Bearing	Length	Cosine	Sine	N	S	E	W
AB	N57°10′E	310.20	0.54220	0.84025	168.19	×	260.65	×
BC	—	—	—	—	—	—	—	—
CD	S43°18′W	234.32	0.72777	0.68582	×	170.53	×	160.70
DA	N82°36′W	406.90	0.12880	0.99167	52.41	×	×	403.51
				Σ =	220.60	170.53	260.65	564.21
					$E_L = 50.07$		$E_D = 303.56$	

Next, compute the length and bearing of the missing side.

$$l_{BC} = \sqrt{(50.07)^2 + (303.56)^2} = \textbf{307.66 ft}$$

$$\text{Tan of bearing } \measuredangle = \frac{E_D}{E_L} = \frac{303.56}{50.07} = 6.6027122$$

$$\text{Bearing} = \text{S80°38′E}$$

13-10 USING THE COMPUTER DISKETTE TO DETERMINE THE LENGTH AND BEARING OF A MISSING SIDE

In this section Example 13-2 is reworked using SURVEY. After the usual input, DOS, BASIC and LOAD″SURVEY the user in answer to the screen prompt selects the specific program desired, which in this case is 2 (*omitted-line computation*). The RETURN key is pressed and the directions given on the screen followed. The computer will then determine the length and bearing of a line from the ending point to the beginning point.

Example 13-3

Using the computer program SURVEY, repeat Example 13-2.

Solution

```
PROJECT TITLE: WILD HOG FARM
--------------
```

```
            ----------------------------------------------------
                                    INPUT
            ----------------------------------------------------
```

```
NUMBER OF SIDES =  ⟨3⟩
```

SIDE	LENGTH	AZIMUTH	BEARING
1 - 2	234.32	223°18' 0.0''	S 43°18' 0.0''W
2 - 3	406.90	277°23'60.0''	N 82°36' 0.0''W
3 - 1	310.20	57°10' 0.0''	N 57°10' 0.0''E

```
TOTAL    951.42
```

```
            ----------------------------------------------------
                                    OUTPUT
            ----------------------------------------------------
```

UNBALANCED		BALANCED	
LATITUDE	DEPARTURE	LATITUDE	DEPARTURE
-170.5320	-160.7000	-182.8600	-85.9400
52.4050	-403.5110	30.9900	-273.6800
168.1900	260.6460	151.8700	359.6200

```
ERROR IN LATITUDE    =    50.06

ERROR IN DEPARTURE   =   -303.57


ERROR OF CLOSURE =  307.666  Feet

Length and Bearing of omitted line 307.666  Feet   S 80°38' 9.7'' E
```

13-11 LENGTH OF ONE SIDE AND BEARING OF ADJOINING SIDE UNKNOWN

Figure 13-5 shows a closed traverse for which the bearing of side *AB* and the length of side *BC* are unknown. (This is the same as saying that the angles at *A* and *B* are missing as well as the length of side *BC*.) The lengths and bearings of the other sides are known. The latitudes and departures of these known sides may be calculated and from them the length and bearing of the dashed line *AC* determined.

The missing information for triangle *ABC*, which is redrawn in Fig. 13-6,

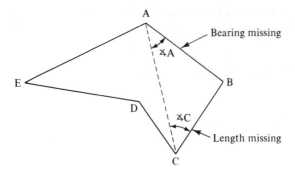

Figure 13-5

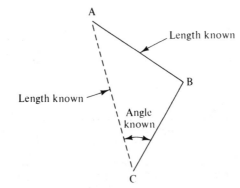

Figure 13-6

may easily be determined. First, the angle at C can be computed from the known bearings of sides CA and CB.

When one angle and the lengths of two sides of a triangle are known, the sine law may be used to find the missing information. For instance, the following equation may be written for the triangle ABC of Fig. 13-6:

$$\frac{\sin \angle C}{l_{AB}} = \frac{\sin \angle B}{l_{AC}}$$

With this equation the angle at B can be determined, thus giving two of the three interior angles of a triangle. Angle A equals $180° - \angle B - \angle C$ and the length of side BC can be determined from the equation

$$\frac{\sin \angle C}{l_{AB}} = \frac{\sin \angle A}{l_{BC}}$$

It is theoretically possible to obtain two sets of answers that will satisfy the problem (see Fig. 13-7). The lines AB and BC shown in Fig. 13-7(a) represent one possible solution, while the lines AB' and $B'C$ shown in part (b) of the figure provide another solution. If, however, the approximate bearing of side AB had

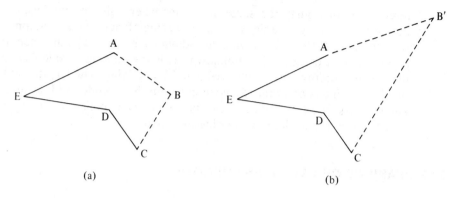

Figure 13-7

been observed in the field the confusion would have been avoided since the correct set of answers would be obvious.

Two further comments need to be made concerning the accuracy of this method. If the angle between the side of unknown bearing and the side of unknown length is close to 90°, the solution may be quite poor. In using the sine law it may be impossible to tell on which side of 90° an angle is located. (The sines of 89° and 91° are the same.) Second, the value of the sine near 90° changes very slowly and thus a small variation in the calculated values can cause a relatively large error in the angle calculated by the sine law. For these reasons the answers may be unsatisfactory.

13-12 LENGTH OF ONE SIDE AND BEARING OF ANOTHER UNKNOWN—SIDES NOT ADJOINING

Should the length of one side and the bearing of another side be missing, the two sides not adjoining, the unknowns can still be determined as they were when the sides were adjoining. The latitude and departure of a line of fixed length and

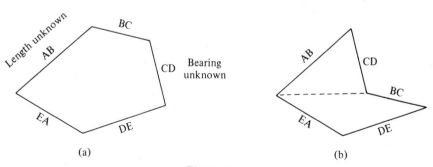

Figure 13-8

direction is obviously the same no matter where the line is placed. Thus it is possible to rearrange the sides of a traverse (their lengths and directions remaining the same) so that the unknown sides adjoin each other. For instance, in Fig. 13-8(a) the length of side *AB* is unknown, as is the bearing of side *CD*. In part (b) the figure is redrawn with sides *AB* and *CD* adjoining. The length and bearing of the dashed line is determined by summing up the latitudes and departures of the known sides *BC, DE,* and *EA*. The sine law is then applied to the triangle on top of the figure to calculate the missing information.

13-13 BEARINGS OF TWO SIDES UNKNOWN

Should the bearings of two sides of a closed traverse be missing (the same thing as saying that three angles are unknown), they can easily be calculated with the use of a common triangle formula. The two sides in question may or may not be adjoining. If they do not adjoin, the figure can be redrawn so that they are adjoining. In Fig. 13-9 it is assumed that the bearings of sides *AB* and *BC* are unknown.

The latitudes and departures of sides *CD, DE,* and *EA* are computed and summed; from the results the length and bearing of the dashed line *AC* is determined. Then the lengths of all sides of the oblique triangle *ABC* are known as well as the bearing of side *AC*. In Fig. 13-10 the letters *a, b,* and *c* represent the

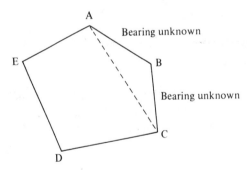

Figure 13-9

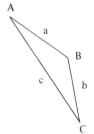

Figure 13-10

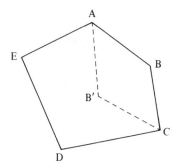

Figure 13-11

lengths of the sides of the triangle and s is a constant for the triangle, which simplifies the use of the angle equation.

Letting $s = \frac{1}{2}(a + b + c)$,

$$\sin \tfrac{1}{2}\angle A = \sqrt{\frac{(s - b)(s - c)}{bc}}$$

After the angle at A is determined, the sine law may be applied to obtain the other missing angles in the triangle.

It is necessary to know roughly the direction of at least one of the sides, since the solution of the trigonometric equation determines the shape of the triangle but not necessarily its position. In Fig. 13-11 it is observed that sides AB' and $B'C$ form another possible solution for the problem of Fig. 13-9.

13-14 LENGTHS OF TWO SIDES UNKNOWN

Sometimes a surveyor may read angles to a point that he or she does not intend to occupy. An angle may be read to a water tank, church steeple, tower, and so on, from two or more points. In others words, we do not want to take the time to measure distances to the point directly, either by taping or with an EDM device. Should this situation occur, the missing lengths can be computed with the sine law. In Fig. 13-12 the lengths of sides AB and BC are assumed to be unknown. The latitudes and departures of sides CD, DE, and EA are computed and summed and the length and bearing of line AC determined. For triangle ABC the bearings of all three sides are known, as is the length of line AC. With this information it is possible to write

$$\frac{\sin \angle B}{l_{AC}} = \frac{\sin \angle C}{l_{AB}}$$

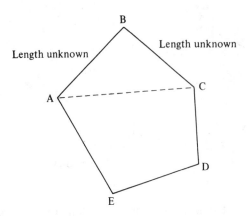

Figure 13-12

In this expression the only unknown is l_{AB}. After its value is determined, the following expression may be written to determine the length of side BC:

$$\frac{\sin \measuredangle C}{l_{AB}} = \frac{\sin \measuredangle A}{l_{BC}}$$

13-15 CUTTING OFF REQUIRED AREAS FROM GIVEN TRAVERSES

A frequent problem faced by the surveyor is that of cutting off required areas from given traverses. For instance, a buyer may purchase 40 acres from a larger tract with two sides of the land made parallel to a given direction—as perhaps a highway—as shown in Fig. 13-13, may buy the land on one side of a line drawn through two points (Fig. 13-14), may obtain a certain number of acres below a line drawn through a certain point (Fig. 13-15), and so on.

We now consider four kinds of area cutoff problems. There are various types in addition to the ones described, but if the reader understands the cases presented

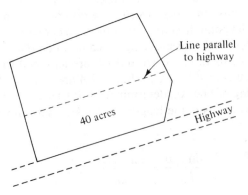

Figure 13-13

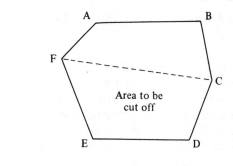

Figure 13-14

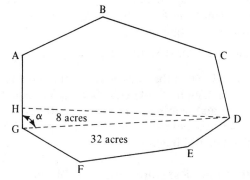

Figure 13-15

here, he or she will be able to develop a reasonable approach to the other situations. In each of the cases that follow it is assumed that:

1. A boundary survey has been made for the entire tract of land and a satisfactory degree of precision obtained.
2. The errors have been distributed throughout the traverse by one of the balancing methods, such as the compass or Bowditch rule.
3. The lengths and directions of the sides of the balanced traverse have been computed from the adjusted values of step 2 and are used for area cutoff calculations.

Cutting Off an Area with a Line Drawn between Two Given Points

For the traverse shown in Fig. 13-14, it is desired to determine the area below a line drawn from *F* to *C*. The latitudes and departures of sides *CD, DE,* and *EF* may be computed and used to determine the length and bearing of line *FC*. (The same value would be obtained if sides *FA, AB,* and *BC* were used.) Once this information is obtained, the area of *CDEF* can be determined with DMDs, coordinates, or some other method.

Cutting Off a Required Area with a Line Drawn through a Given Point

For the traverse of Fig. 13-15, it is desired to draw a line through point D such that it cuts off 40 acres below. As a first step in solving this problem the surveyor might draw a line from point D to another corner such that he or she thinks it cuts off an acreage somewhere in the neighborhood of the desired value. In this case corner G is selected and the length and bearing of line GD is computed with latitudes and departures, after which the area $DEFG$ is computed. It is assumed that the resulting area is 32 acres. It is next necessary to draw a line DH such that the triangle DGH contains 8 acres. Since the bearings of DG and GH are known, the angle between them (α in Fig. 13-15) may be computed and the following equation written for the triangle area:

$$\text{area } DGH = (8)(43{,}560) = \tfrac{1}{2} \sin \alpha l_{DG} l_{GH}$$

In this equation the number 43,560 is the conversion from acres to square feet.

The only unknown is the length of side GH. Solving the equation will yield its value, and then the length and bearing of DH may be determined by the sine law or by latitudes and departures. Finally, the area of figure $DEFGH$ should be calculated to be sure that it equals 40 acres.

Cutting Off an Area with a Line Passing through a Given Point in a Given Direction

For the traverse shown in Fig. 13-16, it is desired to cut off the area below a line drawn through point D parallel to side EF. The length and bearing of a line from point D to a convenient corner such as F can be computed by latitudes and

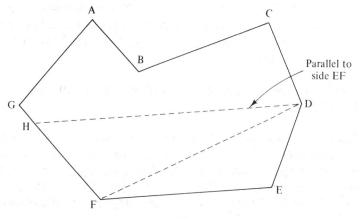

Figure 13-16

departures. Then with the bearings of lines *DF, FH*, and *HD* known as well as the length of line *DF*, it is possible with the sine law to determine the lengths of lines *FH* and *HD*. Once these values are determined, the area *DEFH* can be calculated.

Cutting Off a Required Area with a Line in a Given Direction

For the traverse of Fig. 13-17 it is desired to draw a line parallel to side *EF* so that it will cut off 60 acres below. A line is drawn through a convenient corner *D* parallel to side *EF* until it intersects the line *GA* at point *H*. The corner *D* is selected by the surveyor so that a line through that point will cut off an area below it somewhere in the neighborhood of the desired value. The lengths of lines *GH* and *HD* are computed as just described and then the area *DEFGH* is calculated. Here the result is assumed to be 45 acres.

It is then necessary to draw a line *IJ* parallel to line *HD* such that the area *DHIJ* equals 15 acres. It will be noted in Fig. 13-17 that since all the bearings are known, the angles α and β shown may easily be determined. The area of the parallelogram *DHIJ* can then be calculated as follows:

$$\text{area } DHIJ = \text{area of rectangle of } y \text{ height and length } l_{DH}$$

$$- \text{ area of triangle 1 shown in Fig. 13-17}$$

$$+ \text{ area of triangle 2}$$

$$(15)(43{,}560) = (y)(l_{DH}) - (\tfrac{1}{2})(y)(y \tan \alpha) + (\tfrac{1}{2})(y)(y \tan \beta)$$

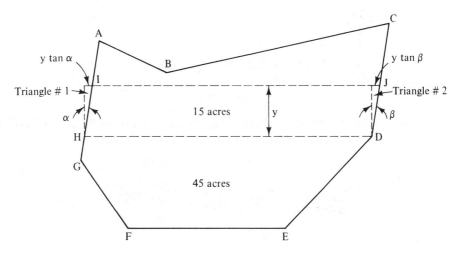

Figure 13-17

By solving this equation the value of y can be determined and then by trigonometry the lengths *HI* and *JD* can be calculated. The length *IJ* equals the length of *DH* − y tan α + y tan β. Having these values the area *JDEFGI* can be calculated to check its intended value of 60 acres.

PROBLEMS

For Problems 13-1 to 13-3, determine the precision and land areas using the computer diskette provided with this book.

13-1. Problem 12-1.

(*Ans.:* E_C = 0.133 ft, precision $\frac{1}{6405}$)

13-2. Problem 12-2.

13-3. Problem 12-3.

(*Ans.:* E_C = 1.259 ft, precision $\frac{1}{5043}$)

13-4. Compute the lengths and bearings of sides *BG* and *CF* for the accompanying closed traverse. The latitudes and departures for the outer traverse (in solid lines) have been balanced.

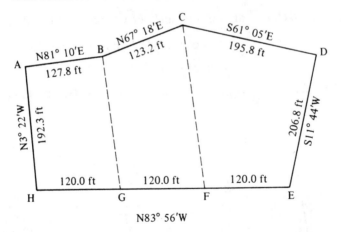

13-5. A random traverse is run from points *A* to *F*, as shown in the accompanying figure. Compute the length and bearing of a straight line from *A* to *F*.

(*Ans.:* N89°45'E, 1148.95 ft)

In Problems 13-6 to 13-8, for each of the closed traverses given in the tables, the length and bearing of one side are missing. Compute the latitudes and departures of the given sides and determine the length and bearing of the missing side.

13-6.

Side	Bearing	Length (ft)
AB	N45°30'E	260.00
BC	S31°30'E	310.00
CD	S41°42'W	212.00
DE		

13-7.

Side	Bearing	Length (ft)
AB	N63°19'E	332.66
BC	S18°03'W	302.54
CA		

(*Ans.:* N55°48'W, 246.02 ft)

13-8.

Side	Bearing	Length (ft)
AB	N48°50'E	211.00
BC	S42°48'E	239.00
CD	S18°10'W	234.50
DE		
EA	N20°32'W	316.00

For Problems 13-9 to 13-12, repeat the problems given using the computer diskette provided with this book.

13-9. Problem 13-5.

(*Ans.:* N89°45'E, 1148.95 ft)

13-10. Problem 13-6.

13-11. Problem 13-7.

(*Ans.:* N55°48'W, 246.02 ft)

13-12. Problem 13-8.

13-13. Using the computer diskette determine the length and direction of the missing side. Azimuths and lengths are provided for the other sides.

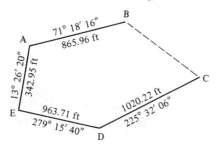

(*Ans.:* S86°12'45"E, 780.99 ft)

In Problems 13-14 to 13-16, compute the missing data in each table.

13-14.

Side	Bearing	Length (m)
AB	S62°24'E	352.3
BC	S39°40'E	398.6
CD		501.4
DA		497.6

13-15.

Side	Bearing	Length (ft)
AB	N49°48'E	
BC	S18°36'W	1122.2
CD		854.9
DA	N43°40'E	736.3

(*Ans.:* Length AB = 919.34 ft, Bearing CD = S85°49' W)

13-16.

Side	Bearing	Length (ft)
AB	N70°42'E	636.2
BC	S36°18'E	
CD	S59°36'W	
DA	N15°10'W	813.4

13-17. For the accompanying sketch of the balanced traverse *ABCDE* it is desired to cut off the tract of land *ABFG* so that *FG* is parallel to *AB* and the distance *GA* is 150.00 ft. Determine the lengths *BF* and *FG*.

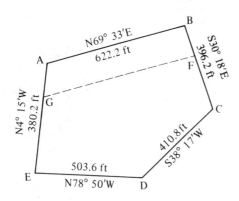

(*Ans.:* BF = 151.35 ft, FG = 664.30 ft)

13-18. For the balanced traverse shown in the accompanying illustration, compute the length *DA* and the area *DEFA*.

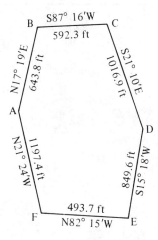

13-19. Repeat Problem 13-18 using the computer diskette provided with this book.
(*Ans.:* 1205.88 ft, 17.68 acres)

13-20. For the traverse of Problem 13-18, it is desired to cut off 6 acres below a line drawn through point *D*. Determine the lengths and bearings of the necessary sides of the figure.

13-21. For the traverse of Problem 13-18, it is desired to cut off 12 acres below a line drawn through point *D*. Determine the lengths and bearings of the necessary sides of the figure.
(*Ans.:* zD = 966.65 ft, N82°21′E)

13-22. For the traverse of Problem 13-18, it is desired to cut off the land below a line drawn through point *D* parallel to line *EF*. Determine the length of the necessary sides of the figure and the area of the figure.

13-23. For the traverse of Problem 13-18, it is desired to cut off 20 acres below a line drawn parallel to line *AD*. Determine the length of the necessary sides of the figure.
(*Ans.:* On side *AB* it's 86.4 ft, and on *DH* it's 110.6 ft)

CHAPTER FOURTEEN

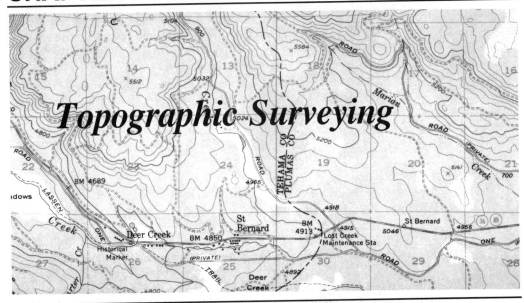

Topographic Surveying

14-1 INTRODUCTION

Topography can be variously defined as the shape or configuration or relief or roughness or three-dimensional quality of the earth's surface. Topographic maps are made to show this information, together with the location of artificial and natural features of the earth including buildings, highways, streams, lakes, forests, and so on. Obviously, the topography of a particular area is of the greatest importance in planning large projects such as buildings, highways, dams, or pipelines. In addition, unless the reader lives in fairly flat country, he or she would probably want to have a topographic map prepared for the land before locating and planning the building of a house. Topography is also important for soils conservation projects, forestry plans, geological maps, and so on.

The ability to use maps is very important to people in many professions other than surveying, for example, engineering, forestry, geology, agriculture, climatology, and military science. In October 1793, at the age of 24, Napoleon Bonaparte received his first promotion because of his ability to make and use maps when he was placed in command of the artillery at the siege of Toulon.

Since the preparation of maps is quite expensive, the surveyor should learn what maps have previously been made for the area in question before beginning a new one. For instance, the U.S. Geological Survey has prepared topographic maps for a large part of the United States. These maps are readily available at very reasonable prices. A large percentage of their old maps were published with a scale of 1:62,500 (1 in. = almost 1 mile). These maps, however, are being phased out and the present goal of the National Mapping Program is to provide maps with a scale of 1:24,000 (1 in. = 2000 ft) for all of the United States except Alaska, where a scale of 1:25,000 has been used on some recent metric maps. The USGS maps have contour intervals (a term defined in Section 14-2) of 10 ft in relatively flat country and up to 100 ft in mountainous country.

Many other government agencies, state and federal, have prepared maps of parts of our country and copies of them are easily obtained. In fact, more than 30 different federal agencies are engaged in some phase of surveying and mapping. To obtain information about the maps available for a particular area, one may write to the National Cartographic Information Center, 507 National Center, 12001 Sunrise Valley Drive, Reston, VA 22092. If the surveyor cannot find a map from these sources that has a sufficiently large scale for the purpose at hand, he or she should nevertheless be able to find one that will give general information and serve as a guide for the work.

14-2 CONTOURS

The most common method of representing the topography of a particular area is to use contour lines. It is thought that contours were first introduced in 1729 by the Dutch surveyor Cruquius in connection with depth soundings of the sea. Laplace used contours for terrain representation in 1816.[1] A contour is an imaginary level line that connects points of equal elevation (see Fig. 14-1). If it were possible to take a large knife and slice off the top of a hill with level slices at uniform elevation intervals, the edges of the cut lines around the hill would be contour lines (see Fig 14-2). Similarly, the edge of a still lake is a line of equal elevation or a contour line. If the water in that lake is lowered or raised, the edge of its new position would represent another contour line.

The contour interval of a map is the vertical distance between contour lines. The interval is determined by the purpose of the map and by the terrain being mapped (hilly or gently sloping). For normal maps, the interval varies from 2 to 20 ft, but it may be as small as $\frac{1}{2}$ ft for relatively flat country and as large as 50 to 100 ft for mountainous country.

The selection of the contour interval is a very imporant topic. The interval must be sufficiently small so that the map will serve its desired purpose while being as large as possible, to keep costs at a minimum. When the maps are to be used for earthwork estimates, a 5-ft interval is usually satisfactory unless very shallow cuts and fills are to be made. For such cases a 2-ft contour interval is probably needed. When the maps are to be used for planning water storage projects, it is usually necessary to use 1- or 2-ft contours.

A sufficient number of contour lines are numbered so that there is no confusion regarding the elevation of a particular contour. Usually, each fifth contour

---

[1] M. O. Schmidt and K. W. Wong, *Fundamentals of Surveying*, 3rd ed. (Boston: PWS Engineering, 1985), p. 350.

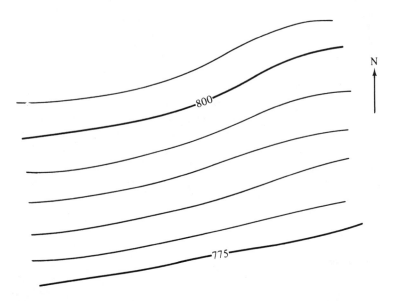

Figure 14-1 Typical contours (5-ft contour interval). Notice that the slope is fairly uniform, for the lines are almost equally spaced.

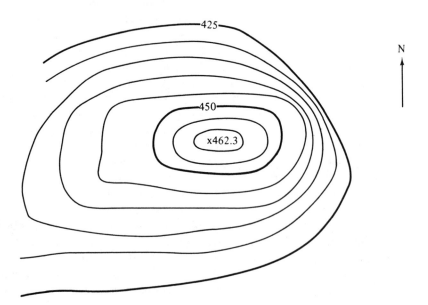

Figure 14-2 Notice that ×462.3 ft indicates the top of the hill and its highest point. The close spacing of the contours on the upper right indicates a steep slope.

line is numbered and shown as a wider and heavier line. Such a line is called an *index contour*. If any uncertainty remains, other contour lines may be numbered.

Figures 14-1, 14-2, and 14-3 present introductory examples of contour lines together with descriptive notes. Figure 14-4 shows a portion of a topographic map prepared by the U.S. Geological Survey.

Other methods that may be used to show elevation differences are relief models, shading, and hachures. Relief models are a most effective means of showing topography. They are made from wax, clay, plastic, or other materials and are shaped to agree with the actual terrain. The Army Map Service once produced and distributed beautifully colored molded plastic relief maps for those parts of the country that have significant elevation differences. These maps are now sold by Hubbard Scientific Co. of Northbrook, Illinois.

An old method used to show relative elevations on maps was shading. In this method an attempt was made to shade in the various areas as they might appear to a person in an airplane. For instance, the steeper surfaces might be in shadows and would be darkly shaded on the drawings. Less steep slopes would be indicated with lighter shading, and so on.

A few decades ago it was common practice to represent topography by means of hachures, but their use today is more infrequent, and they have been replaced by contour lines. Hachures are short lines of varying widths drawn in the direction of the steepest slopes. They give only an indication of actual elevations and do not provide numerical values. By their spacing and widths, these lines produce an effect similar to shading, but they are perhaps a little more effective.

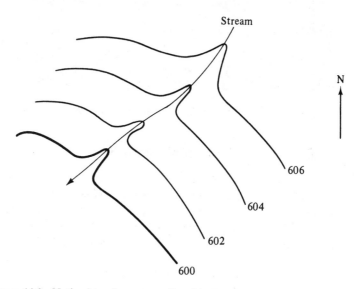

Figure 14-3 Notice how the contour lines bend upstream in crossing creeks or rivers.

Figure 14-4   Sample contour map. (Courtesy of U.S. Geological Survey.)

14-3 METHODS OF OBTAINING TOPOGRAPHY

There are three general methods used to obtain topography. These are the transit-stadia, plane table, and photogrammetric methods. In the transit-stadia method the necessary measurements are made in the field, recorded in the field book, and then plotted on paper in the office. In the plane table method the measurements are made in the same way as those in the foregoing method, but the data are plotted in the field on paper that is attached to a drawing board mounted on a tripod. This device is called a plane table. Both of these two methods are explained in this chapter, and the subject of photogrammetry is presented in Chapter 24.

Today, topographic maps are usually prepared using stadia methods for small areas such as for buildings, small industrial plants, and subdivisions while photogrammetric methods are used for large areas. Stadia surveys can be conducted for almost all weather and vegetation conditions. On the other hand, photogrammetric surveys need to be conducted on clear days and for situations where snow or dense vegetation do not obscure the ground surface. These items can decidedly reduce the accuracy of photogrammetric surveys and may actually prevent their use.

14-4 STADIA

Stadia is commonly used to obtain data needed for preparing topographic, hydrographic, and other maps. It is also useful for making rough surveys and for checking more precise work, but it is not sufficiently precise for property surveying. With stadia the surveyor can with one instrument setup make readings that will provide the distances, directions, and elevation differences to many surrounding points. The usefulness of this procedure is gradually declining, however, in the face of advances made in aerial mapping.

Figure 14-5 Stadia rods. (Courtesy of Keuffel & Esser, a Kratos Company.)

14-5 STADIA EQUIPMENT

The equipment used for stadia surveying consists of an instrument whose telescope is provided with stadia hairs and a regular level rod or a stadia rod. The
stadia hairs are mounted on the cross-hair ring, one a distance above the center
horizontal hair and one an equal distance below. A regular level rod is satisfactory
for stadia readings of up to 200 to 300 ft, but for greater distances, stadia rods
similar to the ones shown in Fig. 14-5 are used. A study of this figure will show
how easily good stadia rods can be made, and for this reason there are numerous
types of rods in use, many of which are homemade. Because stadia rods are large,
unwieldy, and inconvenient to transport in vehicles, a stadia design that is printed
on a strip of coated woven fabric is sometimes used. The fabric can be tacked
onto a board for a particular job and then removed, the board thrown away, and
the fabric folded up and carried to the next job.

14-6 STADIA THEORY: HORIZONTAL MEASUREMENTS

In this section an optics discussion of transit telescopes is neglected. The result
is that the theory is very simple. A detailed theoretical discussion of this matter
is presented in *Elementary Surveying*, 8th ed., by R. C. Brinker and P. R. Wolf.[2]
For this discussion it is assumed that a modern internally focusing instrument is
used. These instruments are manufactured so that if the telescope is horizontal
and the level rod vertical, the distance D from the center of the instrument to the
rod equals the stadia multiplier K times the distance from the rod reading at the
top stadia hair to the rod reading at the bottom stadia hair. This interval between
the stadia hairs is referred to as the *rod intercept* and is represented herein by
the letter s, as shown in Fig. 14-6. The stadia multiplier is 100 in almost all instruments, although it is 333 in a few.

　　Many older instruments are externally focusing. The external focal point of
these instruments is located a distance c (the distance from the center of the
instrument to the center of the objective lens) plus f (the distance from the focal

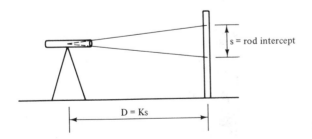

s = rod intercept

D = Ks

Figure 14-6

[2] New York: Harper & Row, Publishers, 1989, pp. 282–285.

point to the objective lens). These values are shown in Fig. 14-7, where it is desired to determine the distance D from the center of the instrument to the level rod.

If we let the distance from the top stadia hair to the bottom stadia hair be i, it is possible to write the following expression and from it determine the distance D:

$$\frac{i}{f} = \frac{s}{d}$$

$$d = \frac{f}{i}s$$

$$D = \frac{f}{i}s + (c + f)$$

In nearly all old transits the value of f/i or K is 100 and the value of $c + f$, called the *stadia interval factor*, varies from approximately 0.8 to 1.2 ft, with an average value of 1 ft. Manufacturers showed on their instrument boxes more precise values for the stadia constant for each instrument. In general, when the telescope is in a horizontal position, the horizontal distance from the center of the instrument to the center of the rod is

$$H = Ks + 1$$

The stadia intervals usually measured are probably a little too large because of unequal refraction and because of unintentional inclination of the rod by the rodman. To offset these systematic errors, the stadia constant was neglected in most surveys.

For modern internally focusing instruments the distance from the center of the telescope to the focal point is almost zero and is ignored. The objective lens of one of these instruments remains fixed in position while the direction of the light rays is changed by means of a movable focusing lens located between the plane of the cross hairs and the objective lens.

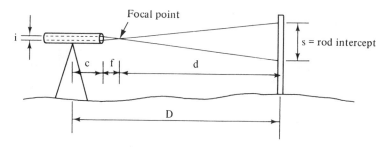

Figure 14-7

14-7 STADIA THEORY: INCLINED MEASUREMENTS

Since stadia readings are seldom horizontal, it is necessary to consider the theory of inclined sights. For such situations it is necessary to read vertical angles as well as the s values or rod intercepts and from these values to compute elevation differences and horizontal components of distance.

If it were feasible to hold the rod perpendicular to the line of sight, as shown in Fig. 14-8, the horizontal and vertical components of distance would be as follows (again, the stadia interval factor is neglected):

$$H = D \cos \alpha = Ks \cos \alpha$$

$$V = D \sin \alpha = Ks \sin \alpha$$

Obviously, it is not practical to hold the rod perpendicular to the line of sight, that is, at an angle α from the vertical; therefore, it is held vertically. This means that the rod is held at an angle α from being perpendicular to the line of sight. As can be seen in Fig. 14-9, the rod intercept reading is too large if the rod is not perpendicular to the line of sight. If the readings are taken with the rod plumbed, they can be corrected to what they should be if the rod had been perpendicular to the line of sight by multiplying them by the cosine of α. The resulting values for H and V may then be written as follows:

$$H = (Ks \cos \alpha)(\cos \alpha)$$

$$H = Ks \cos^2 \alpha$$

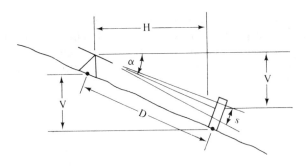

Figure 14-8

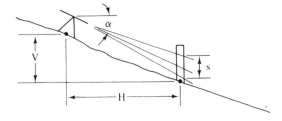

Figure 14-9

$$V = (Ks \sin \alpha)(\cos \alpha)$$

$$V = \tfrac{1}{2}Ks \sin 2\alpha$$

As a practical matter, if the vertical angle α is less than 3°, the diagonal distance may be assumed to equal the horizontal distance, but for such cases the vertical differences in elevation (the V values) definitely cannot be neglected.

14-8 STADIA REDUCTION

The derivations of the stadia reduction formulas were presented in the preceding section. In this section a few remarks are made regarding their application and solution. These formulas may be used directly for calculating horizontal distances and elevation differences, but their direct application for such purposes is infrequent because more efficient methods are available. If large areas are to be surveyed by stadia, for example, for mapping purposes, there may be hundreds or even thousands of readings to be reduced. If each of these were directly solved by the formulas, the task would be very large indeed.

One very common method for making these calculations involves the use of the stadia reduction tables given in Table A of Appendix A. These tables give horizontal distances and elevation differences for different vertical angles where Ks is equal to 100, that is, where s equals 1.00 ft. For s values other than 1.00, the needed values are simply equal to s times the numbers given in the table. These multiplications are usually rounded off to the nearest foot for horizontal distances and to the nearest 0.1 ft for elevation differences.

Other devices for stadia reduction are available, such as the stadia slide rule, which includes the ordinary log scales plus two special scales, one with log $\cos^2 \alpha$ values and another with log $\tfrac{1}{2} \sin 2\alpha$ values. Stadia slide rules are not as accurate as the tables just mentioned, but they are sufficient for most topographical work.

Another method of stadia reduction involves the use of the *Beaman stadia arc*, also called the *stadia circle*. This device is a specially graduated arc that either may be part of the vertical circle of the transit or alidade or may be separate. The arc consists of two scales graduated in percent: (1) one that enables the surveyor to determine the number of feet that must be subtracted from each 100 ft of stadia distance in order to obtain the horizontal distance measured, and (2) one that enables him or her to determine the number of feet of rise or fall of the line of sight for each 100 ft of stadia distance. These devices (stadia slide rules, Beaman stadia arc, etc.) were commonly used in the past but hand-held programmable calculators and electronic computers today have the field.

In Section 14-9 a sample set of stadia reduction notes is presented. The needed calculations were made with a pocket calculator. Then in Section 14-10 the same problem is repeated using a computer and the diskette enclosed with this book.

There are available today theodolites especially designed for stadia readings. These instruments are referred to as *reduction tacheometers*. Instead of having three horizontal cross hairs, they have three curved lines of varying radii. These lines appear to move farther apart or closer together as the telescope is raised or lowered to different vertical angles so that 100 times the rod intercept from the top to the bottom hair automatically provides the horizontal distance. Furthermore, the middle hair appears to move as vertical angles are changed so that the difference from the bottom line to the middle line multipled by a constant (100 for some instruments) equals the difference in elevation.

In Chapter 15 total station instruments are discussed. Topographic mapping is one area in which total stations tremendously increase production and economy.

14-9 FIELD PROCEDURE FOR STADIA MAPPING

The precision obtainable with stadia (averaging approximately 1/300 to 1/500) is sufficient for many preliminary or rough surveys. As such, it provides a quicker and more economical job than may be obtained with a transit-tape survey, particularly in rough country. The procedure used follows that of transit-tape surveys in that the horizontal angles or azimuths are measured with the transit. Distances are measured by stadia. The precision of stadia distance measurements can be improved to about 1/1000 or a little better by calibrating the instrument carefully for its *K* value, repeating readings several times and averaging results and perhaps by using targets.

The most common use of stadia surveying is for obtaining the details necessary for mapping. For topography, it is necessary to measure distances, directions, and vertical angles. For most maps only the details are obtained by stadia. A control traverse is carefully run and differential levels are made to the stations of the control traverse. From these stations stadia is used to obtain the details needed for the map.

The instrument may be set up at each of the traverse stations and a large number of stadia readings taken on points in the surrounding area to determine their position and elevation. If all the required area is not sufficiently covered from the traverse stations, short stadia traverses may be run out from those stations into the required areas.

The instrument is set up at a particular point, its h.i. is measured by holding the stadia rod or a tape up to the center of the telescope, and its value is recorded in the notes. With the middle cross hair set on the h.i. of the rod, the line of sight is parallel to a line drawn from the point over which the transit is centered to the point under the level rod. Thus the difference in elevation between the two points on the ground is the same as the difference in elevation between the center of the telescope and the h.i. up on the level rod. The horizontal circle is oriented by backsighting to the last station. The scale may be set to zero for the backsight or it may be set to the back azimuth of the line. The upper motion is then used for sighting all subsequent points.

The rodman moves to the first point and the instrumentman sets the middle horizontal cross hair on the h.i. and reads the rod intercept s. To do this, he or she may very well move the bottom cross hair up or down to a whole foot because it simplifies the subtraction from the top cross-hair reading. The instrumentman then resets the middle cross hair on the h.i. and lines up the vertical hair. The rodman is waved on to the next point while the horizontal and vertical angles are read.

A typical set of stadia notes is shown in Fig. 14-10. The first four columns identify the points by number and show the three measurements taken for each. The last three columns are used to record computed values. The remaining space on the right-hand page is used to describe the points carefully. Obviously, these descriptions are very important to the map plotter back in the office. In addition, the recorder should make any necessary explanatory sketches in the field book.

Sta.	Azimuth	Rod. Int.	Vert. ∡	Elev. Diff.	Horiz. Dist.	Elev.	
⊼ @ C.	Elev = 207.30,			h.i. = 4.7, BS		on D	
1	27°08'	1.45	-11°05'	-27.4	140	179.9	℄ of creek
2	51°56'	2.60	-3°06'	-14.0	259	193.3	Keller's front corner far side of road
3	77°05'	1.30	-14°20'	-31.2	122	176.1	Pt. under power line
4	95°00'	1.70	-10°50'	-31.4	164	175.9	" " " "
5	127°53'	2.97	-3°41'	-19.0	296	188.3	Keller's back corner
6	161°20'	0.75	-4°50'	-6.3	74	201.0	Edge of cleared field
7	270°00'	1.20	+4°40'	+9.7	118	217.0	P.O.G. (pt. on ground).
8	347°20'	0.80	-4°05'	-5.7	80	201.6	P.O.G.
⊼ @ ℔,	Elev. = 213.60,			h.i. = 4.6, BS		on C	
1	8°20'	1.48	+4°08'	+10.6	147	224.2	Top of creek bank
2	13°50'	0.96	0°00'	0.0	96	213.6	" " " " "
3	56°35'	0.93	+4°48'	+7.8	92	221.4	" " " " "
4	177°00'	2.00	-8°34'	-20.5	196	184.1	℄ of creek
5	196°25'	0.96	-17°38'	-27.7	87	175.9	" " " "
6	266°15'	1.95	-8°15'	-27.7	191	185.9	" " " "

STADIA SURVEY OF WILD HOG FARM

Dec. 5, 1990
Clear, mild, 60°F
L+E Transit # 3K621

J.B. Johnson ⊼, N
N.C. Hanson ∅
A.N. Nelson ∅

J.B. Johnson

Figure 14-10

As an example, suppose that for a particular point with the telescope sighted on the h.i., a vertical angle of $-6°18'$ and a rod intercept of 3.20 are read. Using the stadia reduction tables of Appendix A, the horizontal distance and the difference in elevation can be determined as follows:

$$H = (98.80)(3.20) = 316.2 \text{ ft} \quad \text{say } \textbf{316 ft}$$

$$V = -(10.91)(3.20) = -34.91 \text{ ft} \quad \text{say, } \textbf{-34.9 ft}$$

Should it be assumed that a level with an externally focusing telescope and a stadia constant of 1.25 were used for the readings, the value can be corrected as described below. As mentioned in Section 14-6, it is quite common to neglect the stadia constant for old transits and the resulting corrections as we did in the preceding calculations. If, however, it is desired to make a correction for the constant, it can be handled by using the values listed opposite C (the stadia constant) given at the bottom of the stadia reduction tables. For the example above,

$$H = (98.80)(3.20) + 1.24 = 317.4 \text{ ft} \quad \text{say, } \textbf{317 ft}$$

$$V = -(10.91)(3.20) - 0.14 = -35.05 \text{ ft} \quad \text{say, } \textbf{-35.1 ft}$$

For the stadia notes shown in Fig. 14-10, the stadia constant is neglected.

14-10 STADIA CALCULATIONS WITH THE COMPUTER PROGRAM SURVEY

In this section the computer diskette enclosed with the book is used to make the stadia calculations for the first eight points listed in Fig. 14-10. The number of the applicable program in this case is 3 (stadia computations). Then the computer instructions for inputting azimuths, rod intercepts, and vertical angles are carefully followed. The results are given in Fig. 14-11.

```
PROJECT TITLE: WHITE HORSE ROAD TRACT
---------------

Elevation= (207.30)     Number of points= (8)    K=  100
```

Point	Azimuth	s=Rod Intercept (ft)	Vertical Angle	Elevation Difference (ft)	Horizontal Distance (ft)	Elevation (ft)
1	27° 8' 0.0''	1.45	-11° 5' 0.0''	-27.35	140	179.9
2	51°56' 0.0''	2.60	-3° 6' 0.0''	-14.04	259	193.3
3	77° 5' 0.0''	1.30	-14°20' 0.0''	-31.18	122	176.1
4	95° 0' 0.0''	1.70	-10°50' 0.0''	-31.38	164	175.9
5	127°53' 0.0''	2.97	-3°41' 0.0''	-19.04	296	188.3
6	161°20' 0.0''	0.75	-4°50' 0.0''	-6.30	74	201.0
7	270° 0' 0.0''	1.20	+4°40' 0.0''	+9.73	119	217.0
8	347°20' 0.0''	0.80	-4° 5' 0.0''	-5.68	80	201.6

Figure 14-11 Stadia calculations with SURVEY.

14-11 SELECTION OF POINTS FOR STADIA MAPPING

For mapping purposes it is necessary to determine the elevations of a sufficient number of points in order to picture the contour or relief of the area accurately. Several approaches are possible in selecting these points. One is the so-called *checkerboard method*, in which the desired area is divided into squares or rectangles and the elevation determined at the corners of each of these figures. In addition, elevations are determined at points where slopes suddenly change, as, for example, at valley or ridge lines. This method is best suited to small areas that have little relief or roughness.

The most common method of selecting the points is the *controlling point method*. Elevations are determined for key or controlling points and the contour lines are interpolated between them. The controlling points are usually thought of as those points between which the slope of the ground is approximately uniform. They are such points as the tops of hills, the bottoms of valleys, and tops and bottoms of the sides of ditches, and other points where important changes in slope occur. If there is a uniform slope for a long distance, it is theoretically necessary to obtain only one elevation at the top of the slope and one at the bottom, for the map plotter will merely interpolate between those points in order to obtain contour lines. Practically speaking, however, since the eye is easily fooled by slopes, it is well to take an occasional intermediate point, even though the slope seems constant.

A third method, which makes the plotting of contours quite easy but which is not so simple to apply, is the *tracing contour method*. A number of points are located at the elevation of a desired contour and their locations are plotted on the map and connected by the contour line. To do this, it is usually easier to level the telescope and calculate from the H.I. what the FS rod reading should be when the rodman is on a particular contour line. For instance, if the H.I. is 457.2 ft, the rodman will be on the 450-ft contour line if the FS reading is 7.2. He or she can be waved up or down the slope until the reading is 7.2. After a number of points are located for the 450-ft contour line, the surveying party can work on, say, the 448-ft contour line, by positioning the rodman at points where the FS readings are 9.2 ft, and so on.

Sometimes in sighting to a point the instrumentman is unable to set the center cross hair on the h.i. because there is an obstruction, for example, the bushes shown in Fig. 14-12. If it is not feasible to cut the bushes down, it will be necessary to sight on some point on the rod other than the h.i. It is usually convenient to select a value that is an exact number of feet above or below the h.i. The difference involved is carefully recorded in the notes and must be added (or subtracted) to the calculated vertical distance between the instrument and the point. In Fig. 14-12 the correction is 2.0 ft and the calculated elevation difference V must be increased by that amount.

A similar problem may occur when in sighting to a point the instrumentman is unable to make the readings at both the top and bottom stadia hairs. If, however,

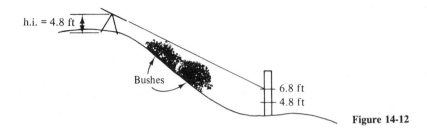

h.i. = 4.8 ft

Bushes

6.8 ft
4.8 ft

Figure 14-12

he or she is able to make the readings at the middle hair and one of the other two, he or she may take this value (which is one-half the desired rod intercept) and double it.

Some stadia rods may be read for distances considerably greater than 1000 ft. It will be noted, however, that because of the inconvenience of handling the bulky stadia rods, they are usually limited to maximum lengths of 10 or 12 ft. If the full intercept between the top and bottom hairs is read, the maximum sight distance then is theoretically limited to 1000 or 1200 ft. Actually, though, the presence of grass, bushes, and so on often keeps the instrumentman from being able to see the full length of the rod. Therefore, it is usually necessary for sights in the range of 1000 ft or greater to be obtained by reading only one-half the intercept and then by doubling it. Some instruments are equipped with a quarter hair between the middle cross hair and the upper stadia hair. By using this hair and taking a one-quarter intercept reading, it is occasionaly possible to take a sign for a distance as great as 4000 ft.

14-12 SAMPLE STADIA WORK

In this section a very brief description of the steps involved in working from the stadia notes to a completed topographic map is presented. In Fig. 14-13 the control traverse for a tract of land is plotted along with the elevation of a group of points taken by stadia.

In Fig. 14-14, 2-ft interval contours are drawn from the elevations that were plotted in Fig. 14-13. These lines were sketched in freehand by "eyeball" interpolation of the proportionate distances between elevation points. For very precise maps, the interpolation may be made with a calculator with the proportionate distances scaled between the points.

Finally, in Fig. 14-15 the control traverse is removed from the drawing, and the locations of important objects, such as roads, buildings, power lines, and so on, are plotted. On many topographic maps, for example, those prepared for house lots or building lots, the property lines and corners are so important for the project in mind that they are included on the map. Final details (the legend, title box, meridian arrow) are not shown in this figure.

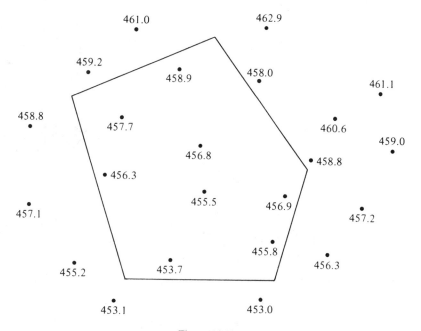

Figure 14-13

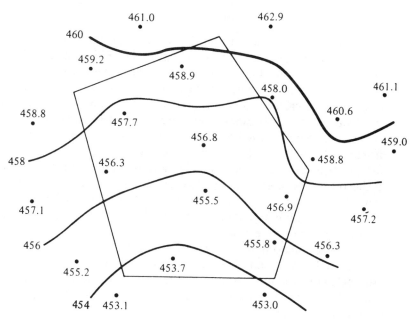

Figure 14-14

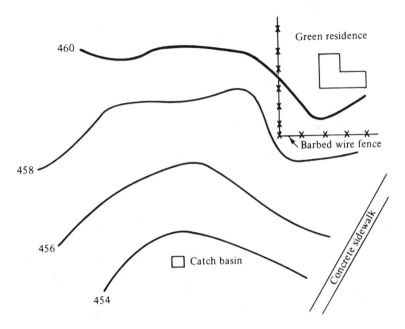

Figure 14-15

There is no doubt but that the work of plotting topographic maps by hand is a very tedious and time-consuming process. Within the last few decades much work has been done in the area of plotting such maps with computers. Originally, computer-aided contouring was available only on mainframe computers, but today programs are available to perform the necessary calculations and plot the contour lines and other details quickly and accurately with personal computers. The information can be inputted to the computer from the field notes, or a total station instrument can be used. The instrument will simultaneously compute the x, y, and z coordinates of the points and record them in an electronic field book. Back in the office the map is plotted by the computer and the contour map can be viewed on the screen. Furthermore, the contour lines can be smoothed or rounded off. Today, computer-generated graphics is a part of a large number of surveying projects in North America.

One serious problem may occur with the computer drafting of topographic maps if we have a point whose elevation is completely wrong. The computer will proceed to draw contour lines indicating a hill or a hole in the ground where none exists. If the map is drawn by hand, we are more likely to discover and discard the erroneous value.

14-13 SUMMARY OF CONTOUR CHARACTERISTICS

To aid the student in learning to read and prepare topographic maps, a summary of contour line characteristics follows:

1. Contour lines are evenly spaced when the ground surfaces are uniformly sloping.
2. For steep slopes contours will be close together, while for relatively flat areas they will be widely spaced.
3. When the ground surface is rough and irregular, contours will be irregular.
4. Contours are drawn perpendicular to the directions of the steepest ground slopes and thus water flow will be perpendicular to the contours.
5. Contour lines crossing streams or ditches point upstream in approximate $\cap$ or $\wedge$ shapes.
6. Contours connect points of equal elevation and thus cannot cross each other except in caves and similar situations.
7. The tops of hills and ground depressions look the same as to contour lines and may not easily be distinguished. For this reason spot elevations may be shown as at the low or high point, some extra contour lines may be numbered and shading or hachure lines may be used.
8. Contour lines are stopped at the edges of buildings.

14-14 PLANE TABLE SURVEYS

A very satisfactory mapping method involves use of a plane table. An alidade (Fig. 14-16) is used to take stadia measurements and the results are plotted on drawing paper attached to a portable drafting board or plane table. The table is mounted on a tripod so that it can be rotated and leveled. Map details are plotted on the paper as soon as they are measured. The contour lines are drawn as the work proceeds. The great advantage of plane table mapping is that the plotter can compare or correlate his or her drawing with the actual terrain as the work progresses, thus greatly reducing the chance of making mistakes.

The alidade consists of a telescope mounted on a brass ruler of approximately $2\frac{1}{2}$ in. by 15 in. The ruler is aligned with the line of sight of the telescope, which enables its user to take readings and scale them off in the proper direction along the ruler's edge. The telescope is similar to those used on transits and has the usual stadia hairs. In addition, the alidade has a trough compass mounted in a small metal box, a bubble tube mounted on the straight edge for leveling purposes, and a vertical arc that permits the measurement of vertical angles needed for determining elevation differences. If the surveyor has his or her position located

Figure 14-16 Plane table and alidade. (Courtesy of Wild Heerbrugg, Inc.)

on the drawing and if the drawing is oriented in the proper direction, he or she can sight on a desired point and determine the distance and elevation difference to that point by stadia and immediately plot the information on the paper.

Plane table paper is subject to rough field conditions, particularly dirt and moisture. Consequently, a special tough paper is used that minimizes expansion and contraction. This paper has been subjected to alternating drying and wetting cycles to make it more resistant to moisture changes. Sharp, hard pencils (6H to 9H) are used for plotting details, and great care is needed in handling the alidade to reduce smearing; it is lifted and moved rather than slid on the paper.

The plane table is set up at approximately waist height so that the surveyor or topographer can bend over it without resting on it. The topographer sets up, levels, and orients the table with great care. For small-scale maps (e.g., those having a scale smaller than 1 in. = 50 ft), it is generally satisfactory to center the table over each traverse station by eye, but for larger-scale maps, it is necessary to center the table more carefully. Each plane table is equipped with an arm under the board from which a plumb bob may be hung and used to center the table over the desired point.

14-15 PLANE TABLE DEFINITIONS

Several terms pertaining to plane table surveying that must be defined in order to enable the reader to understand how the work is accomplished are listed below.

Foresight. Sighting a point with the alidade and drawing a line from the plotted point occupied by the table to the other point is called a foresight.

Orientation.   The process by which the plane table is set up and aligned so that the lines on the drawing paper are parallel to the lines on the ground which they represent is called orientation. Orientation may be accomplished by use of the compass, by backsighting, and by resection.

Backsight.   A sight that is taken from the position of the plane table at a known and plotted point to another known and plotted point is called a backsight. The edge of the alidade scale is placed along the line between the two known points and the table is turned until the telescope line of sight hits the other point. When this is done, the table is oriented in relation to the points on the ground.

Radiation.   The process by which a series of points are located in relation to the plotted position of the table is referred to as radiation. The plane table is set up and oriented, and rays are drawn in the direction of each of the points whose elevation and horizontal position is desired. Figure 14-17 illustrates this process. The orientation of the table should be checked occasionally to see if any table slippage has occurred. This might happen if the topographer has leaned against the table, or bumped it.

Traversing.   In plane table surveying, traversing involves a procedure very close to the one followed with the transit-tape procedure. The plane table is moved successively to each of the traverse stations. At each station the table is oriented, ordinarily by backsighting, after which a foresight is taken to the next station. A line of the proper length is drawn to that station by the radiation method just described. The traversing procedure is illustrated in Fig. 14-18, in which points *A, B, C, D,* and *E* are the traverse stations on the ground and *a, b, c, d,* and *e* are the plotted points for those stations on the drawing. Notice that when the

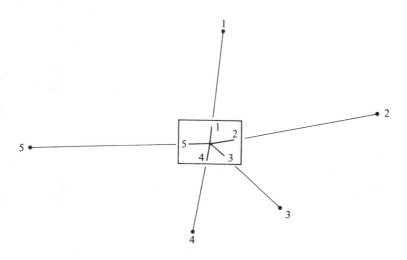

Figure 14-17

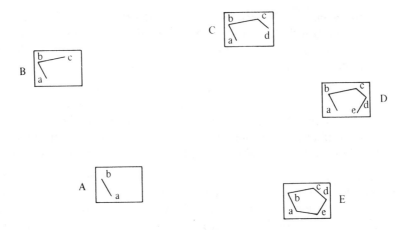

Figure 14-18

table was set up at *E*, the entire traverse was previously completed while the table was at *D*. This means that the setup at *E* can be used for checking.

Resection.   If the table is set up at a point on the ground that has not previously been plotted on the drawing, the process of locating the point is called resection. This topic is also discussed in Section 15-6 with particular reference to total station instruments. Several excellent references are available which discuss resectioning in great detail.[3]

Resectioning is accomplished by sighting on three previously plotted points whose positions are visible from the table and which have previously been plotted on the drawing. On frequent occasions these may be outstanding features such as church steeples, flag poles, individual trees or other prominent objects. From this information the points whose positions are desired may be determined by means of trigonometry or by graphical procedures.

Only a graphical procedure is presented in this section. This method called the *tracing paper-contour method* can be executed very easily and quickly in the field. A sheet of tracing paper is taped down over the main drawing and a point X is selected on the tracing paper about where we think the point will be located. Then a line is drawn from that point towards each of the prominent points in the distance. The sheet of tracing paper is untaped and moved around until the lines just plotted pass through the plotted positions of the prominent reference points. The point X is now the coreect instrument location. That position is marked through the tracing paper onto the map and the table is oriented by taking backsights on the reference points.

[3] R. E. Davis, F. S. Foote, and J. W. Kelly, *Surveying Theory and Practice,* 5th ed. (New York: McGraw-Hill Book Company, 1966), pp. 431–436.

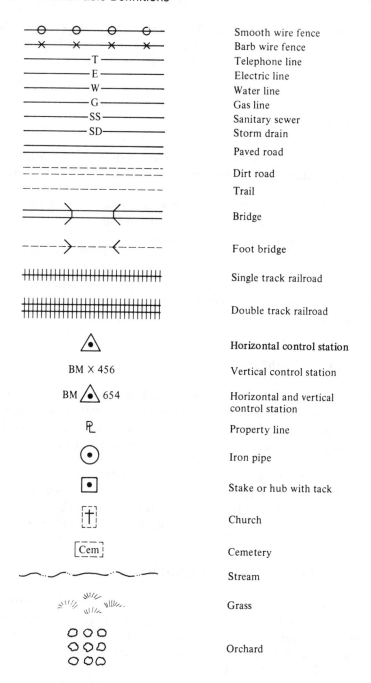

	Smooth wire fence
	Barb wire fence
T	Telephone line
E	Electric line
W	Water line
G	Gas line
SS	Sanitary sewer
SD	Storm drain
	Paved road
	Dirt road
	Trail
	Bridge
	Foot bridge
	Single track railroad
	Double track railroad
	Horizontal control station
BM X 456	Vertical control station
BM 654	Horizontal and vertical control station
PL	Property line
	Iron pipe
	Stake or hub with tack
	Church
Cem	Cemetery
	Stream
	Grass
	Orchard

Figure 14-19 Typical map symbols.

In selecting the points to be observed the surveyor should be aware of one problem which can occur. If the point whose position is desired falls on or near a circle passing through the three sighted points we will not be able to locate the desired point.

14-16 MAP SYMBOLS

It is quite convenient to represent various objects on a map with symbols that are easily understood by the users. Such a practice saves a great deal of space (compared to the lengthy explanatory notes that might otherwise be necessary) and results in neater drawings. Many symbols are standard across the United States, whereas others vary somewhat from place to place. In Fig. 14-19 a sample set of rather standard symbols are presented. Very detailed symbol lists are available from the U.S. Geological Survey, these being the ones they use for their maps.

Occasionally, the surveyor encounters some feature for which he or she does not have a standard symbol. For such a situation a new symbol may be made up, but it must be described in the map's legend.

14-17 COMPLETING THE MAP

The appearance of the completed map is a matter of the greatest importance. It should be well arranged and neatly drawn to a scale suiting the purpose for which it is to be used. It should include a *title, scale, legend, and meridian arrow.* The lettering should be neatly and carefully performed or perhaps lettering sets or press-on letters should be used.

The title box may be placed where it looks best on the sheet but is frequently placed in the lower right-hand corner of the map. It should contain the title of the map, the name of the client, location, name of the project, date, scale, and the name of the person who drew the map and the name of the surveyor.

14-18 SPECIFICATIONS FOR TOPOGRAPHIC MAPS

To describe the accuracy of the horizontal dimensions for a particular topographic map the usual procedure is to say that the average error that results in scaling a distance between two plotted points should not exceed a certain value as compared to the measured value on the ground. The accuracy of contour lines may be described by saying that any elevation taken from the map may not exceed a certain value, such as perhaps one-half of the contour interval used on the map.

The federal agencies involved in mapping have agreed upon certain minimum requirements which when met enable them to print on their maps the following statement: "This map complies with the National Map Accuracy Standards." These requirements are summarized in the paragraphs to follow.

Horizontal Accuracy.   For maps with scales larger than 1:20,000, no more than 10% of the positions of well-defined points tested may be in error by more than 1/30 of an inch. For maps with scales of 1:20,000 or smaller, the value becomes 1/50 of an inch.

Vertical Accuracy.   It is specified that no more than 10% of the points tested can be in error by more than one-half of the contour interval.

14-19 PROFILES FROM CONTOUR MAPS

One of the important advantages of contour maps is the fact that their users can quickly plot the profiles for lines running across the map in any direction. This is illustrated in Fig. 14-20, where the profile for line *AB* is plotted. Such a profile

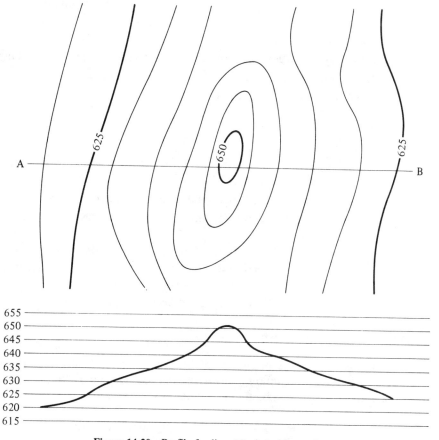

Figure 14-20   Profile for line *AB* plotted from above contour map.

presents information that can conveniently be used for establishing grades for various projects and estimating earthwork quantities.

14-20 CHECKLIST OF ITEMS TO BE INCLUDED ON A TOPOGRAPHIC MAP

A checklist of items that may be needed for a topographic survey is presented in this section. This list should be helpful for surveyors in the field as well as for persons drawing maps in the office. The most important item is for all parties to be aware of the purpose of the map. If a surveyor were to imagine that he or she were preparing a map of his or her own property for some type of development and wanted to list the items that would be of significance, the result would probably be a list similar to the following:

1. Location of property (as to state, county, town, highways, etc.).
2. Direction of north meridian.
3. Available access to property (from highways, railroads, etc.).
4. Information concerning property corners and monuments as well as the lengths and directions of property lines and land area.
5. Sufficient elevations to plot contours and show ridges and valleys.
6. Locations, sizes, and descriptions of buildings on property.
7. Locations and sizes of any roads (in use or abandoned) on or near property. Similar information concerning railroads.
8. Location of power lines, water lines, sewer lines, and other utilities on or near property.
9. Location and sizes of springs, streams, lakes, wells, and drainage ditches—also culverts, bridges, and fences.
10. Position and areas of forest land, cleared land, cultivated land, and so on.
11. Description and location of any horizontal or vertical control monuments on or near the land.
12. Other significant features that the surveyor might think are of importance to the owner. Included in this list might be information concerning the characteristics and development of neighboring property.
13. Title, scale, legend, and names of surveyor and draftsperson.

14-21 DIGITAL TERRAIN MODELS[4]

As the use of digital computers has become more widespread in recent years, a new method of assembling and processing topographic data has developed. The topographic data assembled in this manner form what is known as a digital terrain

---

[4] This section is from an unpublished paper by Donald B. Stafford of Clemson University.

model (DTM) or, more recently, a digital elevation model (DEM). A digital terrain model consists of a series of numerical values that represent the two coordinates of points in a horizontal grid or coordinate system and the corresponding elevation of each point. For instance, if an X–Y coordinate grid system is superimposed on an area and the elevation determined and recorded at each grid intersection point, the series of X and Y coordinate values and elevations for each grid point constitutes a digital terrain model. The elevation values are often considered to be the Z coordinate in a three-dimensional coordinate system. Therefore, the digital terrain model consists of the X, Y, and Z coordinates of a dense system of points within a particular project area. The horizontal grid can be developed using state plane coordinates (see Chapter 21) or any other convenient coordinate system.

The degree to which the digital terrain model accurately represents the three-dimensional shape of the terrain depends on the density of the horizontal grid points for which elevations are determined. For generalized studies of large areas, a sparse grid pattern of 500 or 1000 ft might be utilized. For more detailed studies of small areas, a grid pattern of points at intervals of 50 or 100 ft might be selected. The development of a digital terrain model using a denser grid pattern requires considerably more effort. The digital terrain model can be developed by using photogrammetric techniques or by digitizing an existing topographic map and reading the elevation at each grid point. The digital terrain model is usually stored on punch cards, magnetic tape, floppy disk, or other data storage system.

The primary advantage of the digital terrain model approach to topographic data is that the information can easily be input into a computer for processing, analysis, or manipulation. Over the past several years, a wide variety of computer software systems have been developed to analyze digital terrain models. Some examples include computer generation of contour lines, earthwork computation, route location studies, and similar applications. It is widely believed that digital

TABLE 14-1

Point	X (east)	Y (north)	Z (elevation)
1	0	0	410
2	500	0	420
3	1000	0	425
4	0	500	418
5	500	500	423
6	1000	500	430
7	0	1000	428
8	500	1000	432
9	1000	1000	435

terrain models will become a more common method in the future for assembling, storing, and processing topographic data.

Table 14-1 constitutes a digital terrain model for a small area. Because the distance increments between the points for which elevations are given is so large in this example, this is a rather coarse model.

CHAPTER FIFTEEN

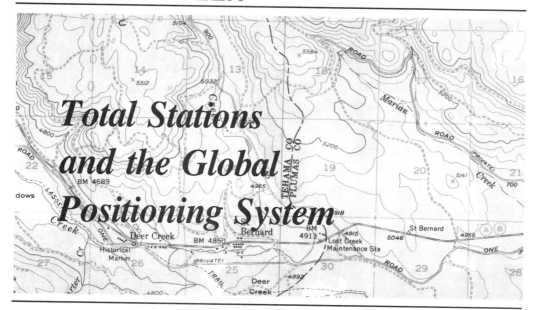

Total Stations and the Global Positioning System

15-1 INTRODUCTION

An increasingly large percentage of surveyors today use total station instruments. These are marvelous devices that can be used to measure slope distances and horizontal and vertical angles, and which may automatically apply the trigonometric values so that horizontal and vertical distances will be generated. The original name used for these instruments was *tacheometer* or *tachymeter* or *electronic tacheometer*, but Hewlett-Packard introduced the name *total station* for their Model 3810A instrument and it caught on in the profession. These devices are actually combinations of theodolites and EDM instruments together with equipment that has capabilities for determining horizontal and vertical distances.

15-2 TYPES OF TOTAL STATIONS

Porter W. McDonnell, Jr. of Metropolitan State College in Denver has placed these instruments into three different classes in his total station surveys in *P.O.B. Magazine.*[1] These are:

1. *Manual total stations.* These instruments (Fig. 15-1) make use of the same telescopes for distance and angle measurements. To obtain horizontal and vertical distances, vertical angles or zenith angles are read optically and

[1] P. W. McDonnell, Jr., "P.O.B. 1988 Total Station Survey," *P.O.B. Magazine*, Apr./May 1988, vol. 13, no. 4, pp. 36–54.

Figure 15-1 Manual-type total station. (Courtesy of Topcon Instrument Corporation of America.)

Figure 15-2 Elta 4 automatic total station. (Courtesy of Carl Zeiss, Inc.)

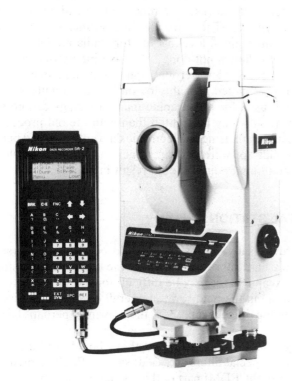

Figure 15-3 DTM-1 automatic total station with data recorder DR-2. (Courtesy of Nikon, Inc.)

calculations for horizontal and vertical distances are made with on-board calculators or with pocket calculators.

2. *Semiautomatic total stations.* These instruments contain a vertical angle sensor for the automatic calculations of vertical and horizontal distances. Horizontal angles are read optically.

3. *Automatic total stations.* With these instruments (Figs. 15-2 and 15-3), horizontal angles and vertical angles are both read electronically for use with slope distances in internal computers or data collectors.

With some total stations it is possible to have considerable data reduction in the field, whereas some instruments are constructed so that the data reduction and the plotting will be with office computers. Some instruments provide for the computation of coordinates, whereas others do not.

15-3 USES OF TOTAL STATIONS

There is today in the surveying profession a very strong movement away from theodolites and EDMIs toward the use of total station instruments. With total stations the surveyor can measure slope distances as well as horizontal and vertical angles. Furthermore, these devices may or may not have the capacity to compute and display horizontal distances, differences in elevations as well as directions, and perhaps coordinates. These wonderful instruments may also be used with superb results for differential leveling and for stadia readings for topographic work. Usually the leveling is not done by measuring vertical distances with respect to a horizontal line. Rather, vertical angles and slope distances are measured and elevations are computed probably by on-board computers.

When total stations are used for topographic maps, readings can be made so quickly that many more points may be sighted than with the old procedures. This is particularly true when we are using an electronic data collector and perhaps autoCAD (a computer graphics software package) back in the office to plot the needed maps. The use of additional points should result in better maps.

15-4 DISADVANTAGES OF TOTAL STATIONS

Total stations do have a few disadvantages that should be clearly understood. Their use does not produce a hard set of field notes of the types we have studied in previous chapters. This means that it may be difficult for the surveyor to look over and check the work while in the field. For an overall check of the survey it will be necessary to return to the office and prepare the drawings (perhaps handled by computers).

Another disadvantage of total stations is that they should not be used for observations of the sun unless special filters are used, such as Roelof's prism (see Chapter 16). If this is not done, the EDMI part of the instrument may be damaged. We will see another disadvantage of total stations described in the next section of this chapter, relating to mistakes when electronic data collectors and office computers are used.

15-5 TRAVERSING WITH A TOTAL STATION

It may have seemed in earlier chapters that for traversing the instrument was set up at every one of the traverse corners and distances and angles measured between them. This procedure was certainly followed for centuries. However, with modern equipment the traverse corners are rarely occupied except on rather large traverses. Rather, the instrument is set up at convenient locations (can be one corner of the transverse or a point inside or outside of the tranverse) and sights made on the corners as described in Section 11-6. This process is called radiation. It is true, however, that the most accurate results will be obtained by working around the traverse and occupying each corner.

Radiation does have one major weakness, in that a mistake may be made in one of the values. You may say "but we are using a total station instrument and the readings are recorded in electronic data collectors and thus there are no

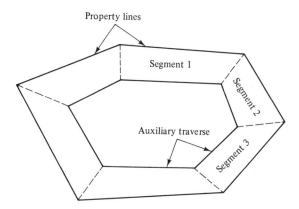

Figure 15-4 Auxiliary traverse and segmental survey.

mistakes." Unfortunately, mistakes can be made with this modern equipment as with older equipment. A sight may be taken on the wrong point for the corner, the electronic signal may be reflected by something other than our reflectors (perhaps a nearby road sign or a piece of reflector tape on a parked vehicle), and so on.

Our radiation measurements for traversing need to be checked for the reasons just described. Many surveyors will very carefully measure one or two sides of the traverse (by EDMI or tape) and compare the results with the computed values. Another way to check the work is to set up the total station at a different point and repeat the measurements to the corners. If the results are close we can average them. If not we will have to repeat the work. As we get into larger traverses, we may run an auxiliary traverse inside the other one as discussed in Section 11-6 and as shown in Fig. 15-4.

An auxiliary traverse may be partitioned into segments as shown in Fig. 15-4. Closure is computed on each segment to see if the segments match up for the entire figure.

15-6 RESECTION

A very useful procedure with total station instruments for many situations is *resection*. This is a method sometimes called *free stationing* by which the location of an unknown point can be determined by setting up the instrument at the point in question and taking sights on at least three points of known location. The position of the desired point is determined by measuring the two angles between the points as shown in Fig. 15-5 and by solving the triangles involved. The detailed trigonometric equations needed are given in the book footnoted here.[2]

---

[2] R. E. Davis, F. S. Foote and J. W. Kelley, *Surveying Theory and Practice*, 5th ed. (New York: McGraw-Hill Book Co., 1966) pp. 413–415.

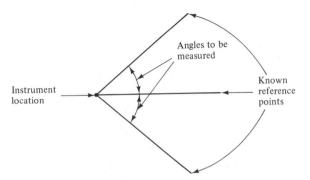

Figure 15-5 Resection.

We can set up the instrument at some convenient point where there is excellent visibility for other work and then sight on at least three points whose positions have previously been determined. Then with our total station which is programmed for this situation we can determine the coordinates of the instrument position. An accuracy check can be made by making a sight to a fourth point whose position is known. If the surveyor sights on just three known points and is careless with his work a bad solution will be the result. Obviously it is desirable to use another point when feasible.

Once the unknown position is determined sights can be made to other points and their positions determined by measuring directions and distances. Resection is a particularly useful procedure for building layout work where the instrument can be set up at any convenient position perhaps out on the building itself. After the coordinates of the instrument are determined the coordinates of desired layout points can be input into the total station and its programs will compute the lengths and directions of ties to those points.

A very simple and quick graphical method of resectioning which is sometimes useful in topographic mapping was previously presented in Section 14-15.

15-7 GLOBAL POSITIONING SYSTEM SATELLITE NETWORK

The surveying profession is on the threshold of a tremendous change in the way in which control points are located on the earth's surface—that is, the use of a global positioning system (GPS) of satellites called NAVSTAR. By the early 1990s the U.S. Department of Defense plans to have a network of 24 orbiting satellites (21 active, 3 spares) 20,000 km above the earth sending out radio signals from which precise horizontal and vertical positions on the earth's surface can be obtained.

The essential purpose of the system, which is today partly operational, is to enable planes, ships, and other military groups to determine their geodetic

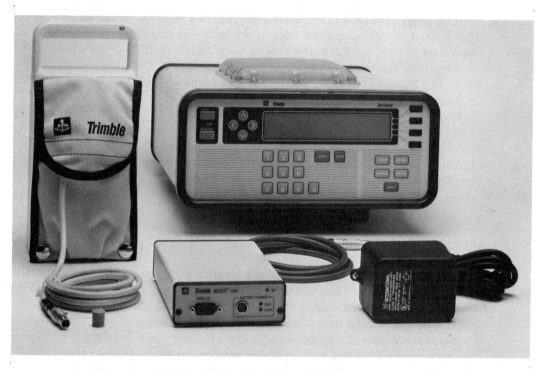

Figure 15-6   4000 ST GPS field surveyor consisting of internal antenna and receiver/datalogger system. (Courtesy of Trimble Navigation, Ltd.)

positions rapidly. Although the system is being developed for military purposes, it will be of great benefit to other governmental organizations (such as the National Geodetic Survey and the U.S. Geological Survey) and to private surveyors as well.

The satellites are being placed on six different orbital planes, three in each orbit. The nearly circular orbital planes are inclined 55° to the equator and are spaced every 60° in longitude. The orbital periods are about 12 hours. An observer at any point on the earth's surface will be able to receive radio signals from at least four of the satellites. This coverage will enable the surveyor to locate his or her position in the $x, y,$ and z directions.

This brings up something of a problem concerning a control system. With GPS receivers work may often be performed at accuracies far exceeding the accuracies of the old control monuments to which the new survey work is tied. As a result, it appears that some states are going to establish control networks of their own, which will have greater accuracies than the ones of the NGS.[3]

Each satellite contains an atomic clock and the radio signals are sent out at

[3] J. Collins, "Shooting for a Full GPS Constellation," *Professional Surveyor*, July/Aug. 1989, vol. 9, no. 4, pp. 40–41.

certain times. Thus the range from a satellite to a receiver on the ground can be determined by measuring the time at which a signal is received. If ranges are observed to four satellites, the precise horizontal and vertical coordinates of a point can be determined from the measured ranges.

The National Geodetic Survey has approximately 240,000 horizontal control monuments located at intervals of 6 to 8 miles in most localities. To establish first-order triangulation points they build towers, run theodolite lines at night, and so on, and spend $6000 to $8000 per point. If the satellite system is used, the establishment of such monuments may eventually be as low as a few hundred dollars apiece and it may be cheaper not to establish the permanent and expensive monuments now used. With satellites it will be possible, quickly and cheaply, to determine the positions of any points desired without the use of fixed monuments.[4]

15-8 APPLICATIONS OF GPS

Satellites are being used today for the location of control points for important government surveys. They will undoubtedly be used much more frequently in the future to provide the location of natural and manmade objects. The information obtained will be used by engineers, tax offices, fire departments, and so on, to better manage their resources. Today a few surveyors are using them to determine coordinates for control for various types of construction and route projects. GPS is extremely well suited for aerial survey control, route surveys, and other projects that have rather isolated points. The method works exceptionally well for surveys that are to be referenced to a state plane coordinate system (see Chapter 21). Software is readily available to convert GPS data to state plane coordinates. Perhaps someday GPS will be used for establishing accurate coordinates for every piece of land in the United States. Surveyors will be able to establish coordinates for every corner of a piece of property, regardless of how far it may be to any reference monuments. A GPS receiver may be set up at one point in a project and then while doing usual distance and direction traversing and stakeout, the GPS receiver will be measuring the precise coordinates of the point.[5]

In this regard of "leaving the GPS receivers working" it should be noted that precision can be increased by having a longer observation period. For instance, if a 1-hour observation results in a precision of 1/100,000, it could probably be increased to 1/250,000 by lengthening the observation two- or threefold.

The same amount of station time is required at points located 30 miles apart as for those located 50 ft apart. As a result, when establishing control points, the farther apart they are, the more economical GPS becomes. Levels do not have

[4] J. Collins, "A Satellite Solution to Surveying," *Professional Surveyor*, Nov./Dec. 1982, vol. 2, no. 6, pp. 13–17.

[5] J. Collins, "Shooting for a Full GPS Constellation," *Professional Surveyor*, July/Aug. 1989, vol. 9, no. 4, pp. 22–23, 40–41.

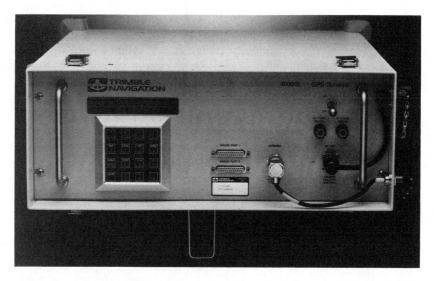

Figure 15-7   4000 SL GPS surveyor consisting of a receiver/processor/datalogger system. (Courtesy of Trimble Navigation, Ltd.)

to be run in from the distant points or the usual traversing done for lengths and directions.

It usually takes about a minimum of an hour or a little more to set up the equipment, collect the data, and take up the equipment. The equipment must not be disturbed during the data collection period. The rather definite time period required for an observation means that it is easier to make cost estimates for a GPS job than for other types of surveys.

There are today a few dozen private firms who make use of GPS technology; there are, however, many dozens more who are making plans to move in this direction if equipment costs come down appreciably. Such a reduction in costs does seem likely.

Not all points are going to be located by the GPS, but rather, certain control points will be located and traditional surveying measurements will be used to measure the location of other points with respect to these control points. One possibility may be to bring in a firm that specializes in GPS work to establish certain control points and then use conventional traversing methods for the remainder of the survey. GPS may be particularly useful where conventional methods cannot be used to cross someone else's land because of their objection.

GPS surveying is far less affected by weather than are most other types of surveying. Since it makes use of microwaves, the work can be done in snow, rain or fog. Even snow building up on the antenna does not affect the accuracy.

To use GPS to locate points it is not necessary to be able to sight between adjacent points, but the system does not work if there are obstructions blocking the satellite signals from the antenna of the equipment. For instance, the method will probably not work very well deep in a forest without quite a bit of wood cutting. It is probably going to be necessary to have unobstructed views of the satellites for vertical angles of greater than about 20° above the horizon. GPS may not be a very economical method when high accuracies are not required or where points are relatively close together and where it is not difficult to traverse between them.

The satellite signals cannot penetrate water, soils, walls, or other obstacles. Thus, they cannot be used for underground positioning or for underwater navigation. Furthermore there can be problems in large cities with many tall buildings.

15-9 GPS PROCEDURES

GPS is used in several ways. Some of these are discussed in the paragraphs to follow.

Relative Positioning

With relative positioning one GPS receiver is placed at an NGS control point and one is placed at a point whose position is desired. (Several other receivers may be used to determine the positions of more than one other point at a time.) The receivers are used to observe the same satellites for minimum periods of 45 minutes. As mentioned previously, precision can be improved by lengthening the observation periods. There must be at least four satellites visible. (If only three are visible, then only the horizontal coordinates of a point are obtainable.) With four satellites the latitude and longitude of each point, the azimuth between them, and their elevations are obtained.

At the present time there are only certain hours of the day or night when at least four satellites are available. Furthermore, the periods of visibility are constantly changing (about 4 minutes per day). When all planned satellites are in orbit in the early 1990s, surveyors will be able to determine position by GPS at any time of day or night at any point on the earth's surface.

After the observations are made, the user can take the receiver from the original known station and leapfrog it over those that have been used to determine positions of new points. Depending on the visibility of the satellites and upon the accuracy required, the surveyor can make one to four observations per day.[6]

Distance Measuring

If it is desired just to measure the distance between two points it isn't necessary to start at a known position. A GPS receiver is placed on each point and the

[6] J. P. Reilly, "Practical Surveying with GPS," *P.O.B. Magazine*, Apr./May 1988, vol. 13, no. 4, pp. 12–22.

satellites are observed for the same time periods. The position obtained will not be accurate, but the distances between the points will be. It is not even necessary to be able to see between the points and they can be any distances apart.

Point Positioning

The surveyor can roughly determine his or her position within about 100 ft with only one GPS receiver. Even though this accuracy is not acceptable for surveying work, the method is very useful. The receiver can be placed in a vehicle and used to quickly find monuments or other points of certain coordinates.

A very important use of GPS observations is with open traverses. If a long open traverse is run along a highway, power line, or railroad, by the usual methods a check is not available upon reaching the other end. Thus it is usually necessary to tie into some NGS monuments (if they happen to be reasonably close) or more likely to run a new set of lines back to the starting point and check the closure and precision of the resulting closed figure. Such a procedure is very time consuming and there are often precision problems if the lines are run back along approximately the same route as the initial ones. That is because the end angles can be very small, and slight errors in their values can greatly distort the traverse. In Section 20-6 this is referred to as a very weak figure.

With GPS the end points can be located and perhaps a few intermediate ones along the route and the traverse can be tied into them. The results will be considerable saving in time and probably appreciable increases in accuracy.

15-10 ACCESS TO THE SATELLITE SYSTEM

Although the Department of Defense has control of the GPS system, they have stated that they have no plans to deny users access to the system except in times of emergency. They do, however, fear that an extremely accurate system available to every one might be used by enemies of the United States for harmful purposes. As a result, they may decide someday to degrade the broadcast so that users cannot locate a single position to better than ±100 m.

If the Department of Defense does establish this "denial of accuracy," it would not really affect the work of surveyors with two GPS receivers who are working in a differential survey mode. In a differential survey mode, one receiver is placed at a point whose coordinates have previously been located, and the other receiver is placed at a point whose coordinates are desired. It is possible in this way to determine the differences in coordinates between the two points to great accuracy (about ± 2 ppm).[7]

---

[7] J. Collins, "GPS Equipment Survey: What Does It All Mean?" *P.O.B. Magazine*, June/July 1987, vol. 12, no. 5, pp. 12–22.

CHAPTER SIXTEEN

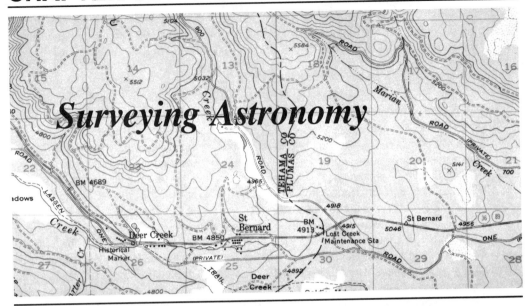

Surveying Astronomy

16-1 INTRODUCTION

Up to this point in the book an appreciable amount of time has been spent describing methods for accurately measuring distances, elevations, and angles. Almost nothing has been said, however, about accurate directions. This chapter was written with the objective of presenting accurate methods for determining directions by making observations of the sun and other stars.

To present a complete theoretical background and explanation of astronomical observations would require an entire book in itself. Fortunately, however, it is possible to present in a single chapter sufficient information to enable the surveyor to work effectively in this area.

16-2 ASTRONOMIC AND GEODETIC NORTH

Sometimes surveyors use the term *true north*, yet that term could possibly mean astronomic north or it could mean geodetic north. Unfortunately, the two terms do not mean exactly the same things. *Astronomic north* is based on the direction of gravity and the axis of rotation of the earth. It is determined from observations of the sun or other stars.

Geodetic north is determined from a mathematical approximation of the earth's shape. The difference between astronomic north and geodetic north, which is referred to as the Laplace correction, is rather small in the eastern and central parts of the United States but can be as large as 20 arc-seconds in the western

322

parts of the country.[1] (For comparison, the width of the cross hair of a theodolite is about 2 or 3 arc-seconds.) Because this difference may be significant, the term *true north* should probably not be used. To be specific the term astronomic north or geodetic north should be used, whichever is applicable. Astronomic directions will be used in this book and explored in more detail in this chapter.

16-3 ADVANTAGES OF ASTRONOMICAL DIRECTIONS

The use of astronomical directions has several advantages over the use of assumed or magnetic directions. These include the following:

1. Permanence is given to the direction of boundaries in land surveying compared to magnetic directions, where the magnetic directions are forever changing. If only one corner of a piece of property surveyed on the basis of astronomical directions can be located, all the corners can be reestablished assuming that the distances were measured accurately.
2. As described previously, astronomical directions are quite useful for correlating surveys and for checking the angles for long open or closed traverses.
3. Astronomical directions are useful for orienting important maps and charts as well as for orienting radio and radar antennas.

16-4 CELESTIAL BODIES TO BE CONSIDERED

Astronomical directions may be obtained by sighting on the sun or on one of several thousand stars if their positions are known. (Actually, the sun is also a star, but that name is not frequently applied to it.) In this book only observations on Polaris (the north star) and the sun are presented because it is felt that they are the most practical for the surveyor located in the northern hemisphere. In the southern hemisphere there is not a bright star located near the south pole, and if observations are not made on the sun, they are probably made on the Southern Cross. The hour-angle method to which a large part of this chapter is devoted is equally applicable to the northern and southern hemispheres.

16-5 METHODS OF OBTAINING ASTRONOMIC DIRECTIONS CONSIDERED HEREIN

Three methods are presented in this chapter for obtaining astronomic directions:

1. Hour-angle method of observing the sun

[1] R. C. Brinker and R. Minnick, editors, *The Surveying Handbook* (New York: Van Nostrand Reinhold Company, Inc., 1987), pp. 541–542.

2. Altitude method of observing the sun

3. Hour-angle method of observing Polaris

It now appears that the hour-angle method of observing the sun is going to be the common procedure used for determining astronomic directions in the immediate future. As a result, it is the method emphasized in this chapter. The azimuths to any celestial bodies, including the sun, Polaris, other stars, and even artificial satellites, can be determined using the hour-angle method. To determine an azimuth it is necessary to have the declination of the body (which is its angle above or below the equator), the latitude and longitude of the observer's position, and the time of the observation.

16-6 ASTRONOMICAL TABLES USED BY THE SURVEYOR

An *ephemeris* (plural *ephemerides*) is an astronomical almanac that shows the position of the sun, the planets, the moon, and various stars at certain time intervals. The position of any one of these celestial bodies at times other than those listed in the tables can be determined by interpolation. In the following discussion, reference is made to the tables of the *Lietz 1990 Celestial Observation Handbook and Ephemeris*. Some of its tables are reproduced in the Appendix of this book with the permission of Elgin, Knowles & Senne, Inc. Included is the information necessary to solve the example and exercise problems contained in this chapter. For practical work it is necessary to use the ephemeris of the year in question, although the tabular values for a particular date of the year change rather slowly year by year.

Other ephemerides readily available include *The Nautical Almanac* published by the U.S. Naval Observatory and available from the U.S. Government Printing Office in Washington D.C. 20402 and *The Apparent Place of Polaris and Apparent Sidereal Time* which can be obtained from the U.S. Department of Commerce, National Geodetic Information Center, Rockville, Maryland, 20852.

Although only a discussion of observations of Polaris and the sun is presented in this chapter, a little additional study will enable the reader to determine astronomical directions by making observations of other stars for which information is given in the ephemerides.

16-7 INTRODUCTORY DEFINITIONS

For the purposes of surveying astronomy, it is assumed that the stars are fixed on the inside of a gigantic sphere with an infinite radius. This tremendous sphere, called the *celestial sphere*, includes all of the heavenly bodies and is assumed to rotate around the earth (its speed of rotation being 360°59.14' per 24 hours, as described in Section 16-9). Since it is of infinite size, in comparison the earth seems a mere dot. The observer is assumed to be at the center of this sphere, the radius of the earth being negligible compared to the distance to the stars. When this assumption causes an appreciable error (as it may for observations of the sun, the moon, or the planets), a correction can be made as described in Section 16-19.

The assumption of the existence of this tremendous sphere is perfectly consistent with the view seen by an observer standing on the earth's surface. As he or she looks toward the heavens, the stars are so far away that they all appear to be equidistant and seem to lie on the inside of a gigantic sphere.

Several definitions relating to the celestial sphere follow.

Celestial Poles.   If the axis of rotation through the north and south geographic poles of the earth is extended in both directions until it intersects the celestial sphere, the intersection points are referred to as the celestial poles.

Great Circle.   If a plane were passed through the center of the celestial sphere (i.e., the center of the earth) and extended outward in all directions until it intersected the celestial sphere, its line of intersection would be a great circle.

Celestial Equator.   A great circle perpendicular to the polar axis of the celestial sphere is called the celestial equator. It is the same as a plane passing through the earth's equator extending outward until it intersects the celestial sphere.

Zenith and Nadir.   If a plumb line at the observer's position on earth were extended upward until it intersected the celestial sphere, that point would be the zenith. If the plumb line were extended in the other direction until it intersected the celestial sphere, that point would be the nadir.

Ecliptic.   The sun goes around the celestial sphere once each year and its path across that sphere is a continuous curved line called the ecliptic. The sun moving northward on this line crosses the equator on March 21 and reaches a maximum northerly point $23\frac{1}{2}°$ above the equator on June 21. Similarly, moving southward it crosses the equator on September 22 and reaches a maximum southerly point $23\frac{1}{2}°$ below the equator on December 21.

Vernal Equinox.   On its northward journey each spring the sun crosses the celestial equator on March 21, and that imaginary point on the celestial sphere is called the vernal equinox. It is referred to in ephemerides as "the first point of Aries" or just Aries and is often represented by the zodiacal symbol ♈.

Horizon.   A great circle passing through the observer's position and perpendicular to a plumb line at the observer's position is called the horizon. This circle is halfway between the observer's zenith and nadir and is the plane in which azimuth is measured.

Vertical Circle.   A great circle passing through the observer's zenith and any celestial body is referred to as a vertical circle. Such a circle is perpendicular to the horizon.

Reference Meridian. The meridian passing through the Royal Observatory at Greenwich, England, is generally taken as the reference meridian. Longitude is measured east or west from the reference meridian from 0 to 180°.

Celestial Meridian. The celestial meridian at any point on the earth's surface is the great circle that passes through the point in question, the celestial poles, and the zenith and nadir for that point.

16-8 LATITUDE AND LONGITUDE

Positions on the earth's surface are described in terms of latitude and longitude. Brief definitions of these terms follow.

Latitude (which is represented by ϕ is normally defined as the angular distance (0 to 90°) that a point is above or below the equator. It is also equal to the angle between the plane of the earth's equator and a plumb line held at the point in question.

As defined previously, *longitude* (represented by λ herein) is the east or west angular distance (0 to 180°) measured from the Greenwich meridian. It can be expressed either as an angle or in terms of time. If the sun is considered to travel around the earth in 24 hours, that would be equal to 15° per hour or 1° in 4 minutes. A longitude of 78°20' west could be expressed in time as $5^h13^m20^s$ west (where the abbreviations h, m, and s represent hours, minutes, and seconds, respectively).

For astronomical observations it is possible to scale latitudes and longitudes from a good map. The topographical maps published by the U.S. Geological Survey are usually handy. If these or other good maps are not available, the Director of the National Geodetic Survey, National Oceanic and Atmospheric Administration (NOAA), Rockville, MD 20882, can provide values. The person making the request will need to describe the location as carefully as possible, such as giving the name of the nearest post office and in Public Land Survey areas giving the section number and perhaps the name of the township. (See Chapter 19.)

It will be noted that 1 statute mile on the ground is equal to a little less than 1' of latitude. If we are to obtain equivalent azimuth accuracies for our various observations, we need to determine latitudes and longitudes more accurately for celestial bodies located near the equator (e.g., the sun) than we do for celestial bodies located near the poles (e.g., Polaris).

16-9 APPARENT MOTION OF THE CELESTIAL SPHERE

Here on earth it appears that the sun and all the other stars, supposedly plastered on the inside of the celestial sphere, rotate around us once each day. However, these other stars rotate around us one more time in 365 days than does the sun.

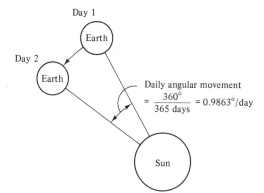

Figure 16-1 Movement of earth around sun in one day.

This is due to a combination of the earth's rotation about its own axis and the apparent angular movement of the sun along its path or ecliptic. The latter movement is shown in Fig. 16-1 and equals 0.9863° per day.

For the stars other than the sun to rotate around us apparently in a complete circle takes 24 hours—the time required for a 0.9863° movement: that is, 24 hours $- 0^h3^m56^s = 23^h56^m04^s$. Thus a star day is 3^m56^s shorter than a mean solar day. You can see that when we work with Polaris or one of the other stars, we have to account for the shorter star days.

16-10 TIME

There are several systems for reckoning time with which the reader needs to be familiar in order to understand astronomical observations. These include (1) apparent time or true solar time, (2) mean solar time, and (3) sidereal time. The conversion of one kind of time to another is of the greatest significance in all phases of astronomical observations. The following paragraphs present a brief description of these various times.

Apparent Time or True Solar Time

An apparent solar day is the time required for one apparent revolution of the true sun about the earth. The earth moves along an elliptical orbit about the sun and thus the apparent movement of the sun around the earth is not constant. The sun that we see is called the *apparent sun* or the *true sun*. During four periods of the year its velocity across the sky is greater than its average velocity, and during four periods its velocity is less than its average velocity. Thus the days given by the apparent sun are not of uniform length.

Mean Solar Time or Civil Time

So that our solar days will be of equal length, astronomers created the *mean sun*, which is a fictitious sun that moves at a constant rate around the earth. This sun apparently makes one complete circuit from west to east among the stars and around the earth in one year. A *mean solar day* is the time required for one revolution of this fictitious or mean sun. This is the same as a *civil day*.

Civil time is the same as *mean solar time* and is the *standard time* generally used. *Local civil time* is the time based on the central meridian of an observer. Civil time based on any other meridian is designated by name: for instance, *Greenwich Civil Time*.

The various ephemerides provide the time intervals between times as determined from the apparent sun and the mean sun. This interval is referred to as the *equation of time* and can be written as follows:

$$\text{equation of time} = \text{apparent solar time} - \text{mean solar time}$$

Sidereal Time

As we have seen, an apparent solar day is the amount of time required for the sun to make one complete revolution of the earth. Thus apparent solar time may simply be called sun time. Sidereal time, on the other hand, is star time. A sidereal day is the time required for the vernal equinox to make one complete revolution of the earth. A mean solar day is 3 minutes 56 seconds longer than a sidereal day, which is 23 hours 56.1 minutes. Sidereal time at any point is equal to the hour angle of the vernal equinox at that point. As described previously, the equation of time can give a plus answer (if the real sun is ahead of the mean sun) or a negative answer (if the real sun is behind the mean sun), and the interval can be as large as 16 minutes.

The prime meridian for figuring longitudes anywhere in the world passes through the observatory at Greenwich, England. The standard time for Greenwich, which is commonly used for astronomical observations, is called *Greenwich Civil Time* (GCT) or *Universal Time* (UT) or *Greenwich Mean Time*. As we have noted, the sun makes one apparent revolution of the earth in 24 hours or that is it apparently moves 15° per hour. The longitude of the Greenwich time zone is 0° or 0^h.

The United States is divided into four time zones: Eastern, Central, Mountain, and Pacific times. Through each of these zones there is a central meridian at 75°, 90°, 105°, and 120° west of the Greenwich meridian, respectively. These central meridians differ from each other by 1 hour of time.

A person might think that noon should coincide with the appearance of the sun at its highest point over his or her position. If such a practice were followed in every town, there would be an endless number of time zones throughout the country. Therefore, within a given time zone all the watches are set to the time which applies to the central meridian in that time zone.

Since the sun's apparent daily motion is from east to west, it reaches its highest point at locations to the east of the central meridian in a particular time zone before it does at the central meridian. In a like manner it reaches noon later for the western part of the time zone.

To determine GCT at any point in the United States, the observer must add to watch time the number of hours of longitude from Greenwich to the central meridian of the time zone. If daylight saving time is being used in a particular zone, standard time will be the same as daylight time for the time zone just to the east. For instance, 8 A.M. Pacific Daylight Time is the same as 8 A.M. Mountain Standard Time.

16-11 MEASUREMENT OF TIME

When the altitude method of observing the sun is used, a very accurate time measurement is not required. Watch time is generally satisfactory but the observer must compare his or her watch against a standard clock or time signal before the observation is made.

For the hour-angle type of observation, an extremely accurate time is absolutely essential. It is, in fact, desirable to have time to within about $\pm\frac{1}{2}$ second. One source of such accurate time measurement is the National Bureau of Standards. They broadcast Coordinated Universal Time (UTC) on radio station WWV (WWVH in Hawaii) on 2.5, 5, 10, 15, and 20 MHz to the nearest 0.01 second. This time can also be obtained with a shortwave radio or by calling (303) 499-7111.

UTC is the uniform time for general use while the time needed by surveyors for hour-angle observations is called mean universal time or UTl. It is based on the actual rotation of the earth. The very small difference between UTC and UTl is called the DUT correction.

In addition to its time signals, station WWV provides the DUT correction. This correction, which changes about 0.1 second every two months, cannot be greater than ± 0.7 second. Immediately after WWV gives the minute tone they give the DUT correction by broadcasting a series of audible double ticks with a computerized voice. This correction in leap seconds accounts for the gradual change of the earth's rotation rate.

If we are tuned to WWV, we can count the number of double ticks which they broadcast during the first 15 seconds of each minute. Each double tick represents a 0.1-second correction of time. The first seven double ticks are positive; beginning with the ninth second, each double tick is negative. In other words if we hear double ticks in the first 5 seconds after the minute tone we add them to UTC. If double ticks are given only between the 9th and 15th seconds they are subtracted.

We will make this correction and use UT1 rather than UTC because its use can on occasion provide appreciably better astronomical directions. We can actually neglect the DUT correction for observations of Polaris or other high declination stars because its effect on azimuths is extremely small.

Some calculators (e.g., HP-41CX or the HP-41C/CV) have a time module that can be set to UTC time corrected by DUT to UTI time. Another excellent way to obtain UTI time is to make use of the WWV broadcast times and a stopwatch.[2] If we are unable to obtain a very accurate time measurement as described herein, we probably should consider a night observation of Polaris or perhaps the altitude method for a sun observation.

16-12 OBSERVATIONS OF THE SUN

The sun presents a large target that can be observed during regular working hours. It can be seen even if there is considerable haze and/or if partly cloudy conditions exist. **The sun should not be sighted directly through a normal telescope because injury to the eyes may result.** It is necessary to use a solar attachment such as a dark filter. Alternatively, the telescope can be pointed toward the sun and a field book page or white card held several inches away from the eyepiece and the telescope moved until the sun is seen on the page or card.

If we make an observation with a total station instrument we must use an objective lens filter because some of the components of the electronic distance-measuring equipment may be injured. If we have a separate electronic distance-measuring instrument mounted on the equipment, it should be removed or a lens cap should be used to protect its parts when the sun is being observed.

There are two basic methods commonly used for making sun observations: the hour-angle method and the altitude method. With each of these methods a horizontal angle is measured from a line on the ground to the sun. With the altitude method it is also necessary to measure the vertical angle (or altitude) of the sun. Then the azimuth to the sun is determined (as described in the sections to follow) and the azimuth of the line on the ground is computed using the horizontal angle.

With the hour-angle method we make a very accurate time measurement for each observation of the sun. In addition, we need the latitude and longitude of the point. With this method we can probably obtain an azimuth that is accurate within about 10″.

With the altitude method we accurately measure the vertical angle to the sun (the altitude) and obtain an approximate time for the observation. In addition, our latitude is needed. When this method is used carefully, we can probably obtain an azimuth accurate to 1 or 2′.

The hour-angle method is definitely the better procedure to use, but the altitude method has been commonly used in the past. Until recently the measurement of accurate time in the field (an absolute necessity for the hour-angle

---

[2] D. R. Knowles, "Accurate Time for Celestial Observations Using an HP-41," *P.O.B. Magazine*, Dec. 1985/Jan. 1986, vol. 11, no. 2, pp. 26–30.

procedure) was impractical. In the past few years, however, tremendous progress has been made in the development of accurate timepieces, such as digital watches, time modules for calculators, and so on. As a result, the hour-angle method, which is much more accurate, is quicker and easier to learn, and has fewer limitations as to time of day when the observations can be made and as to geographic locations, is definitely the preferred method. Thus it is strongly recommended that the surveyor learn and use this method.

16-13 HOUR-ANGLE OBSERVATIONS OF THE SUN

Hour-angle observations of the sun can be made at any time of the day from sunrise to sunset. We will find in Section 16-18, however, that when the altitude method is used for sun observations, we are practically limited to periods of about 2 hours in the middle of the morning and about 2 hours in the middle of the afternoon.

Observations of the sun, frequently called sun shots, are not as easily made as are those on Polaris or other stars. The stars make small targets and move slowly, enabling the surveyor to sight directly on them with little difficulty. The sun's image, on the other hand, is large (about 32′ diameter) and moves swiftly.

It is to be remembered that if a person sights the sun directly through a telescope, permanent injury to the eyes may result. Observations must be made using some type of dark filter attached to the eyepiece or by observing the sun's image on a light surface (such as a page in the field book or a white card) held several inches behind the eyepiece. Initially, it is assumed that we will make our observations directly on the sun using a filter. Observations made indirectly using a card or paper are discussed in Section 16-17.

If a dark glass is placed in front of the objective lens, it must be made plane and perfectly parallel to the objective lens or it will introduce an error in the sighting. There are available on the market several special devices or reticules which are specifically made for sighting directly on the sun. If the surveyor is going to make a large number of sun shots, he or she should seriously consider purchasing one of these devices. Use of these devices improves the accuracy of telescope pointings.

One type of glass reticule is shown in Fig. 16-2. As can be seen in the figure, the regular vertical and horizontal cross hairs are each augmented with a pair of tick marks. These marks are spaced so that a circle of radius 15′45″ can be inscribed within them. The diameter of such a circle is slightly smaller than that of the sun, and as a result we can center the sun's image very accurately, although it will overhang the tick marks just a little bit.

Another device for making observations of the sun is Roelof's solar prism. This device is fitted to the objective lens of the telescope. The viewer looking through the telescope sees four overlapping images of the sun, as shown in Fig. 16-3. If the cross hairs are centered on the little diamond-shaped area between

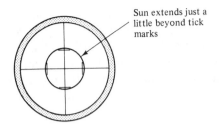

Figure 16-2   Reticule for directly viewing the sun.

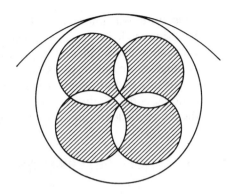

Figure 16-3   View of sun's image with Roelof's solar prism.

these images, the telescope will be pointed at the sun's center with a high degree of precision. In addition to these attachments, several other useful devices are on the market for observing the sun.

16-14 HOUR ANGLES

An hour angle is the angular distance along the equator from a particular meridian to the great circle passing through the heavenly body in question. For surveying astronomy, three types of hour angles are particularly important: the Greenwich hour angle (GHA), the hour angle of a heavenly body with respect to the Greenwich meridian; the local hour angle (LHA), the hour angle of a heavenly body with respect to our local meridian; and the sidereal hour angle (SHA), the hour angle of a heavenly body with respect to the vernal equinox.

Greenwich Hour Angle

The ephemerides provide the Greenwich Hour Angle (GHA) of the sun for 0 hours UTl each day. In the Appendix of this book these values are provided for a few months in 1990. The tables are copied with the permission of Elgin, Knowles & Senne, Inc. from their ephemeris which is published by the Leitz Company.

The GHA for a particular UTl time can be determined from the tables by interpolation using the following equation:

$$\text{GHA} = \text{GHA } 0^h + (\text{GHA } 24^h - \text{GHA } 0^h + 360)\frac{\text{UTl}}{24}$$

Local Hour Angle

The local hour angle (LHA) at the time of the observation (UTl) of the sun is needed for use in the azimuth formula given in Section 16-16. It is equal to the Greenwich hour angle minus the west longitude of the observer's position.

$$LHA = GHA - \text{west longitude} = GHA - W\lambda$$

or for east longitude,

$$LHA = GHA + \text{east longitude} = GHA + E\lambda$$

Once this value is obtained, it is adjusted if necessary to a value between 0 and 360° by adding or subtracting 360°.

16-15 DECLINATIONS

The declination (δ) of the sun, Polaris, or some other star is the angular distance measured north or south of the equator to the celestial body in question. Declinations are given at 0^h UTl for each date in the ephemerides, and we can determine its value at other times by linear interpolation:

$$\text{declination at some UTl} = \text{declination } 0^h + (\text{decl. } 24^h - \text{decl. } 0^h)\frac{UTl}{24}$$

It is true, however, that linear interpolation of a function that is changing non-linearly results in some error. This error is usually negligible in its effect on the final azimuth, but a little better value can be obtained by using the following equation, which amounts to three-point interpolation. In this expression declination is substituted in decimal degrees:[3]

$$\text{declination} = \text{decl. } 0^h + (\text{decl. } 24^h - \text{decl. } 0^h)\frac{UTl}{24}$$
$$+ (0.0000395)(\text{decl. } 0^h) \sin (7.5UTl)$$

The preceding equation is to be used only for determining the declination of the sun. It must not be used for observations of Polaris or any other stars.

16-16 ASTRONOMIC NORTH FROM THE SUN USING THE HOUR-ANGLE METHOD

If we have accurately timed the instant at which the sun's center was sighted and if we have obtained our latitude (probably from a USGS map), we can compute the north azimuth of a line from our instrument to the sun at the time of mea-

---

[3] R. C. Brinker and R. Minnick, editors, *The Surveying Handbook* (New York: Van Nostrand Reinhold Company, Inc., 1987), pp. 538–571.

surement using the following equation. The application of this equation using a pocket calculator is very tedious; however with computers the work is quite simple. The enclosed computer diskette contains such a program and its use is illustrated with Example 16-2 later in this section.

$$\text{azimuth} = \tan^{-1} \frac{-\sin \text{ local hour angle of sun}}{\cos \text{ latitude } \tan \text{ declination} - \sin \text{ latitude } \cos \text{ LHA}}$$

$$= \tan^{-1} \frac{-\sin \text{ LHA}}{\cos \phi \tan \delta - \sin \phi \cos \text{ LHA}}$$

Using our pocket calculator to solve this equation, we will get a value between $-90°$ and $+90°$. We then make a correction to this value to get it between 0 and 360°, as shown in Table 16-1.

TABLE 16-1

	Corrections	
Value of LHA	If computed Az is plus	If computed Az is minus
0 to 180°	180°	360°
180 to 360°	0°	180°

Example 16-1 which follows illustrates the calculations involved for a single hour-angle observation of the sun. As is discussed in the next section, we should make several observations, compute the azimuth for each, and average the results. It is assumed in this example that a sighting was made directly on the center of the sun using a special reticule.

It might appear to the reader that this is a very tedious calculation when made with a pocket calculator as is done in this example. If we use a programmable calculator or a computer, the calculations become very simple. The ephemeris referred to in these pages provides programs that can be used to make these calculations.

In the example problem it is assumed that we set a stopwatch (with a split-time feature) by the WWV minute tone. It is wise to check it a minute later to see if we get a minute's difference. If there is an appreciable difference from 1 minute it will be necessary to try it again. After the observations of the sun are made, we should again check the watch against station WWV.

Example 16-1

For the data given for an hour-angle observation on the center of the sun, determine the azimuth of a line from station *A* to station *B*. This information is the first observation shown in the field notes of Fig. 16-4 for the telescope in its normal or direct position.

Date = September 11, 1990

Latitude $= \phi = 37°52'20''$N

Longitude $= \lambda = 119°12'43''$W

Instrument at station A, BS on station B

Clockwise angle to sun $= 265°24'22''$

UTC from WWV when we set stopwatch $= 14^h32^m03^s$

DUT correction from WWV $= +0^s.4$

Stopwatch time elapsed from time of setting
until time of sun obervation $= 15^m22^s.1$

Solution

1. Derive information from ephemeris tables in Appendix.

$$\text{GHA } 0^h = 180°46'53''.8$$

$$\text{GHA } 24^h = 180°52'09''.8$$

$$\text{Declination } 0^h = 4°44'26''.3$$

$$\text{Declination } 24^h = 4°21'37''.1$$

2. Determine UTl at time of observation.

$$\text{UTl} = \text{UTC from WWF} + \text{DUT} + \text{stopwatch time}$$

$$= 14^h32^m03^s + 0^s.4 + 15^m22^s.1$$

$$= 14^h47^m25^s.5$$

3. Determine GHA.

$$\text{GHA} = \text{GHA } 0^h + (\text{GHA } 24^h - \text{GHA } 0^h + 360)\frac{\text{UTl}}{24}$$

$$= 180°46'53''.8 + (180°52'09''.8 - 180°46'53''.8 + 360)\frac{14^h47^m25^s.5}{24}$$

$$= 402°41'31''.1 = 42°41'31''.1$$

4. Determine LHA.

$$\text{LHA} = \text{GHA} - W\lambda$$

$$= 42°41'31''.1 - 119°12'43'' = -76°31'11''.9$$

Adding 360° LHA $= 283°28'48''.1$.

5. Determine declination of sun.

$$\text{Declination} = \delta = \text{decl. } 0^h + (\text{decl. } 24^h - \text{decl. } 0^h)\frac{\text{UTl}}{24}$$

$$+ (0.0000395)(\text{decl. } 0^h)\sin(7.5\text{UTl})$$

$$= 4°44'26''.3 + (4°21'37''.1 - 4°44'26''.3) \frac{14^h47^m25^s.5}{24}$$

$$+ (0.0000395)(4°44'26''.3) \sin [(7.5)(14^h47^m25^s.5)]$$

$$= 4°30'23''.1$$

6. Compute azimuth to sun.

$$AZ = \tan^{-1} \frac{-\sin LHA}{\cos \phi \tan \delta - \sin \phi \cos LHA}$$

$$= \frac{-\sin 283°28'48''.1}{\cos 37°52'20'' \tan 4°30'23''.1 - \sin 37°52'20'' \cos 283°28'48''.1}$$

$$= -85°14'42''$$

Since the LHA is between 180 and 360° and since the computed AZ is positive, the correction is 0°:

$$AZ = \text{azimuth to sun} = 94°45'18''$$

$$AZL = \text{azimuth of line on ground}$$

$$= AZ + 360° - \text{Angle to right}$$

$$= 94°45'18'' + 360°00'00'' - 265°24'22''$$

$$= 189°20'56''$$

Example 16-2 which follows illustrates the solution of Example 16-1 using the enclosed computer diskette.

Example 16-2

Repeat Example 16-1 using the enclosed computer diskette

Solution

```
PROJECT TITLE: JONES GAP SUBDIVISION
---------------


--------------------------------------------
            INPUT
--------------------------------------------

Latitude = 37°52'20.0''
Longitude = 119°12'43.0''
UT when stopwatch set = 14h 32m  3.0s
DUT correction = 0.4 sec.

GHA at 0 hours UT = 180°46'53.8''
GHA at 24 hours UT = 180°52' 9.8''
Declination at 0 hours UT = 4°44'26.3''
Declination at 24 hours UT = 4°21'37.1''
```

Observation number	Elapsed Time from stopwatch setting	Clockwise angle to Sun
1	15m 22.1s	265°24'22.0''

RESULTS

Observation	Azimuth to Sun	Azimuth of line
1	94°45'17.2''	189°20'55.2''

Azimuth of line (average) 189°20'55.2''

It is desirable to make sun shots with a repeating theodolite so that several readings may conveniently be taken. Two general procedures are commonly used: the single foresight and multiple foresight procedures.

With the *single foresight procedure* we sight on a mark on the ground, sight on the sun with the telescope in its normal or direct position, invert the telescope

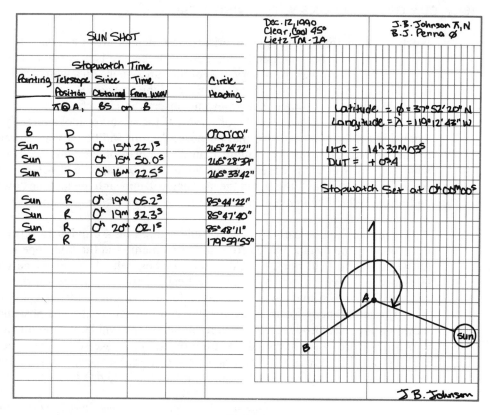

Figure 16-4

and sight on the sun, and then with the telescope still inverted, sight back to the ground mark. The four horizontal angles and the times for the two pointings to the sun are read. This group of readings is referred to as one *data set*. An observation consists of one or more data sets, with a minimum of three sets being desirable.

With the *multiple foresight procedure* we sight on the mark, take several direct readings on the sun (a recommended minimum of three), take the same number of readings with the telescope inverted, and then make a reading on the mark with the telescope inverted.

After the observations are made, the azimuths to the sun and the azimuths of the ground line are computed. If the computed azimuths of the ground line are acceptably close to each other, they are averaged. A typical set of notes for an hour-angle observation of the sun using the multiple foresight method is shown in Fig. 16-4.[4]

16-17 SUN OBSERVATIONS WITH A WHITE CARD

We have seen that when special prisms or reticules are not available for the observer to sight the sun directly, an indirect procedure is necessary. This is the procedure that involves the observation of the sun's image on a white card, paper, or field-book page held behind the eyepiece of the telescope.

Initially, it is assumed that a morning observation is to be made and further, that the sun is to the south of the observer. If the observer looks directly at the sun using a special reticule for that purpose, the sun will be moving upward and to the right if the observer is more than 23.5° north of the equator. However, when the sun's image is observed on a white card or paper held behind the eyepiece of the instrument, the sun in the morning will appear to be moving downward and to the right. *The image will move in the same direction as a shadow cast by the sun.*

The telescope is carefully leveled and sighted on a previously set point on the ground some distance away from the instrument, and the telescope is turned to the sun. The field book is held several inches behind the eyepiece and the telescope is moved horizontally and vertically until its shadow on the card is circular. The sun's image will move much faster than the reader might expect.

The telescope is focused to make the sun's image as clear as possible. The image on the card produced by an erecting telescope will be upside down as we look at the card. As a result, the bottom of the image we see is the upper limb of the sun and the top of the image is the lower limb. Furthermore, the right edge of the image is the western limb of the sun and the left edge is the eastern limb.

---

[4] R. L. Elgin, D. R. Knowles, and J. H. Senne, *1990 Celestial Observation Handbook and Ephemeris* (Fayetteville, Ark.: Elgin, Knowles & Senne, Inc., 1989), p. 40.

Using a card, we will generally find that for accurate measurements we cannot make our observations on the center of the sun but instead, make them tangent to the sun, as described herein. Nevertheless, as a starting point for this discussion the author assumes that we will sight on the center of the sun.

The first procedure described (it is a rather poor one) is to draw a circle on the card or paper the size of a nickel or a dime. Dashed cross-hair lines are then drawn through the center of the circle. The image of the sun is focused until it falls into the circle and the cross hairs of the telescope coincide with and fill in the cross hairs in the circles (see Fig. 16-5).

We can do a much more accurate job by making observations with the sun's image tangent to the cross hairs instead of attempting to center it in a circle. Pointings to the sun for the hour-angle method are made by lining up the vertical cross hair on one edge of the sun's image with no consideration of the horizontal cross hairs, as we need only a horizontal angle. If we take readings on only one edge of the sun, we will have to make a correction for the sun's semidiameter. On the other hand, if we sight on both edges, compute azimuths of the ground line, and average results, semidiameter errors will theoretically be eliminated.

If a sight is taken tangent to the sun, it will be necessary to correct the horizontal angle to the center of the sun using the equations presented here. First the altitude of the sun can be computed from the following equation:

$$h = \sin^{-1}(\sin \phi \sin \delta + \cos \phi \cos \delta \cos LHA)$$

Then the correction to the horizontal angle dH is determined from the following expression, in which the sun's semidiameter appears (its value is given in ephemerides A).

$$dH = \frac{\text{sun's semidiameter}}{\cos h}$$

Notice that the trailing edge of the sun is always the left edge if we are working at latitudes greater than 23.5° north of the equator, while the left edge will always be the leading edge if we are working at latitudes greater than 23.5° south of the equator. Should we be pointing at the trailing edge of the sun we need to add dH to our horizontal angle, and vice versa for the leading edge.

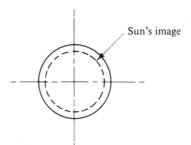

Sun's image

Figure 16-5

Should a white card be used for sighting the sun for an hour-angle observation the surveyor will focus the telescope for infinite distance, point the telescope in the general direction of the sun, hold the card behind the telescope and then move it horizontally until the shadow caused by the standards and telescope appears symmetrical on the card. Then the telescope is moved vertically until the sun's image becomes clearly visible.

To sight on the leading edge of the sun the vertical cross hair is turned forward with the slow motion screw until it becomes tangent to the sun's image on the card. To sight on the trailing edge of the sun the vertical cross hair is set so the sun's image will move onto the vertical cross hair. It is well to show by sketch in the field notes the edge which was sighted and the direction the sun's image was moving for each observation.

16-18 OBSERVATION OF THE SUN: ALTITUDE METHOD

For many years surveyors made altitude observations of the sun to obtain astronomic north. This method is perhaps a little more difficult to perform than the hour-angle procedure because both horizontal and vertical angles need to be measured. Furthermore, the accuracy obtained is not nearly as good.

In the discussion that follows the author keeps referring to the vertical angle measured. With a very large percentage of modern instruments we actually measure the zenith angle. For such cases the vertical angle equals 90° minus the zenith angle.

Altitude observations of the sun are best made when its altitude or vertical angle is at least 20° above the horizon and at least 2 hours before or after noon. Thus the most desirable times are 8 to 10 A.M. and 2 to 4 P.M. (9 to 11 A.M. and 3 to 5 P.M. if daylight saving time is being used). Should the sun's altitude be less than 20°, the effect of atmospheric refraction on the measurement of the altitude becomes significant and difficult to estimate. Observations should not be taken between 10 A.M. and 2 P.M. standard time because during those hours the rate of change of altitude of the sun with the lapse of time approaches zero. Very small errors in vertical angle measurement will cause relatively large errors in the calculated azimuth to the sun. (In addition, vertical angles greater than approximately 50° cannot be read with many instruments because the instrument bases are in the way of the telescope lines of sight.)

When a single observation of the sun is made, the surveyor will probably be able to achieve an accuracy of about ±2 or 3'. Should several observations be made, and the azimuths computed and averaged, the results may be close to ±1'.

As it is quite difficult to measure the horizontal and vertical angles to the sun accurately at the same instant, and as the accuracy of the vertical angle is extremely important, it is suggested that several foresights be taken on the sun for each backsight taken to the point on the ground. It is often recommended that one backsight be taken, then three foresights to the sun with the telescope in the direct or normal position, then three sightings on the sun with the telescope reversed, and then another sight on the mark. A single azimuth is computed between the points on the ground for each set of observations and these values averaged.

If the tangent method is used, half of the observations should be made with the sun in one quadrant and the other half with the sun in the opposite quadrant. Thus the observation would be made as shown in parts (a) and (b) of Fig. 16-6.

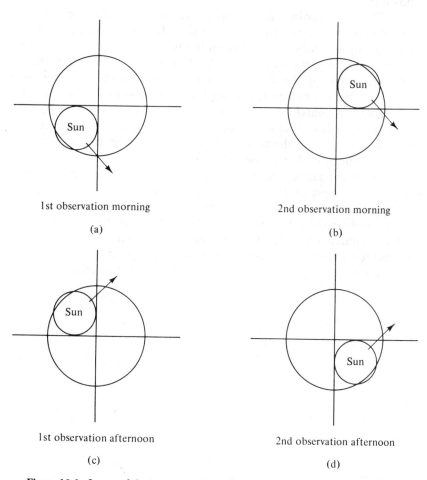

1st observation morning

(a)

2nd observation morning

(b)

1st observation afternoon

(c)

2nd observation afternoon

(d)

Figure 16-6 Image of the sun as seen in a field book. The sun is to the south of the observer. The vertical hair is at the same azimuth for the two morning observations. Similarly, the azimuth for the two afternoon observations (although different from the morning azimuth) is the same.

If the two measurements made with such a procedure are averaged, the results will be the same as though the observation had been made with the cross hairs centered on the sun. For the procedure described in the next few paragraphs it will be noted that the horizontal angle is the same for both the first and second observations of parts (a) and (b) of Fig. 16-6. Should afternoon observations be made, they would be as seen in parts (c) and (d) of the figure.

There are several procedures that may be followed to observe the sun with the cross hairs tangent to its image on the paper. One such procedure is described here with respect to Fig. 16-7. It is assumed for this discussion that the observations are being made in the morning with the sun to the south of the observer's position.

The observer first moves the telescope so that the upper edge of the sun's image is above the horizontal cross hair. Then he or she moves the vertical hair with the upper motion until it is on the western edge of the sun. The image will be as shown in part (a) of Fig. 16-7. The observer then clamps the upper motion and keeps moving the vertical hair to the right with the upper motion tangent screw so that the hair remains tangent to the leading edge of the sun.

The horizontal cross hair is left fixed in position until the sun apparently falls far enough so that its upper edge is tangent to the horizontal cross hair, as shown in part (b) of the figure. Then the time is recorded and the vertical angle is read if the altitude method is being used.

Next the vertical hair is left where it is and the horizontal hair is moved until it is tangent to the bottom of the sun's image, as shown in part (a) of Fig. 16-8. The surveyor keeps moving the tangent screw for vertical motion so as to keep the horizontal hair tangent to the bottom of the sun until finally the sun, which is moving from left to right, becomes tangent to the vertical hair as shown in part (b) of the figure. The surveyor then records the time and the vertical angle (the horizontal angle did not change) if the altitude method is being used.

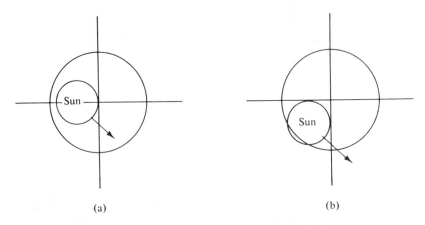

(a)   (b)

Figure 16-7   Morning observation of sun with horizontal hair stationary.

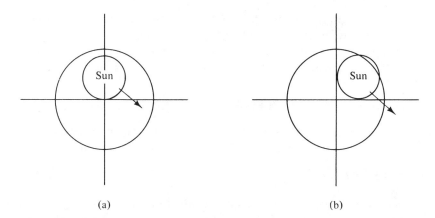

(a)   (b)

Figure 16-8   Afternoon observation of sun with vertical hair stationary.

In the afternoon a similar procedure is followed except that the image of the sun in the field book is apparently rising from left to right (if it is south of the observer) and it is necessary to make the observations as shown in parts (a) and (b) of Fig. 16-9.

If readings are taken by a tangent method *only* to the top or bottom edges of the sun, it is necessary to make a semidiameter correction to the vertical angle or altitude read for the altitude method. A similar adjustment would be required if the readings were made only with the vertical hair tangent to the leading or trailing edge of the sun's image. If the vertical angle is read to the lower edge of the sun's image, it is necessary to add the vertical angle.

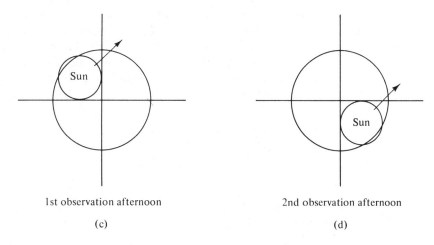

1st observation afternoon   2nd observation afternoon

(c)   (d)

Figure 16-9   Afternoon observations of sun.

As the vertical angles or altitudes of the sun are rather large, it is extremely important to level the instruments very carefully. It is also quite important to use an equal number of direct and reverse pointings of the telescope. This procedure is used in an attempt to balance out the systematic errors.

16-19 ERRORS INVOLVED IN MEASURING VERTICAL ANGLES

Several errors need to be considered when vertical angles are being measured: the index error and errors due to parallax and refraction.

Index Error

A critical aspect in making observations of the sun is the measurement of the vertical angle. Very few old transits have a vertical scale that reads accurately to zero when the instrument is leveled. This is not a problem for modern equipment with its compensators that index vertical circles automatically. The result for older instruments is that there is an index error. If only a single observation is made, a correction will then have to be made to the vertical angle. However, if one observation is made with the telescope normal and one with the telescope inverted and the results averaged, the index error is theoretically canceled. Strictly speaking, the sun moves along a curved path; unless readings are taken within a very small time interval (say, about 3 minutes), it will be advisable to calculate the astronomic azimuth to the point in question from each observation and average the results, rather than averaging the angle and time readings.

Parallax

In discussions thus far we have pictured the earth as being a mere dot in the center of the celestial sphere with a radius so small as to be insignificant with regards to the observations made. From this viewpoint a vertical angle measured from the earth to a star would be the same as if it were measured from the center of the earth. However, the actual distance to the sun or the moon or one of the planets is relatively small compared to the other stars, and as a result it is necessary to make a correction for parallax as shown in Fig. 16-10. From the figure

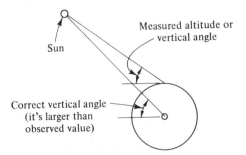

Figure 16-10

we can see that the parallax correction should be added to the measured vertical angle. The value of the correction can be determined from the expression to follow, in which C_p is the correction (always +) in decimal degrees and h is the vertical angle:

$$C_p = +8.794 \cos h$$

Refraction

The rays of light coming from the sun or stars are bent downward by the earth's atmosphere. As a result, celestial objects seem to be higher than they actually are and it is necessary to make a negative correction to the vertical angles for refraction. The magnitude of the correction is dependent on the temperature and barometric pressure of the atmosphere and on the altitude of the rays. The effect of refraction is much larger for lower-altitude rays.

The value of the refraction correction (*always negative*) can be estimated with the following expression, in which b is the absolute barometric pressure in inches of mercury (not corrected to sea level), F is the temperature in degrees Fahrenheit, and h is the measured vertical angle:

$$C_r = -\frac{0.27306b}{(460 + F) \tan h}$$

If a pressure is not immediately available, it can be estimated with the following formula:

$$b = \text{inverse log} \frac{92,670 - \text{elevation}}{62,737}$$

16-20 AZIMUTH ANGLE EQUATION: ALTITUDE METHOD

There are several forms of the azimuth angle equation. One very convenient one follows:

$$Z = \cos^{-1} \frac{\sin \text{declination} - \sin \text{altitude} \sin \text{latitude}}{\cos \text{altitude} \cos \text{latitude}}$$

$$= \cos^{-1} \frac{\sin \delta - \sin h \sin \phi}{\cos h \cos \phi}$$

In this expression δ is the declination of the sun, h is the vertical angle (corrected for the errors discussed in the preceding section), and ϕ is the latitude of the observer's position. In applying the expression it is important to notice that the declination is + if the sun is north of the equator and − if it is south of the equator. Z will be clockwise from astronomic north for morning observations

and will be measured counterclockwise from astronomic north for afternoon ob-
sevations.

16-21 EXAMPLE OBSERVATION OF THE SUN: ALTITUDE METHOD

In Example 16-3 that follows, the observer sights on a point away from the instrument and turns the angle clockwise to the center of the sun. At that time the
horizontal angle and the vertical angle (or perhaps more frequently, the zenith
angle) are measured and the time is recorded. UTC is used in this example, as
the use of the DUT correction and UTl does not significantly improve the accuracy
of the results. The necessary calculations are shown to determine the astronomic
azimuths from the observer's position to the sun and from the instrument to the
other point on the ground.

Example 16-3

The following information was obtained for an observation on the center of the sun.
The instrument was located at point 1 and a BS was taken to point 2. It is desired
to determine the astronomic azimuth for a line from point 1 to point 2.

Date = November 8, 1990

Latitude = 34°34′00″N = ϕ

Longitude = 83°20′12″

Zenith angle to sun's center = 64°49′40″

Observed time = 3:05 P.M. EST

Clockwise angle or angle to right from point I to sun's center = 154°16′20″

Temperature = 60°F

Elevation = 800 ft

Solution

1. Calculate UT or GCT of observation.

$$\text{Watch time} = 3^h05^m \text{ P.M. EST}$$

$$\text{Time on 24-hour basis} = 15^h05^m$$

$$+ \text{ Zone correction} = 5^h$$

$$\text{UT} = \overline{20^h05^m}$$

2. Determine the apparent declination of the sun on November 8, 1990, at 20^h05^m
(using Table D in Appendix A).

$$\text{Decl.} = \text{decl. } 0^h + (\text{decl. } 24^h - \text{decl. } 0^h) \frac{\text{UT}}{24^h}$$

$$= -16°25'51''.1 + [(-16°43'16''.6) - (-16°25'51''.1)] \frac{20^h05^m}{24^h}$$

$$= \mathbf{-16°40'26''} = \delta$$

3. Determine vertical angle = altitude = 90° − zenith angle.

$$\text{Vert. angle} = 90° - 64°49'40'' = 25°10'20''$$

4. Correct vertical angle for parallax and refraction.

$$C_{\text{parallax}} = +8.794 \cos h = +(8.794)(\cos 25°10'20'') = +8''$$

$$b = \text{inverse} \log \frac{92,670 - \text{elevation}}{62,737}$$

$$= \text{inverse} \log \frac{92,670 - 800}{62,737} = 29.13 \text{ in. Hg}$$

$$C_{\text{refraction}} = -\frac{0.27306b}{(460 + F)\tan h} = -\frac{(0.27306)(29.13)}{(460 + 60)(\tan 25°10'20'')}$$

$$= -0.0325479° = -1'57''$$

Corrected vertical angle = 25°10'20'' + 8'' − 1'57''

$$= \mathbf{25°08'31''} = h$$

5. Calculate azimuth to sun.

$$Z = \text{angle from astronomic meridian to sun}$$

$$= \cos^{-1} \frac{\sin \delta - \sin h \sin \phi}{\cos h \cos \phi}$$

$$= \cos^{-1} \frac{\sin - 16°40'26'' - \sin 25°08'31'' \sin 34°34'00''}{\cos 25°08'31'' \cos 34°34'00''}$$

$$= 135°05'38'' \text{ measured counterclockwise from north}$$
since it is an afternoon observation

6. Determine Azimuth from point 1 to point 2. With reference to Fig. 16-11, we subtract 135°05'38'' from 360° and get the north azimuth from point 1 to the sun = 224°54'22''. Then with further reference to the same figure we can see that the north azimuth from point 1 to point 2 = 224°54'22'' minus the observed angle to the right of 154°16'20''.

$$\text{Azimuth from pt 1 to pt 2} = \mathbf{70°38'02''}$$

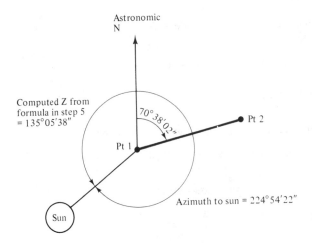

Figure 16-11   Sun-shot information for Example 16-3.

16-22 CIRCUMPOLAR STARS

Sun shots made by the hour-angle method described in earlier sections of this chapter provide sufficient accuracy for most surveying projects. Should an accuracy better than about ± 10″ be needed, however, it will be necessary to make a star observation.

If the axis of the earth were extended upward until it intersected the celestial sphere, that point (as previously defined) would be called a celestial pole. Unfortunately, that point is not marked in the sky and we therefore sight on a nearby star, usually Polaris, the north or polestar. Polaris is located approximately 1° from the celestial pole and as the earth rotates about its axis toward the east, Polaris appears to move in an orbit or small counterclockwise circle around the celestial pole. The radius of this circle (called the polar distance) is approximately 1°.

A star that rotates around the celestial north pole and never goes below the observer's horizon is referred to as a *circumpolar star*. Polaris is such a star and it rotates very close to the celestial north pole with an angular distance of approximately 1°. When a circumpolar star is directly above the celestial pole it is said to be at *upper culmination* (UC) and when directly below it is said to be at *lower culmination* (LC). When the star is at its most easterly point it is said to be at *eastern elongation* (EE), and when at its most westerly point it is said to be at *western elongation* (WE).

Figure 16-12 shows the apparent motion of a circumpolar star such as Polaris as it is seen from an observer's position X on the earth's surface. It will be noted that two times a day the star will be lined up with astronomic north or the celestial north pole shown in the figure.

If the observer is able to sight on Polaris at UC or LC, it will only be necessary to depress the telescope and set a point on the ground to have an astronomic north direction from the instrument to the point on the ground. If more

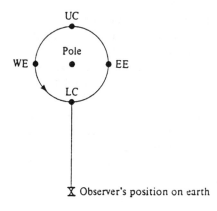

Figure 16-12   Movement of a circumpolar star.

X̲ Observer's position on earth

convenient, an angle from some other line to the line to Polaris can be measured and the azimuth of that other line can be calculated.

Actually, however, there are some distinct disadvantages to making sights on a circumpolar star such as Polaris at UC or LC. The star moves quite rapidly west to east or east to west at culmination and the observer will have to sight on the star at a very precise time. A little error in time can cause an appreciably large error in azimuth. Furthermore, the observer will probably be able to make only one sighting on the star at each culmination. As a result, the measurement cannot be checked and if the instrument is out of adjustment that fact will cause an error when the telescope is turned downward and used to set a point on the ground.

In its rotation about the celestial pole, Polaris moves on an almost vertical path for about 10 minutes before and 10 minutes after WE and EE. For all practical purposes, therefore, the star remains at elongation for about 20 minutes, enabling the observer to make numerous observations (some with the telescope in its normal position and others with it inverted). The azimuth of Polaris does not change more than 5″ during the 10 minutes before and after elongation. With an ephemeris the surveyor can determine the time of WE and EE.

The hour-angle method, however, provides the most convenient and practical method of observing Polaris. It is not necessary to calculate the time of UC, WE, and so on, and be there at that exact time, as was the common practice in the past. The hour-angle method can be used any time we can see the star.

16-23 LOCATING POLARIS

In the northern hemisphere, the most commonly observed star for surveying purposes is Polaris. This star is the last star in the tail of the constellation Ursa Minor (Little Dipper). To locate Polaris correctly, it is necessary to also be familiar with the constellations Ursa Major (whose seven brightest stars are known as the Great or Big Dipper) and Cassiopeia. These are sketched roughly in Fig. 16-13.

In this figure it will be noted that a line drawn through the two end (or pointer stars) in the Big Dipper points almost exactly to Polaris. On the opposite side of Polaris from the Big Dipper is the W-shaped constellation Cassiopeia, which is also shown in the figure. As a further aid in locating Polaris it will be noted that its approximate location can be determined with a compass (perhaps making a correction for magnetic declination). Its altitude will approximately equal the latitude of the point where the observation is being made.

The *declination* of a star is defined as the angular distance the star is above or below the celestial equator. The declination of the sun varies during the year

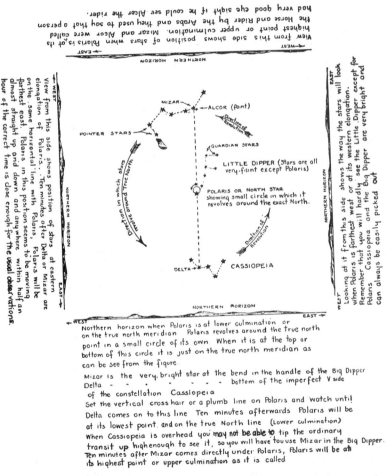

Figure 16-13 [From K. W. Leighton, *Gilbert Civil Engineering* (*Surveying*) *for Boys* (New Haven, Conn.: A. C. Gilbert Company, 1920), p. 88.]

from about $23\frac{1}{2}°$ south to about $23\frac{1}{2}°$ north. For Polaris the declination is in the range of 89°, and for such a star it may be more convenient to refer to its *polar distance* instead of its declination. The polar distance of a star is the angular distance of the star from the north pole. The polar distance of Polaris would then be about 1°. Thus the declination of a star plus its polar distance is equal to 90°. From an ephemeris the surveyor can obtain the polar distance of Polaris for the date of the observation. If the declination is given instead of the polar distance, it can be subtracted from 90° to obtain the polar distance.

A star may be sighted at dusk through the telescope without a light, but after dark it is necessary to use some kind of light so as to see the cross hairs. A flashlight pointed diagonally into the objective lens is satisfactory. The observer has to adjust the flashlight's position so that with the indirect light the cross hairs can be seen, but not have so much light that the star cannot be seen. One thing that will help reflect the light into the telescope is replacing the sunshade with a 3- or 4-in. piece of rolled-up paper held in place with a rubber band. Notice that many theodolites are equipped with lighting systems.

16-24 CHOICE OF SIGHTING ON POLARIS OR THE SUN

A few remarks are presented in this section concerning the desirability of using Polaris for determining astronomic north. Polaris provides a small excellent slowly moving target that can be sighted clearly and directly with the telescope. Polaris observations do have the disadvantage that they must be made at times other than normal working hours. Also, just a little bit of haze or just a few clouds may obscure the star so that it cannot be seen. The errors and mistakes which result while working in the dark can be significant. Furthermore, instrumental errors can be significant for Polaris observations, particularly where the vertical angle to the star is more than 40 or 50° (at latitudes of more than 40 or 50°). For such large vertical angles, accurate leveling of the instrument is of the greatest importance, as it can significantly affect the horizontal angles measured.

If we make an observation of Polaris at any time, the following steps need to be taken to determine the azimuth from the instrument to another point on the ground. In these steps you will note that it is not necessary to correct UTC to UTl by the DUT correction because that correction will not affect the computed azimuths by more than about ±0.2″ at the most. It will also be noted that we need for our computations both the latitude and longitude of the instrument's position. These will probably be taken from a USGS 7.5-minute quadrangle map.

1. Record UTC for the time of the observation.

2. Determine $\text{GHA} = \text{GHA } 0^h + (\text{GHA } 24^h - \text{GHA } 0^h + 360)\dfrac{\text{UTC}}{24}$.

3. Compute LHA = GHA − Wλ.

4. Determine declination = decl. 0^h + (decl. 24^h − decl. 0^h) $\dfrac{UTC}{24}$.

5. Calculate AZ to Polaris = $\tan^{-1} \dfrac{-\sin LHA}{\cos \phi \tan \delta - \sin \phi \cos LHA}$. Should the
value of this expression be +, Polaris will be east of north, and if the result
is −, Polaris will be west of north. For the latter case it will be necessary
to add 360° to normalize its value between 0 and 360°.

6. Determine the azimuth of the line on the ground from the angle to the right.

We will desirably take two or three sets of observations, determine the
azimuth of the line on the ground for each, and average the azimuths resulting if
they are reasonably close to each other ("reasonably close" means within a few
seconds).

The enclosed computer diskette may be used to make these calculations for
Polaris observations just as it was for hour-angle observations of the sun.

PROBLEMS

In Problems 16-1 to 16-5, it is assumed that the instrument is set up at point 1 and backsighted to point 2 and then a clockwise angle is measured to the center of the sun. For
each case determine the astronomic azimuth from point 1 to point 2 using the hour-angle
method.

	Date	Latitude	Longitude	Location	Clockwise to sun	Time of observation (UTI)
16-1.	4/17/90	40°36′10″	89°38′15″	Peoria, Ill.	42°10′32″	$15^h48^m30^s0$
					(Ans.: 85°07′34″)	
16-2.	4/22/90	35°21′30″	118°58′20″	Bakersfield, Calif.	98°16′40″	$13^h24^m16^s2$
16-3.	5/11/90	40°42′00″	81°27′00″	Canton, Ohio	192°32′45″	$20^h33^m20^s8$
					(Ans.: 62°55′35.8″)	
16-4.	12/26/90	32°46′20″	97°20′00″	Ft. Worth, Tex.	316″44′06″	$20^h30^m25^s5$
16-5.	11/12/90	42°47′00″	84°30′30″	Lansing, Mich.	145°40′32″	$16^h42^m36^s0$
					(Ans.: 23°33′31″5)	

In Problems 16-6 to 16-9, for each of the sun shots described a sight was taken from the
instrument at point A to a point B on the ground and a clockwise angle was turned to the
center of the sun. Using the altitude method determine the bearing of line AB.

	Date	Time	Latitude	Longitude	Temp. (°F)	Elev. (ft)	Vertical angle	Horizontal angle
16-6.	11/20/90	3^h10^mP.M.	35°03′	106°32′	60	5500	26°13′00″	61°10′00″
			(Albuquerque, N.M.)					
16-7.	12/9/90	10^h16^mA.M.	33°42′	84°22′	50	1200	25°08′00″	54°16′00″
			(Atlanta, Ga., Eastern Standard Time)					
16-8.	5/16/90	10^h33^mA.M.	42°25′30″	71°12′	60	250	35°16′30″	181°53′40″
			(Boston, Mass.)					
16-9.	11/14/90	3^h42^mP.M.	47°08′50″	122°25′	50	350	20°10′20″	131°16′10″
			(Tacoma, Wash.)					

(*Ans. to 16-7:* Bearing *AB* = S88°25′15″E)
(*Ans. to 16-9:* Bearing *AB* = N76°04′30″E)

In Problems 16-10 to 16-12, a sight was taken from the instrument at point 1 to point 2 on the ground and a clockwise angle was turned to Polaris. Using the hour-angle method determine the astronomic azimuth from point 1 to point 2 in each case.

	Date	UTC	Location	Latitude	Longitude	Clockwise ∡ to Polaris
16-10.	4/27/90	$2^h16^m30\overset{s}{.}0$	Wichita, Kans.	37°42′20″	97°14′00″	34°16′30″
16-11.	5/14/90	$4^h18^m22\overset{s}{.}3$	Denver, Colo.	39°41′00″	105°02′30″	111°20′00″
						(Ans.: 248°15′46.4″)
16-12.	5/16/90	$3^h26^m22\overset{s}{.}0$	Great Falls, Mont.	47°26′30″	111°23′10″	215°36′10″

In problems 16-13 to 16-16 use the computer diskette to determine the azimuth of the ground lines.

16-13.   Prob. 16-2   (Ans.: 331°03′10.6″)
16-14.   Prob. 16-4
16-15.   Prob. 16-10   (Ans.: 324°51′10.2″)
16-16.   Prob. 16-12

CHAPTER SEVENTEEN

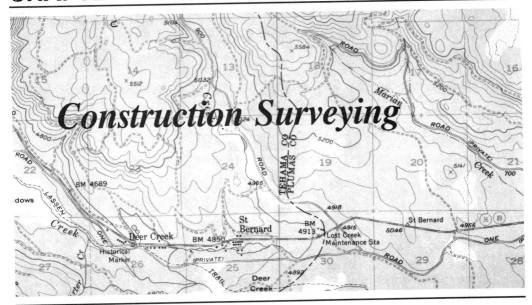

Construction Surveying

17-1 INTRODUCTION

The construction industry is the largest in the United States and surveying is an essential part of that industry. In fact, more than one-half of all surveying relates to the construction industry. To give the reader a feeling for the vast number of construction projects that need appreciable services from surveyors, the following partial list is presented:

1. Highways
2. Drainage ditches
3. Storm and sanitary sewers
4. Culverts
5. Bridges
6. Landfills
7. Pipelines
8. Railroads
9. Canals
10. Airports
11. Transmission lines
12. Buildings
13. Reservoirs
14. Water and sewage treatment plants

15. Dams
16. Mines
17. Quarries

17-2 WORK OF THE CONSTRUCTION SURVEYOR

A boundary survey and the preparation of the necessary topographic maps are the first steps in the construction process. From these maps the positions of the structures are established and when the final plans for the project are available, it is the duty of the surveyor to set the required horizontal and vertical positions for the structures. In other words, construction surveying involves transfer of the dimensions on the drawing to the ground so that the work is done in its correct position. This type of surveying is sometimes called "setting lines and grades." The work of the surveyor for construction projects is often referred to as *layout work* and the term "layout engineer" may be used in place of the term "surveyor."

Of necessity, construction surveying begins before actual construction and continues until the project is completed. Surveying is an essential part of the construction process and must be carried out in coordination with other operations in order to have an economical job and to prevent serious mistakes. The reader can readily understand the expense and inconvenience caused if one reinforced concrete footing for a building is placed in the wrong position. As another example, just imagine the problems involved if a few thousand feet of a sewer line are set on the wrong grade.

Construction drawings show the sizes and positions of structures that are to be erected, such as buildings, bridges, roads, parking lots, storage tanks, pipelines, and so on. The job of the construction surveyor is to locate these planned features in their desired positions on the ground. This is done by placing reference marks (such as construction stakes) sufficiently close to the planned work so as to permit the masons, carpenters, and other tradesmen to position the work properly using their own equipment (folding rules, mason's levels, string lines, and so on).

Construction surveyors will find that their work is quite varied. One day they may be doing topographic work for a proposed building, while on another they may be setting stakes for pipeline excavation. It is often necessary for them to make measurements before and after some types of work (e.g., computing quantities of earthwork moved by a contractor). At other times, they will set stakes to guide the construction of foundations; they may be aligning the columns of a steel frame building, or checking a completed structure to see that it is correctly positioned; and so on.

A construction project requires four kinds of surveys for its completion:

1. A property or boundary survey by a registered land surveyor to establish the location and dimensions of the property.

2. A survey to determine the existing conditions, such as contours, man-made and natural features, streams, sewers, power lines, roads, nearby structures, and so on. This work may also be done by the land surveyor along with the boundary survey.

3. The construction surveys, which determine the position and elevation of the features of the construction work. These surveys include the placing of grade stakes, alignment stakes, and other layout control points. This work is often performed by the contractor.

4. Finally, there are the surveys that determine the positions of the finished structures. These are the "as-built" surveys and are used to check the contractor's work and show locations of structures and their components (water lines, sewers, etc.) which will be needed for future maintenance, changes, and new construction. These measurements should be performed by a registered surveyor.

Only a few of the many possible construction surveying or layout problems can be presented in this chapter. The reader will get a general idea of the problems involved by confronting a range of topics in construction surveying. Included are introductory discussions on building, highway, bridge, and pipeline surveys. Despite the variety, however, the reader will notice that no new surveying principles are involved. All of those needed for construction surveying were presented in the earlier chapters.

17-3 TRADE UNIONS

In most states trade unions do not claim jurisdiction over construction surveying. Nevertheless, surveyors involved in layout work can become involved in disputes where the unions claim that surveying work should be done by union members. For instance, the unions feel that their carpenters are supposed to nail batter boards and lay out building partitions. They further claim that their ironworker foremen are supposed to check the positioning and alignment of structural steel and that their concrete finishers are to set screeds for concrete slabs. In certain areas of the country, the union operating engineers have jurisdiction over field survey work done by the contractor. If, however, the surveyor is employed as an office engineer and part of the job is to do field surveying, only the field surveying work would be under the jurisdiction of the operating engineers.[1]

[1] K. Royer, *Applied Field Surveying* (New York: John Wiley & Sons, Inc., 1970), p. 124.

17-4 PROPERTY SURVEY FROM THE CONTRACTOR'S VIEWPOINT

The contractor for a construction job should be furnished with a property survey made by a registered surveyor. In fact, the construction contract signed by the contractor should require that the property survey be furnished by the owner. Furthermore, the contractor should be able to prove that the survey used was indeed furnished by the owner.

Even if the contractor or one of his or her employees is a registered surveyor, he or she should not make the property survey, nor should the contractor even employ a surveyor to make it. If one of the employees make the survey, the contractor will personally be taking the responsibility of locating the structure, and if a mistake is made, the payment for the entire building project may be placed in jeopardy. Numerous lawsuits occur each year when buildings are placed in the wrong positions—frequently being located wholly or partly on other people's land. Even if the structure is only a small distance out of position, the result can be a disastrous lawsuit. This kind of mistake could possibly bankrupt a contractor.[2]

17-5 PRELIMINARY SURVEYS

To prepare the plans for a building the architect needs a map of the site so that the building will be located carefully. These maps are typically drawn to a large scale, such as 1 in. = 10 ft or 1 in. = 20 ft. The information to be included are property lines; elevations for the preparation of contour lines; locations, sizes, and materials of existing buildings on the site or adjacent to it; location of any immovable objects; locations of existing streets, curbs, and sidewalks; locations of fire hydrants; sizes and locations of gas and water lines and storm and sanitary sewers, including manhole locations and invert elevations (these are low points on the inside circumference of the pipes); and locations of power lines, telephone lines, light poles, trees, and other items.

Before the design of the structure can begin, the above-listed information needs to be furnished to the engineer-architect group. These data will ordinarily be provided on the site map; for buildings this drawing is frequently called the *building plot plan*.

The layout of the building will be based on the information presented on the plot plan and the proposed building design will be superimposed on that plan. The final drawing will show the location of the building with respect to the property lines and with respect to streets, utilities, and so on. It will probably also show the final contour lines that are to exist after the construction is complete.

The preliminary survey may very well include a survey of existing structures

---

[2] K. Royer, *Applied Field Surveying* (New York: John Wiley & Sons, Inc. 1970), pp. 110–111.

on the site and perhaps structures on adjacent property that might be affected by the new construction. Such surveys should be performed prior to the start of construction. They should include measurement of the vertical and horizontal positions of the foundations of these buildings so that it can be determined later if any lateral or vertical movement has occurred during construction. In this regard it should be noted that settlements continue for quite a few years after a building is constructed, and if the existing building or buildings on or adjacent to the site are relatively new, they may still be settling. As a result, settlement that occurs in an existing building when a new one is erected may or may not be due entirely to the new construction.

The surveys of existing buildings should also include examinations of both the exterior and interior of the buildings to record their condition. For instance, such items as the location and sizes of wall cracks should be noted.

17-6 GRADE STAKES

If a project is to be constructed to certain elevations, it is necessary to set stakes to guide the contractor. Once the rough grading has been completed, it is necessary to place these stakes in order to control the final earthwork. A grade stake is a stake that is driven into the ground until its top is at the elevation desired for the finished job or until the elevation of the top has a definite relation to the desired elevation. Grade stakes are necessary for sewers, street pavements, railroads, buildings, and so on.

When appreciable cuts are to be made, it may not be feasible to drive stakes into the ground until their tops are at the desired final elevations because they may have to be driven below the ground or buried. If appreciable fills are to be made, the tops of the grade stakes may have to be located above the ground level. Here it is the custom to drive stakes to convenient elevations above the ground and mark them with the cuts or fills necessary to obtain the desired elevations. It is very helpful to the contractor when the stakes are placed at heights such that cuts or fills are given in whole numbers of feet.

For a particular point the surveyor determines the required elevation and subtracts it from the HI of the level. This tells him or her what the rod reading should be when the top of the stake is at the desired elevation. Then, more or less by trial and error, the stake is driven until the required rod reading is obtained when the rod is held on top of the stake.

When earth grading operations are near their desired values, it is possible to drive stakes until their tops are at the specified final elevations. For these situations it is customary to drive the stakes to grade and color their tops with blue lumber crayons called *keel*. These stakes, known as *blue tops*, are commonly used for grading operations along the edges of highways, railroads, and so on. Some people require that grade stakes be set at road center lines instead of along shoulders.

17-7 REFERENCING POINTS FOR CONSTRUCTION

All survey stakes, even the very stoutest ones, are vulnerable to disturbance during construction. As a result, it is necessary to reference them to other points so that they may be reestablished in case they are displaced.

In construction or layout surveying, the terms *stakes* and *hubs* are frequently used. A *stake* is usually thought of as an approximately 1 × 2 × 18 in. or longer piece of wood sharpened at one end to facilitate driving in the ground. A *hub* is an approximately 2 × 2 in. piece of wood of variable length driven flush or almost flush with the ground, which has a tack driven into it to mark the precise position of the point. Usually, one or more stakes on which identification of the hubs is written are driven partly into the ground by the hubs. The stakes may also be flagged.

For highway construction it is usually common practice to reference every tenth station (1000-ft points) as well as the starting and ending points of curves and the points of intersections of the tangents to curves. For building construction it is conventional to reference building corners and even property corners if they are in danger of being obliterated.

The references used should be fairly permanent. Markers may consist of crosses or marks chiseled into concrete pavements or curbs or sidewalks, or nails and bottle caps driven into bituminous pavements. Other types are tacked wooden stakes, nails in trees, or preexisting features such as the corners of existing buildings. One disadvantage of the latter two types of points is that the tripod may not be set over them. Sometimes references may be marked on a curb or pavement or wall with keel. In addition, directions are often written on pavements describing how to find the marks whether the marks themselves are on the pavement or wall or in some other location.

It is desirable to set a sufficient number of reference points so that if some of them are lost, the referenced point can still be reset. It is advisable that important points be recorded and described in the notes in case the markings are obliterated or the points themselves disturbed.

If the distances between survey points and their references can be kept to less than 6 ft, the points may be reestablished quickly and easily with a carpenter's rule and a plumb bob. Should reference points be set at greater distances, it will be necessary to use a tape and perhaps a transit to check and reset construction points. It is probably true that almost all construction points will be disturbed at least once and perhaps many times during the construction process. As a result, the surveyor needs to position the reference points carefully so that if they are disturbed, they can easily be reestablished. It is always desirable to put guard stakes and flags around important construction points. Despite these precautions, it is still necessary to make continual position and alignment checks. Otherwise, some of the construction work may be improperly positioned, resulting in possible removal and replacement at potentially great expense.

One very common method of referencing survey points is illustrated in Fig.

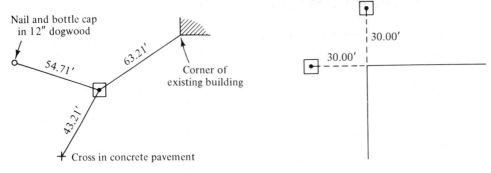

Figure 17-1 Three reference points.   **Figure 17-2**

17-1. In this figure, three reference points are used; if the survey point is disturbed, it can be reestablished by swinging an arc with a steel tape (held horizontally) from each of the reference points. The desirability of keeping the reference points within one tape length of the survey points is obvious (6 ft is even better, as discussed previously). To protect the hubs and make it easier to locate them, it is desirable to drive slanted stakes over their tops.

A convenient and frequently used method for referencing building corners is shown in Fig. 17-2. It is desirable, where possible, to set reference points the same distance away from the survey point being referenced so that no notes have to be used to reset it.

It is quicker to set reference points at random as shown in Fig. 17-3, to make use of relatively permanent objects such as trees or buildings or pavement or poles in the vicinity. To reestablish survey points from these references, it is convenient to use three workers so that simultaneous measurement can be made from both reference points.

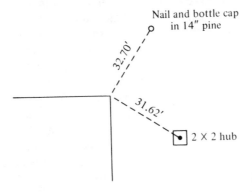

Figure 17-3

17-8 BUILDING LAYOUT

The first step in laying out a building is to locate the building properly on the lot. Many cities and counties have building ordinances that provide certain minimum permissible setback distances from the front edge of lots to the fronts of buildings. The purposes of these regulations are to promote organized growth and to improve appearances and to enhance fire protection. The survey crew sets alignment stakes with distances and directions measured from existing buildings, or streets or curbs, so that the structure will be erected in its desired position.

After the building is located properly, it is necessary for it to be laid out with the correct dimensions. This is always important but is particularly so when some or all of the building components are prefabricated and must fit together. For most residential buildings surveying layout may consist only of surveying for footings because the contractor will be able to obtain all other measurements from the footings. Commercial construction requires detail layout and control stakes primarily due to the length of time the project will be under construction and the size of the project.

It is obviously critical for a construction job that the various parts of the structure be placed at the desired elevations. To accomplish this goal the construction surveyor will establish one or more bench marks in the general vicinity of the project. These bench marks are placed away from the immediate vicinity of the building so they will not be destroyed by construction operations, and are used to provide vertical control for the project. Once these are set, the surveyor will establish a good many less permanent but more accessible bench marks quite close to the project (say, within 100 to 200 ft). The location of these less permanent points should be carefully selected so that turning points will ideally not be needed when elevations have to be set at the project. Such careful selection of the points may result in critical time saving, which is so important on construction projects.

For large construction projects it is desirable to establish elevations based on sea-level values (National Geodetic Vertical Datum of 1929). This datum is particularly important for underground utilities. Should an assumed elevation datum be used, its elevations probably should be sufficiently large that no negative elevations will occur on the project.

Another idea that may save considerable time involves establishing a permanent position for setting up the level so that the instrument can be set up in the same position every day. One way of doing this is to have actual notches for the tripod shoes cut into a concrete sidewalk or pavement or into a specially placed concrete pad. In this way the h.i. of the level will always be the same if the same instrument and tripod are used, and the rod reading for any particular desired location (say, a floor level) will be constant.

Sometimes a special concrete pier can be justified for a large job. Another case where a platform may be advisable is where it is necessary constantly to check nearby existing buildings for possible settlement during the progress of the new construction. This is particularly important during excavation, especially if there is much vibration or blasting. A set of targets can be established on nearby buildings, the level set up at the fixed point, and the telescope used to check quickly for settlements by sighting on the targets. In this regard it is necessary to use the same instrument and tripod (fixed leg, not adjustable leg) each day. It is further necessary to check the elevation of the permanent bench mark frequently against surrounding bench marks that are away from the influence of the construction activity.

Later during construction, it may be convenient to establish a bench mark at a certain point in the walls or other part of the building and use that point to measure or set other elevations in the structure. In addition, other reference points may be set within the building from which machinery or other items may be properly located once the building is enclosed. This is often done by setting brass reference points or disks in the floors or other points in the building. Once the walls are erected, the surveyor will be unable to take long sights from reference points outside the building.

17-9 BASE LINES (LAYOUT PERFORMED BY SURVEYORS)

Before the actual layout measurements can begin, it is necessary for reference lines or base lines to be carefully established. For large construction projects, the usual procedure is to set a main base line down the centerline of the structure and to set stout stakes or hubs (preferably 2 × 2 in. or larger) with tacks in them at intervals not exceeding 100 ft. It is anticipated that these hubs will stay in position for some time during construction. If station numbers are assigned to the points, they should preferably be large numbers so that no negative stations will be needed on the extreme ends of the system.

In addition, monuments are set along the centerline at each end beyond the area of the construction work. The ends of the line are best marked with heavy cast-in-place concrete monuments with metal tablets embedded in their tops. The monuments along the ends of the line may be occupied by the surveyor and will enable him or her to check and reset, if necessary, points within the construction area.

A central base line will be so often disturbed during construction that it is common practice to set another or secondary base line parallel to the central one but offset some distance from it. In fact, sometimes two base lines, one in front of the building and one in the back, may be established.

Figure 17-4 shows the case where a base line is set to one side of the building. Along such a base line hubs are set where needed to enable the surveyor to set or align corners or other important features of the building. After the corners for a building are set it is essential that their locations be carefully checked. One check that can be used for rectangular buildings is to measure their diagonals to see if they are equal (shown by dashed lines in Fig. 17-4).

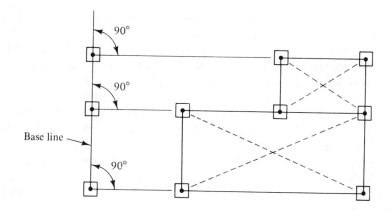

Figure 17-4   Layout of a simple building.

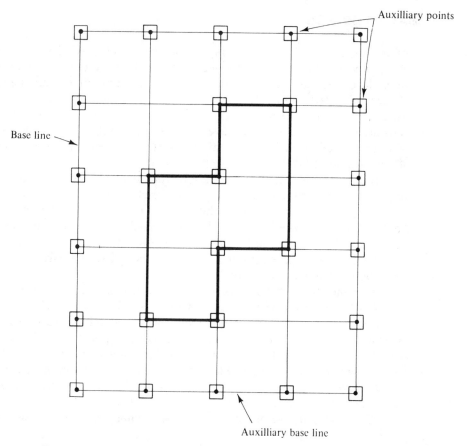

Figure 17-5   Layout of a building with base line, auxiliary base line, and auxiliary points.

When it is necessary constantly to provide alignments for points in a building where reference markers have been disturbed (a typical situation for the construction of most buildings), the procedure described in the preceding paragraphs and illustrated in Fig. 17-4 has the disadvantage that it is necessary to set up the instrument and lay off 90° angles every time a point is needed. It is preferable to use a system where the important points in the building can be checked or re-established merely by sighting between two points or by stretching a string or wire between them. Such a system is illustrated in Fig. 17-5. For this building both a base line and an auxiliary base line were established, together with the auxiliary points shown. When feasible, it is desirable to attach or paint targets on the walls of existing buildings instead of setting the auxiliary points in the ground. Such targets are much less likely to be disturbed than are stakes in the ground.[3]

17-10 BATTER BOARDS

After hubs are set for the corners of a building, they may be secured by means of references set just outside the work area in the form of batter boards. If hubs were placed only at the proposed corners of a building, they would be in the way of excavation and construction operations and would not last long. For this reason, batter boards are ordinarily used. They consist of wooden frameworks that have nails driven into their tops (or have notches cut into them) from which strings or wires are placed to outline the position of the building lines and perhaps the outside of the foundation walls.

Batter boards are used not only for building corners but also for the construction of culverts and sewer lines, for bricklaying, and for many other construction jobs. Examples of batter boards are shown in Figs. 17-6 and 17-7. Batter boards should be placed firmly in the ground and should be well braced. Further, they should be placed a sufficient distance from the excavation so that they will be located in undisturbed ground, yet close enough so that strings may conveniently be stretched between them.

To build the batter boards, several stout timbers are driven in the ground probably 4 to 6 ft from the desired excavation. As wire lines are often used between batter boards, they have to be quite stout to withstand the pulling. The vertical posts are 2 × 4s and 1 or 2 in. boards are nailed across them. The cross boards are set between the timbers with a level rod and adjusted to the desired elevation with a level. To set the nails, which are driven vertically on the cross boards to hold the strings for alignment, the transit is used or perhaps strings are stretched across the corner point using plumb bobs as needed.

The main purpose of batter boards is to enable workers to measure values

[3] B. A. Barry, *Construction Measurements* (New York: John Wiley & Sons, Inc., 1973), pp. 149–155.

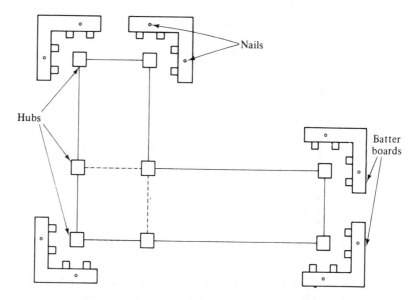

Figure 17-6   Batter boards.

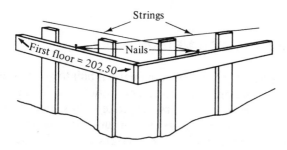

Figure 17-7   Batter boards.

from readily accessible reference points without the necessity of having a surveyor always on the spot. Since these boards are used to provide both alignment and elevation, they must be set very carefully. It is common to set the batter boards at a prearranged height as a certain number of feet above the foundation or the finished floor grade. A controlling elevation such as that of the first-floor level should be clearly marked on the batter boards, as illustrated in Fig. 17-7.

Once the batter boards are set and checked, the strings or wires can be stretched between the nails (or notches) as shown in Fig. 17-7. The strings can be taken down or replaced as often as desired during construction. For instance, they can be put in place and the corner of the structure reset at any time by hanging a plumb bob at the intersection of the strings. The amount of excavation needed at a particular point for the foundation can be determined by measuring down from the string. The strings can be taken down when it is necessary for

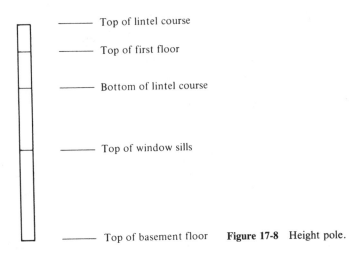

————— Top of lintel course

————— Top of first floor

————— Bottom of lintel course

————— Top of window sills

————— Top of basement floor   **Figure 17-8** Height pole.

workers or machines to excavate for the foundation. They can be put back up for use by the brick masons or other workers. In masonry structures a height pole is often used to give the elevations of various points or courses in the walls. Such a pole is shown in Fig. 17-8.

17-11 BUILDING LAYOUT—CONTRACTOR METHOD

For residential construction and some small commercial buildings the foundation is often subcontracted to a footing contractor. The duties of this subcontractor are to lay out, excavate, place reinforcing bars, set grade stakes, call for inspection and place the concrete. These subcontractors are not surveyors but they use some very ancient and proven layout methods. Levels, lasers or transits are used only for the determination of elevations. Angles are usually not turned with transits or theodolites but are established with tapes.

A rather typical procedure followed by these subcontractors follows:

1. The property corners are located.
2. A 1 × 4 stake is placed at each corner and nylon strings are stretched along the property lines.
3. The setback distances as obtained from the site and floor plans are laid off thus establishing one of the building corners and two of its sides.
4. From the corner and sides established in step 3 the building is laid out as a large rectangular box. The 90° corners for this box are established with a tape using one of the methods described in Section 4-13.
5. The building corners are marked on the ground and batter boards are driven with offsets of about 6 to 9 ft.

6. Nails are driven into the batter boards and nylon string lines are placed over the corners previously marked on the ground making use of plumb bobs.

7. Tapes are used to measure the exterior lines and the diagonals are carefully checked. Offsets are measured from the building lines.

8. Nylon lines are carefully laid out to the outside of the foundation wall (not the outside of the footing). The earth beneath the lines is scratched with some type of rod or a reinforcing bar.

9. The batter boards are set to the desired finished floor elevations.

10. The nylon lines are then removed and the footing excavations made.

11. The placing of the reinforcing bars followed by the inspection and then the placement of the concrete finish the footing subcontractor's work.

17-12 LAYOUT FOR STRUCTURAL STEEL BUILDINGS

The footings or piers for structural steel buildings can be aligned from the base lines as described in Section 17-8. Batter boards are often used to enable workers to excavate for and construct the footings.

Steel columns are set on steel base plates and connected to footings or piers with anchor bolts cast into the footings. The anchor bolts will pass through holes drilled in the base plates. The anchor bolts are set into the footing and carefully aligned vertically. Plywood templates are used to hold the bolts in their desired positions. These templates are set on top of the footing forms and must be aligned vertically and horizontally by the surveyor.

One of the most important aspects in erecting a structural steel building successfully is proper positioning of the anchor bolts and base plates. Except for some very small columns (which may have their base plates welded to their bases in the shop) the base plates are shipped to the job loose and set to their correct elevations very carefully.

The concrete for the footings is placed from $\frac{1}{2}$ to 2 in. lower than its final elevation. The plates are then placed at their correct elevations using shim packs. These packs consist of 3- to 4-in. metal squares with thicknesses varying from $\frac{1}{16}$ to $\frac{3}{4}$ in. each. They are placed under the corners of the base plates and their thicknesses adjusted until the tops of the base plates are at the desired elevation. (An alternative to shim packs involves the use of leveling nuts placed on the anchor bolts under the base plates. They can be adjusted up or down until the plate is at the desired elevation.) At this point a final check of the base plate position must be made using the transit or the batter board strings.

When the plates are in their final position, a small wood form is placed a couple of inches away from the edge of the plates and grout is poured or pumped in to fill in the space under the plates. The grout or mortar is preferably of a nonshrinking type made with a metallic aggregate.

As the columns are set in the building they must be properly aligned. The distances between the columns are carefully checked with a tape, and their positions may be adjusted by means of jacking or hammering. The columns must be kept vertical by using the transit or plumb bobs. Diagonal bracing cables are very useful here and are used to pull the columns into position with turnbuckles.

17-13 LAYOUT FOR REINFORCED CONCRETE BUILDINGS

The layout work needed for reinforced concrete buildings follows very closely the alignment work used for structural steel buildings. The necessary alignment may be accomplished with transits or strings. It is necessary to offset these distances so as to be at the correct positions for setting the concrete forms. The reinforcement that is placed in the footings and runs up into the columns or walls must be carefully aligned, as are the anchor bolts used for structural steel columns.

The snapping of chalk lines onto concrete floors or footings or on the forms is a very useful method of establishing reference lines from which measurements can be taken. Chalk lines can often take the place of mason's strings or wires, and they do not present tripping hazards or problems of resetting when the lines are broken.[4]

It is necessary to do a great deal of spirit leveling in a large reinforced concrete building to set reference marks on concrete floors or columns. As the building is constructed, tapes are often used to control the vertical distances. Because of obstructions, many of these measurements are made in elevator shafts or stairwells or on the outsides of buildings.

To maintain the vertical alignment of walls and columns it is usually necessary to use transits or plumb bobs. Carpenter's levels held against the formwork do not usually provide sufficient accuracy for maintaining vertical alignment.

17-14 LAYOUT NOTES FOR CONSTRUCTION

No particular standard forms are used for keeping field notes for construction layouts, but a sketch should be made showing the actual angles and lengths used. Often the construction drawings themselves can be used and the dimensions thereon circled, checked, or marked in some way. Should the actual construction dimensions be different from the values shown in the drawings, the dimensions used should be noted carefully. For instance, if a distance is shown on the drawing as being 320.00 ft and the actual field measurement is 320.04 ft, the actual value should be recorded. If every field dimension measured agreed exactly with the planned dimension, there could very well be a suspicion that the surveyor did

---

[4] B. A. Barry, *Construction Measurements* (New York: John Wiley & Sons, Inc., 1973), p. 159.

not bother to check the values. A complete set of "as-built plans" is often required or should be furnished as part of a construction job.

17-15 BRIDGE SURVEYS

For short bridges with no offshore piers, the first step is to lay out the centerline of the roadway. Then the station of the abutment is staked on the centerline and the angle of the abutment face across the centerline is established. This should be done by establishing points at the end of the cross line and beyond and referencing them carefully. Should the abutment be out in the water, the cross line will be offset on the shore. In a similar manner the governing lines for the wing walls are established at and beyond the limits of the excavation, preferably at both ends of each wing wall. When the faces of the walls are battered, it is customary to stake one line for the bottom of the batter and another for the top.

Once the foundation concrete has been set, the surveyor will need to provide the lines needed for setting forms. As the walls are constructed, elevation marks are established on the forms or on the hardened concrete. In addition, the alignment is marked on the completed parts of the structure.

For long bridges the measurement of distances is extraordinarily important. They should be measured with a precision even better than that of first-order work because of the necessity of having the parts of the superstructure fit together within very small limits or tolerances. The pier locations are established with a somewhat lower precision than that used for fitting the superstructure. When the entire structure is completed, permanent survey points are usually established at various points so that any lateral and vertical movements of the bridge may be detected.[5]

17-16 PIPELINES

Pipelines can be of two types, pressure or gravity. In the pressure pipeline the liquid or gas that fills the pipes moves along uphill or down under pressure. In gravity pipelines (such as sanitary sewers or storm drains) the liquid flows along by gravity and the grades of the lines have to be controlled very carefully. The layout for pressure lines does not have to be nearly as accurate as it does for gravity lines. Vertical accuracy of ± 0.01 ft may be required for gravity lines, whereas ± 0.1 ft may be sufficient for pressure lines.

The excavation for pipelines is usually handled with a trenching machine and the machine is usually guided across the ditches by horizontal wires or strings. These lines are placed on top of stakes set by the contractor. In addition, guide

[5] R. E. Davis, F. S. Foote, and J. W. Kelly, *Surveying Theory and Practice*, 5th ed. (New York: McGraw-Hill Book Company, 1966), pp. 734–736.

lines are placed parallel to the ditch. They are usually placed by the contractor on the stakes set by the surveyor at certain preset elevations above the pipe inverts. The pipe inverts are used to control the elevations and grades of the lines. Because of manufacturing imperfections there is some variation in pipe dimensions, and therefore the pipe invert is the logical point for elevation control because any dimensional irregularities will be thrown to the top of the pipe, where they will not affect the flow. The old method of controlling pipeline excavation and construction by using batter boards and stringlines is described in this section; more modern procedures are presented in Section 17-17.

To stake out a sewer line, the stakes are set at the desired stations (commonly 50-ft intervals, although 25-ft intervals give much better control) and offset from the planned pipe centerline and their elevations are determined. The station number, offset distance, and cut required are marked on each stake. Alignment is handled with the transit, but the very important elevations are normally obtained with a level. After excavation is completed, the stakes will be used to give alignment and grade for the pipe-laying operation.

The offset stakes are normally driven to a depth such that their tops will be at certain desired elevations above the pipe inverts. During excavation the depth of the trench is often checked from these stakes to ensure that the proper grade is being obtained. The bottom of the trench will obviously be a few tenths of a foot below the pipe invert because of pipe thickness, gravel or sand beds for the pipes, and so on. These values must be considered in the elevations. As shown in Fig. 17-9, it is easy to check trench depths from the offset stakes by using a board, a carpenter's level, and a level rod or some type of height pole.

After the trench excavation is completed, a stringline is placed directly above the desired centerline of the pipe and is made parallel to the desired grade at the inverts. To accomplish this, a cross batter board is placed at each elevation stake, as shown in Fig. 17-10. The top of each batter board will be the same vertical distance above the desired position of the inverts. A nail is placed (as shown in the figure) at the top of the batter board at the desired centerline of the pipe. A string run between these nails and tightly stretched will have a grade equal to that desired for the pipe inverts.

Starting at the lower end of the trench, the pipe is laid joint by joint. The invert elevation of both ends of the first section and its alignment under the string

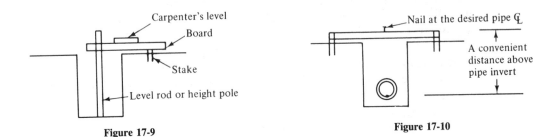

Figure 17-9   **Figure 17-10**

are checked. As the pipe is laid along the line, each additional section of the pipe is set by using a height pole of the proper length from the string to check the elevation of the forward end.

17-17 TRANSIT AND LASER METHODS OF LAYING PIPE

A transit or theodolite can be set up on a platform across the trench and the telescope inclined at the desired grade. With the line of sight set a convenient distance above the proposed bottom of the trench, the grade can be controlled for excavation and later for laying the pipe, as shown in Fig. 17-11. It will be noted that the instrumentman will have to make frequent readings with the transit and will often have to check to see if the instrument has settled or been otherwise disturbed.

Another useful device for controlling pipeline excavations and laying pipe is the laser (Fig. 17-13). So many applications are being found for the laser that some construction people believe that eventually it will be the only tool needed for the layout and control of construction projects. It can be used quickly, accurately, and economically for many purposes, such as distance measurement, alignments for tunnel borings, setting of pipes with desired grades, and setting line and grade for many types of construction.

The laser is an intense light beam that can be concentrated into a narrow ray containing only one color (red) or one wavelength of light. The resulting beam can be projected for short or long distances and is clearly visible as an illuminated spot on a target. It is not disturbed by darkness, wind, or rain, but it will not penetrate fog. A laser can be set up on a bracket or even attached to a transit telescope. The beam is lined in the proper direction at the desired grade and can be left relatively unattended.

Today, instead of using batter boards and strings, the laser can be used to control the alignment for excavating a trench and setting a pipe. The laser can be set so that it shines on the boom of a backhoe so that the operator can clearly see the illuminated spot. By its position the operator can closely control the depth to which he or she is digging. For laying the pipe the laser is set in the proper direction at the desired distance above the pipe invert. With the aid of an L-shaped pole or template, as shown in Fig. 17-12, the workers can control the invert elevation. It may also be possible to direct the laser beam from the inside

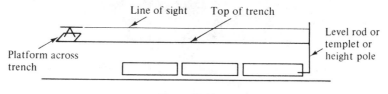

Figure 17-11

Laser beam

Figure 17-12

of manholes through the pipes being laid and control the grade without interference from the backfill operations. This can be done even if the pipes are too small for human access. The units are usually completely watertight. The use of lasers for pipe-laying operations provides several advantages: less labor required, quicker work, better alignments and grades, and no batter boards to be erected and later removed.

It is true that lasers replace guidelines and save a lot of time in the setting and replacing of stakes. To get them in the correct positions and set at the correct slopes, however, requires use of an appreciable amount of conventional surveying. Laser lines are adversely affected by dust and humidity. As a result, when working in trenches blowers are often used to get rid of the dust and reduce the humidity.

Figure 17-13 AccuBeam pipe laser that can be used inside or outside pipes. (Courtesy of CLS Industries.)

17-18 AS-BUILT SURVEYS

As-built surveys are made after a construction project is complete, to provide the positions and dimensions of the features of the project as they were actually constructed. These surveys not only provide a record of what was constructed but also provide a check to see if the work was done according to the design plans. The monuments set as control for the project are checked and readjusted or replaced if necessary. It is essential to protect this control system as much as possible in case there is a future modification or expansion of the project.

From the vertical and horizontal control points a detailed map is prepared showing all changes made during construction. The usual construction project is subject to numerous changes from the original plans due to design modifications and to problems encountered in the field, such as underground pipes and conduits, unexpected foundation conditions, and other unforeseen situations.

Laser levels (discussed in the preceding section) are particularly useful for conducting parts of as-built surveys. An example is the checking of grades for drainage. After setting up a laser level, one person can walk around with a field book and check the elevations.

The reader should note that it is necessary to survey pipelines, sewers, and other underground structures before they are backfilled so that their correct positions will be determined both horizontally and vertically. In addition, these surveys must show the locations of unexpected pipes, structures, and other features, which are encountered in the excavations. The as-built survey is a very important document and must be preserved for use in future repairs, modifications, and expansions.

CHAPTER EIGHTEEN

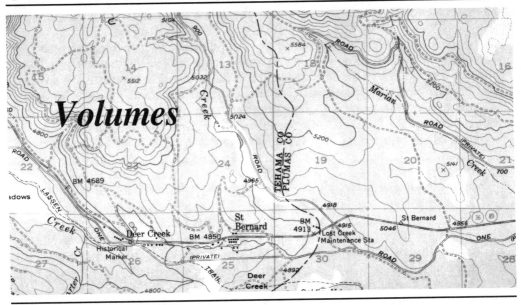

Volumes

18-1 INTRODUCTION

A tremendous volume of earthwork needs to be moved for the construction of highways, railroads, canals, foundations of large buildings, pipelines, and other projects. The surveyor is often directly involved in the determination of the amounts or quantities of this earthwork. He or she is concerned not only with quantities but also with the setting of the grade stakes needed to carry out the earthwork required to bring the ground to the desired grades and elevations.

Before a construction project involving earthwork is begun, the surveyor needs to determine the shape of the ground surface. This is necessary in order that the volume of materials to be added or removed may be determined. In speaking of earthwork it is the custom to refer to excavations as *cuts* and to embankments as *fills*. The quantities of cuts and fills in the types of construction projects described herein are frequently of such magnitude as to make up appreciable percentages of the total project costs.

The principles involved in volumetric calculations are applicable not only to earthwork but also to other materials, such as to volumes of reservoirs and to stockpiles of sand, gravel, and other materials. Volumes of masonry structures may be computed directly from the dimensions on plans but it is also quite common to check pay quantities by computing the volumes from the measurements of the completed structures made by the surveyor in the field.

It should also be noted that maps produced by photogrammetric methods (described in Chapter 24) enable the surveyor to estimate earthwork quantities quite well. Either topographic maps prepared from photographs or stereoscopic models may be used.

18-2 SLOPES AND SLOPE STAKES

In working with cuts and fills for highway construction, the side slopes are generally based on the material involved. The slope is given as a ratio of so many units horizontally to so many units vertically. For instance, a 2-to-1, or 2:1, slope means that the bank in question goes 2 ft horizontally for each 1 ft vertically (see Fig. 18-1). Perhaps a better way to indicate this, with less chance of misinterpretation, is to say 2 on 1. The 1½ on 1 slope is probably the most commonly used, but if the material consists of solid rock, the slopes may be much steeper. For very loose material, such as sand, they will be much flatter (perhaps as much as 8 or 10 or even 12 on 1 in extreme cases).

After the grade of the road has been established and the material slopes decided for cuts and fills, it is necessary for the surveyor to stake out the work. The cut or fill to be made at a particular station along the center line of the road equals the difference between the ground elevation there and the final elevation from the grade line. The surveyor will need to stake the road center line and set *slope stakes* at the intersection of the natural ground line and the side slopes of the cut or fill as shown in Fig. 18-2.

A few descriptive comments are made here regarding the trial-and-error problem of setting slope stakes, because it may give the student a little trouble. If the ground surface is horizontal, the distance from the road center line to the intersection of the ground surface and the cut or fill can easily be computed, as shown in Fig. 18-3.

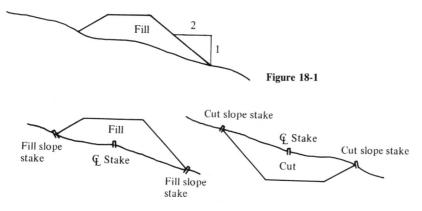

Figure 18-1

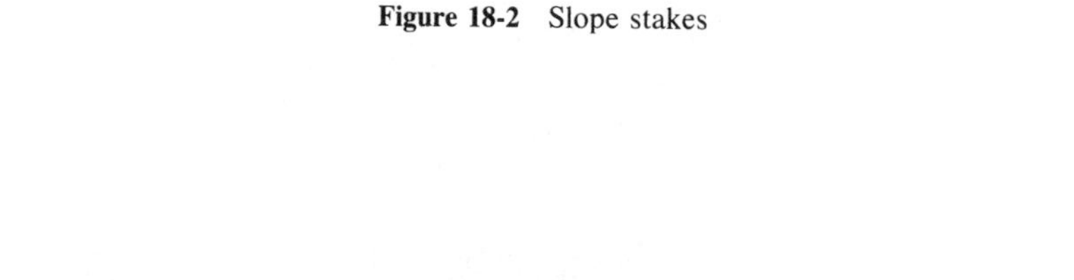

Figure 18-2   Slope stakes

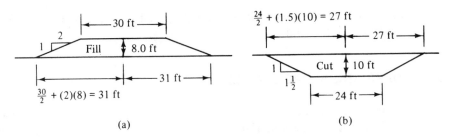

Figure 18-3

If the ground surface is sloping, as is the case in Fig. 18-4, the problem becomes a trial-and-error one because the vertical amounts of cut or fill as measured from the slope stake positions are unknown, as are the horizontal distances from the center line to the slope stakes.

One common practice is to set slope stakes with the 50-ft woven tape and an engineer's level. The zero end of the tape is placed at the center line. The instrumentman roughly estimates the distance to the slope stake and sends the rodman that distance from the center line. The instrumentman then takes a reading on the level rod and calculates to see if the rodman is at the desired point. For this discussion, see Fig. 18-5, in which the instrumentman estimated the fill at the lower slope stake as 8.0 ft and then computed the horizontal distance from the center line as $(30/2) + (2)(8) = 31$ ft.

The rodman was sent out 31 ft from the center line and the level rod reading was taken. It was found that the difference in elevation from the top of the fill to

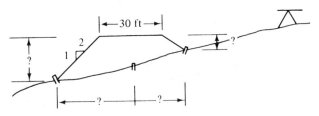

Figure 18-4

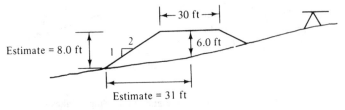

Figure 18-5

the point in question was 10.0 ft. The horizontal distance from the center line for that difference in elevation should have been (30/2) + (2)(10) = 35 ft, so the rodman was not at the correct point.

Next, the rodman was sent out to the 40-ft point and the elevation difference was found to be 12.0 ft. For this point, the horizontal distance should have been (30/2) + (2)(12) = 39 ft. The rodman moved to the 39-ft point, where the elevation difference was 11.9 ft. The calculated horizontal distance equaled (30/2) + (2)(11.9) = 38.8 ft, which was just about right.

Once the trial horizontal distance is within one or two tenths of a foot of the computed horizontal distance, the slope stake is driven. An experienced person usually sets the stake within two or three trials at most, but the beginner may require a few more attempts.

Slope stakes are normally set sloping outward from the road for fills and inward for cuts. The station numbers are usually written on the outside of the stakes and the cuts or fills are given on the inside, often with the distance to the center line of the road given. Sometimes after the correct position of the slope stakes is determined, they are offset by 2 to 5 ft to preserve them during grading operations.

18-3 BORROW PITS

During the construction of roads, airports, dams, and other projects involving earthwork, it is often necessary to obtain or *borrow* earth from surrounding areas in order to construct embankments. These excavations are commonly referred to as *borrow pits*. The quantity of borrow material is very important because the contractor's pay is usually computed by taking the number of yards of borrow times the bid price per yard. In addition, the adjacent areas where the borrow pits are located often belong to other people, who are paid on a quantity basis. Volumes of stockpiled materials such as sand, gravel, coal, and so on, can also be calculated using the borrow pit method.

To determine the data needed for volume calculations, elevations may be obtained at certain points before and after the earth is removed. One or more reference base lines and two or more bench marks are established at convenient and protected locations. Then some type of grid system is normally established (say, 50-ft squares) and the elevations are determined at each of the corners. When the excavation is completed, the levels are run again at the same points and from the differences between the original and final readings the cut made at each point is obtained. Volumes may be approximately calculated by multiplying the average cut for a particular figure times the area of the figure.

Figure 18-6 shows a plan view of a typical borrow pit divided into convenient squares and rectangles and also into some triangles because of the irregular shape of the borrow pit. The numbers shown on the figure represent the dimensions of the figure and the cuts in feet at the corners. The value of earthwork under one

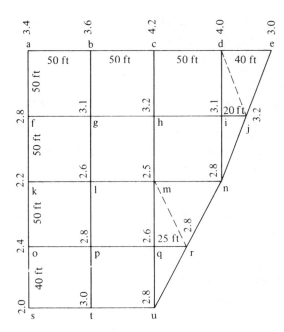

Figure 18-6

of the areas can be estimated as being equal to the average of its corner cuts times the area of the figure. For instance, the estimated cut for figure *opst* is equal to

$$(40)(50)\left(\frac{2.4 + 2.8 + 2.0 + 3.0}{4}\right) = 5100 \text{ cu ft}$$

A similar calculation for figure *pqtu* follows:

$$(40)(50)\left(\frac{2.8 + 2.6 + 3.0 + 2.8}{4}\right) = 5600 \text{ cu ft}$$

The total volume for the two figures is 10,700 cu ft.

Suppose that since these two rectangles have the same horizontal dimensions (40 × 50 ft) we would like to combine their calculations into one set. Looking back at the preceding computations and also at Fig. 18-6, we can see that to determine the sum of the corner cuts for the two figures the values at corners *o, q, s,* and *u* appear one time, while the values at *p* and *t* (2.8 and 3.0, respectively) each appear twice. Thus the calculations for the two figures can be handled in one step as follows:

$$(40)(50)\left[\frac{2.4 + (2)(2.8) + 2.6 + 2.0 + (2)(3.0) + 2.8}{4}\right] = 10,700 \text{ cu ft}$$

In a similar fashion all eight figures within *adnmqo* are squares (50 × 50 ft). To sum the corner cuts for all these figures the one at *a* appears in one square,

TABLE 18-1

Figure	Sum of corner cuts	Area of figure (sq ft)	Multiplier	Volume (cu ft)
oqus	21.4	2000	× ¼	10,700
adnmqo	95.7	2500	× ¼	59,800
uqr	8.2	500	× ⅓	1,370
qrm	7.9	625	× ⅓	1,650
rmn	8.1	1250	× ⅓	3,380
nij	9.1	500	× ⅓	1,520
ijd	10.3	500	× ⅓	1,720
jed	10.2	1000	× ⅓	3,400

Total volume = 83,540 cu ft
= 3094 cu yd

the ones at *b* and *c* each appear in two squares, the ones at *g* and *h* each appear in four squares, and so on.

$$(50)(50) \left[\frac{3.4 + (2)(3.6) + (2)(4.2) + 4.0 + (2)(2.8) + (4)(3.1)}{4} \text{ etc.} \right]$$

$$= 59,800 \text{ cu ft}$$

The calculations for the entire borrow pit can be recorded very compactly as shown in Table 18-1. Groups of figures that have the same horizontal dimensions are combined. Some of the computed volumes shown have been rounded off a little, as is only reasonable for earthwork calculations.

A special comment should be made about the figures on the side of the borrow pit, as, for example, trapezoid *deij*. The volume of this prism may be determined by taking the sum of the four corner cuts divided by 4 and multiplied by the area of the trapezoid. It will be noted that this does not give the same value as that obtained by breaking it into the two triangular figures, *ijd* and *jed*. If the ground surface across the trapezoid is not fairly constant in terms of slope (as where there is a hump or dip out in the figure), a more accurate volume can be obtained by breaking the figure into triangles. For such a case the surveyor should show dashed lines on the drawing (such as *dj* and *mr* in Fig. 18-6) to indicate that triangular figures should be used.

18-4 CROSS SECTIONS

For the purpose of this discussion the construction of a highway will be considered. It is assumed that a longitudinal grade line has been selected as well as the roadway cross section. It is further assumed that cross sections of the ground surface have been taken at each station along the route by the method described in Section 8-6.

A cross section is a section normal to the center line of a proposed highway, canal, dam, or other construction project. The cross sections of the ground surface are plotted for each station and the outline of the proposed roadway as traced from a template is superimposed on the cross sections, with the elevation of the center line of the roadway obtained from the longitudinal grade line. Examples of such cross sections are shown in Fig. 18-7.

18-5 AREAS OF CROSS SECTIONS

To determine earthwork quantities as described later in this chapter, it is first necessary to determine the areas of the cross sections. This may be done either by means of computations or with planimeters, as described in the following paragraphs.

Computer Programs

The calculation of cross-sectional areas and earthwork volumes are very tedious and repetitious for large projects. As such they are ideally suited for computer applications. There are available numerous "canned programs" with which the calculations can quickly be made. Most highway departments today use computers to determine their required areas and volumes for earthwork. Nevertheless, the surveyor should be familiar with the theory behind the calculations described in the remainder of this chapter.

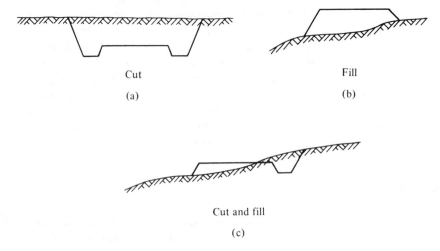

Cut   Fill

(a)   (b)

Cut and fill

(c)

Figure 18-7 Typical cross sections.

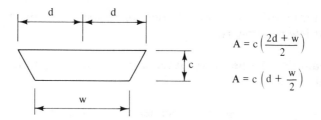

$$A = c\left(\frac{2d + w}{2}\right)$$

$$A = c\left(d + \frac{w}{2}\right)$$

Figure 18-8

Areas of Level Cross Sections

The area of a level cross section such as the one shown in Fig. 18-8 may be calculated by multiplying the average of the top and bottom widths of the cross section by its depth, as shown in the figure. Level cross sections are quite commonly encountered for highways, railroads, ditches, and so on.

Areas of Three-Level Sections

When a three-level section is involved, such as the one shown in Fig. 18-9, its area may be determined by breaking the figure down into triangles and summing up their areas as shown in the figure.

Areas of Five-Level or More Sections

When a section of five or more levels is encountered it is possible to compute its area by summing up the areas of the triangles and/or trapezoids that make up the

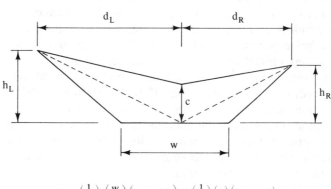

$$A = \left(\frac{1}{2}\right)\left(\frac{w}{2}\right)\left(h_L + h_R\right) + \left(\frac{1}{2}\right)(c)\left(d_L + d_R\right)$$

$$A = \frac{w}{4}\left(h_L + h_R\right) + \frac{c}{2}\left(d_L + d_R\right)$$

Figure 18-9

figure, by using a coordinate rule, or by using a planimeter, as described in the next few paragraphs.

Measuring Areas of Irregular Cross Sections with a Polar Planimeter

An irregular cross section is defined as one where the ground surface is so irregular that neither a three-level cross section nor a five-level cross section will provide sufficient information to describe the area sufficiently. Intermediate elevations at irregular intervals are necessary between center line and slope stakes to specify the cross section adequately.

When irregular cross sections are involved it may be possible to break them down into convenient figures such as triangles and trapezoids and compute their areas. Sometimes a form of the coordinate method may be used. These methods, however, are rather tedious to apply and it is more common to use computer programs or to plot the cross sections on cross-section paper (particularly if the sections are very irregular and also if some of the sides are curved). Once this is done the areas are determined by traversing their perimeters with polar planimeters. In almost every case the area of a cross section can be determined by planimeter with a precision as good as is justified by the precision used in making the field measurements with which the cross section was drawn. Of course, in this age of computers, use of this method has drastically declined.

18-6 COMPUTATION OF EARTHWORK VOLUMES

Earthwork volumes may be computed from the areas of cross sections by two methods, as described in this section. The distance between cross sections is dependent on the precision required for the volume calculations. Obviously, as the price per cubic yard goes up it becomes more desirable to have the cross sections closer together. For instance, for rock excavation or for underwater excavation, costs are so high that cross sections at very close intervals are required, perhaps no more than 10 ft. For ordinary road or railroad earthwork the sections are probably taken at 50- or 100-ft intervals. In addition to the cross sections taken at regular stations it is also necessary to take them at the beginning and ending points of curves, at locations where unusual changes in elevations occur, and for points where ground elevations coincide with natural grades. The latter points are called *grade points*.

The earth between two cross sections forms an approximate *prismoid*. A prismoid is a solid that has parallel end faces (or bases) and sides that are plane surfaces. In this section two methods are presented for estimating the volume of these prismoids: the average-end-area method and a method using the prismoidal formula.

Average-End-Area Method

A very common technique used for computing earthwork volumes is the average-end-area method. In this approach the volume of earth between two cross sections is assumed to equal the average area of the two end cross sections times the distance between them (see Fig. 18-10). The volume is computed as

$$V = \left(\frac{A_1 + A_2}{2}\right)\left(\frac{L}{27}\right)$$

where A_1 and A_2 are the end areas in square feet, and L is the distance between the cross sections in feet. The expression has been divided by 27 to give an answer in cubic yards.

Should cross sections be taken at 100-ft stations, the expression may be reduced to the following form, the volume again being in cubic yards:

$$V = 1.85(A_1 + A_2)$$

The average-end-area method is very commonly used for computing earthwork quantities because of its simplicity. It is not, however, a theoretically exact method unless the two end areas are equal, but the errors are not usually significant. Should one of the areas approach zero, as on a hillside where the cross section is running from cut to fill, the error will be rather large. For this case it may be well to calculate the volume as a pyramid, with $V = \frac{1}{3}$ the area of the base times the height.

Although the average-end-area method is approximate, the precision obtained is fairly consistent with the precision obtained using the field measurements made for the cross sections. The costs per cubic yard for earthwork are usually low, and thus it is not normally justifiable economically to make refinements in earthwork calculations. Also, in most cases the method gives volumes on the high side, which is in favor of the contractor. To improve the accuracy of the average-end-area method, it is necessary to decrease the length between stations. This is particularly desirable if the ground surface is very irregular. Sometimes when a road has a very sharp curve and large cuts or fills, adjustments may be made for curvature in making volume calculations, but usually such adjustments are not considered significant.

Example 18-1 illustrates the computation of earthwork between two stations using the average-end-area method.

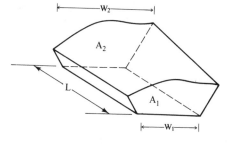

Figure 18-10

Example 18-1

Determine the earthwork volume between stations 100 and 101 using the average-end-area method if the road width is 30 ft. The cut side slopes are 2 horizontally to 1 vertically. Figure 18-11 shows a sketch of the cross section at station 100 + 00.

Station	Left	Center	Right
100 + 00	$\dfrac{C8.0}{31.0}$	$\dfrac{C6.0}{0.0}$	$\dfrac{C5.0}{25.0}$
101 + 00	$\dfrac{C6.0}{27.0}$	$\dfrac{C4.0}{0.0}$	$\dfrac{C2.0}{19.0}$

Solution

$$\text{Area at station } 100 + 00 = \frac{w}{4}(h_\text{L} + h_\text{R}) + \frac{C}{2}(d_\text{L} + d_\text{R})$$

$$= \left(\frac{30}{4}\right)(8.0 + 5.0) + \left(\frac{6.0}{2}\right)(31.0 + 25.0)$$

$$= 265.5 \text{ sq ft}$$

Area at station 101 + 00

$$= \left(\frac{30}{4}\right)(6.0 + 2.0) + \left(\frac{4.0}{2}\right)(27.0 + 19.0) = 152.0 \text{ sq ft}$$

$$V = 1.85(A_1 + A_2) = (1.85)(265.5 + 152.0) = \textbf{772.4 cu yd}$$

Volume by Prismoidal Formula

Should the ground surface be such that two adjacent end areas are quite different from each other or should a high degree of precision be desired in the calculations, as where rock quantities or volumes of concrete are being determined, the average-end-area method may not be sufficiently accurate.

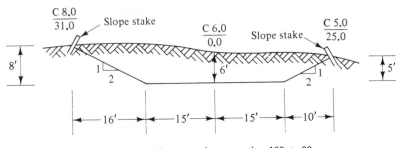

Figure 18-11   Cross section at station 100 + 00.

Actually, most earthwork volumes with which the surveyor deals are prismoids. A *prismoid* is a solid figure that has parallel end faces, and those end faces have the same number of sides. The prismoidal formula that follows was developed by applying Simpson's one-third rule (discussed in Section 12-12):

$$V = \frac{L}{27}\left(\frac{A_1 + 4A_m + A_2}{6}\right)$$

In this expression A_1 and A_2 are the areas of the cross sections at the ends or bases of the prismoid, while A_m is the area of a section halfway between the two ends and L is the distance between the two end cross sections. These values are shown in Fig. 18-12. *The area of the middle section A_m is determined by taking cross sections there or by averaging the dimensions of the end cross sections and using these values to calculate the area. It is not determined by averaging the end areas.*

The prismoidal formula usually yields smaller volumes than does the average-end-area method. Its use is probably justified only when cross sections are taken at very short intervals and where the areas of successive cross sections are quite different. When earthwork is being contracted for, the method to be used for computing volumes should clearly be indicated in the contract. Should no mention be made of the method to be used, the contractor will in all probability be able to require that the owner use the average-end-area method—whatever the owner's intentions may have been.

Example 18-2 illustrates the application of the prismoidal formula.

Example 18-2

Using the prismoidal formula, compute the earthwork volume between stations 100 and 101 using the data of Example 18-1.

Solution   We determine cross-section dimensions at station 101 + 50 by averaging dimensions at stations 100 and 101:

Station	Left	Center	Right	Area by Three-Level Section Formula
100 + 00	$\dfrac{C8.0}{31.0}$	$\dfrac{C6.0}{0.0}$	$\dfrac{C5.0}{25.0}$	265.5 sq ft
100 + 50	$\dfrac{C7.0}{29.0}$	$\dfrac{C5.0}{0.0}$	$\dfrac{C3.5}{22.0}$	206.2 sq ft
101 + 00	$\dfrac{C6.0}{27.0}$	$\dfrac{C4.0}{0.0}$	$\dfrac{C2.0}{19.0}$	152.0 sq ft

$$V = \frac{L}{27}\left[\frac{A_1 + 4A_m + A_2}{6}\right) = \left(\frac{100}{27}\right)\left(\frac{265.5 + 4 \times 206.2 + 152.0}{6}\right)$$

$$= 766.9 \text{ cu yd}$$

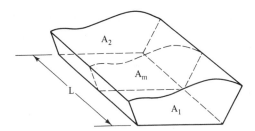

Figure 18-12

Although the prismoidal formula does provide the true volumes of prismoids, the average-end-area method is more commonly used because the difference between the two methods is usually quite small except where abrupt changes in cross sections occur. Furthermore, it is somewhat tedious to apply the prismoidal formula because of the extra work involved in computing the average dimensions and the area of the center cross sections. Unless rock excavation or concrete quantities are being computed, use of the prismoidal formula is not justified anyway because of the low precision of the work usually obtained in cross sectioning. Finally, if it is desired to determine the prismoidal volume, it is easier to use the average-end-area method and correct the results obtained with the *prismoidal correction* formula. This correction expression is accurate for three-level sections and is reasonably accurate for most other cross sections.

In the expression to follow for C_V, the prismoidal correction, C_1 and C_2 are the center cuts or fills at the end cross sections A_1 and A_2, while w_1 and w_2 are the distances between slope stakes (see Fig. 18-10) at these sections. The correction is in cubic yards.

$$C_V = \frac{L}{12 \times 27} (C_1 - C_2)(w_1 - w_2)$$

The correction calculated is subtracted from the volume obtained by the average-end-area method unless the formula yields a minus answer, in which case the correction is added. Example 18-3 shows the application of the prismoidal correction expression to the calculations of Example 18-1, which employed the average-end-area method.

Example 18-3

Correct the average-end-area solution of Example 18-1 to a prismoidal volume using the prismoidal correction formula.

Solution

$$C_V = \frac{L}{12 \times 27} (C_1 - C_2)(w_1 - w_2)$$

$$= \frac{100}{(12)(27)} (6.0 - 4.0)(56.0 - 46.0) = 6.17 \text{ cu yd}$$

Prismoidal volume $= 772.4 - 6.17 = \textbf{766.2 cu yd}$

18-7 MASS DIAGRAM

For highway and railway construction it is desirable to make a cumulative plot of the earthwork quantities (designating cuts plus and fills minus) from one point to another (perhaps the beginning to the ending points). Such a plot, illustrated in Fig. 18-13, is called a *mass diagram*. It is usually plotted directly below the profile of the route. The ordinate at any point on the diagram is the cumulative volume of cut and fill to that point, while the abscissa at any point is the distance in stations along the survey line from the starting point.

From a mass diagram it can readily be seen whether the cuts and fills balance throughout the job and how far the earth will have to be hauled. If there is an excess of fill needed beyond the cut quantities, it will be necessary to obtain earth from other sources, such as borrow pits. If there is more cut than fill, it may be necessary to dump the extra fill as waste, or perhaps widen the planned fills or make the valleys shallower.

When material is excavated and placed in trucks or pans or other earth-moving equipment, it will normally occupy a larger volume than it had in its original position. When solid rock is broken up, it may take up as much as two times its original volume.

When the excavated material is rolled down in thin layers in an embankment at an optimum moisture content such that the greatest compaction is achieved, the resulting fill volume will be appreciably less than the cut quantities. Thus for most fills (other than those consisting of rock) it takes more cut volume than fill

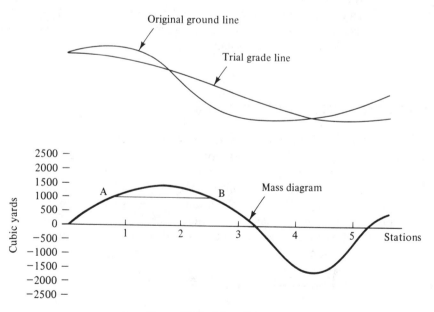

Figure 18-13   Mass diagram.

volume. The excess may vary from 5 to 20% depending on the character of the material involved. If a fill is made in a marshy area, the original material underneath will settle appreciably, thus requiring even more fill. As a result, it is necessary to make use of the so-called *balance factor* in figuring cut and fill quantities. A value of about 1.2 is frequently used: that is, about 1.2 yards of cut make 1 yard of fill. For rock excavation the balance factor will be less than 1.0, to allow for swell from cut to fill. This factor needs to be accounted for in constructing the mass diagram.

Referring to Fig. 18-13, the existing or original ground profile is plotted and a trial grade line (shown in the figure) is drawn. In drawing the trial grade line an effort is made to balance the cut and fill quantities visually. Earthwork volumes are calculated as described in previous sections of this chapter and the mass diagram is drawn taking into account the estimated balance factor. If the cuts and fills do not balance well or if the earth has to be transported or hauled long distances, the grade line will have to be adjusted and the process repeated.

Earth grading contracts refer to a certain distance as being a *free haul* distance. It may be specified as being 500 ft, 1000 ft, 2000 ft, or some other value. If the earth is not moved more than this distance, the contractor will be paid the standard price per cubic yard. If it is moved more than this distance, the contractor will be paid extra as specified in the contract. The extra hauling is referred to as *overhaul*. The unit of measurement usually used for overhaul is the *station yard* where one station yard indicates the hauling of 1 cubic yard of material for one station.

From the figure it will be noted that the peaks on the diagram show where there is a change from cut to fill while the valleys show a change from fill to cut. If a horizontal line *AB* is drawn as shown in the figure, the cuts and fills between the two points will exactly balance.

18-8 VOLUMES FROM CONTOUR MAPS

Should an accurate contour map be available for an area being studied it can be used for computing earthwork volumes as described here. It is quite practical to estimate earthwork quantities from such a map. For this discussion, reference is made to the 5-ft contour interval map of Fig. 18-14. It is assumed that the hill shown is to be graded off to elevation 515. The areas within the 525, 520, and 515 contour lines can be easily determined with a planimeter. From these values the volume of earth to be removed can be calculated from the following average end-area expression:

$$V = (5) \left(\frac{A_{525} + A_{520}}{2} \right) + (5) \left(\frac{A_{520} + A_{515}}{2} \right)$$

$$+ \text{ earth volume above the 525 contour line}$$

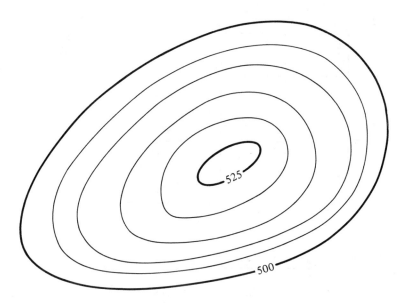

Figure 18-14

PROBLEMS

18-1. For the accompanying sketch and readings with the level placed at the center line of the road, would no. 1, no. 2, or no. 3 be at the correct location for the slope stake?

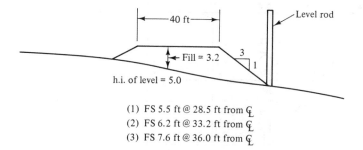

    (1) FS 5.5 ft @ 28.5 ft from $\mathcal{C}_L$
    (2) FS 6.2 ft @ 33.2 ft from $\mathcal{C}_L$
    (3) FS 7.6 ft @ 36.0 ft from $\mathcal{C}_L$

(*Ans.:* no. 2)

In Problems 18-2 to 18-5, find the volume in cubic yards of the excavation for the borrow pits shown. The numbers at the corners represent the cuts in feet.

18-2.

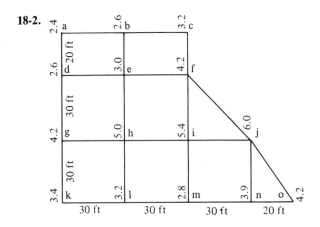

18-3.

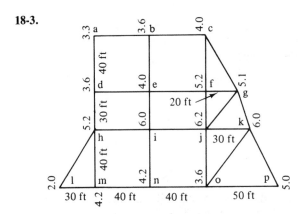

(*Ans.:* 2106.6 cu yds)

18-4.

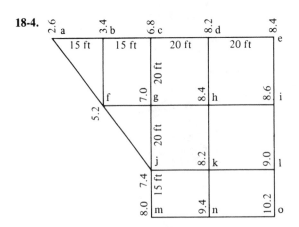

18-5.

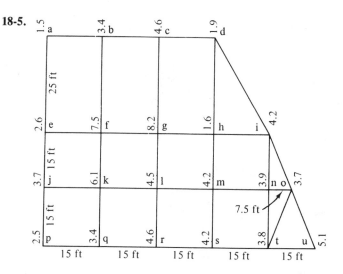

(*Ans.:* 548 cu yds)

18-6. An area has been laid out in 40-ft squares as shown in the accompanying illustration and the ground elevations are as indicated on the sketch. How many cubic yards of cut are required to grade the area level to elevation 40.0 neglecting the balance factor?

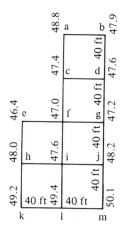

18-7. An area has been laid out as shown in the accompanying illustration. The elevations of the corners are indicated on the sketch. Determine how many cubic yards of cut are required to grade the area level to elevation 60.0 neglecting the balance factor.

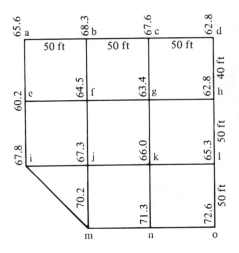

(*Ans.:* 4458 cu yds)

18-8. An area has been laid out as shown in the accompanying illustration and the elevations of the corners are as indicated on the sketch. How many cubic yards of cut are required to grade the area level to elevation 60.0, neglecting the balance factor?

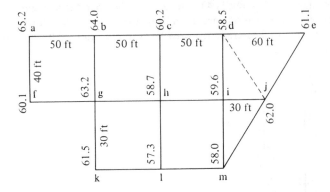

18-9. The following notes are for cross sections at stations 67 and 68. If the width of the roadbed is 30 ft, calculate the area of the two cross sections.

Station	Cross Sections		
67	$\dfrac{C6.2}{27.4}$	$\dfrac{C3.4}{0.0}$	$\dfrac{C1.3}{17.6}$
68	$\dfrac{C10.8}{36.6}$	$\dfrac{C4.8}{0.0}$	$\dfrac{C2.6}{20.2}$

(*Ans.:* 132.75 sq ft, 236.82 sq ft)

18-10. Compute the volume of excavation between stations 67 and 68 of Problem 18-9 using:
(a) The average-end-area method.
(b) The prismoidal formula.

18-11. The cross-section areas along a proposed dike as obtained with a planimeter are as follows:

Station	End Area (sq ft)
46	622
47	1466
48	962

Calculate the total volume of fill in cubic yards between stations 46 and 48 using the average-end-area method.

(*Ans.:* 8355 cu yd)

18-12. For a proposed highway the following planimetered areas in square feet were obtained. Using the average-end-area method, determine the total volumes of cut and fill between stations 84 and 87.

Station	Cut	Fill
84	86	
85	65	
85 + 50	0	0
86		38
87		64

CHAPTER NINETEEN

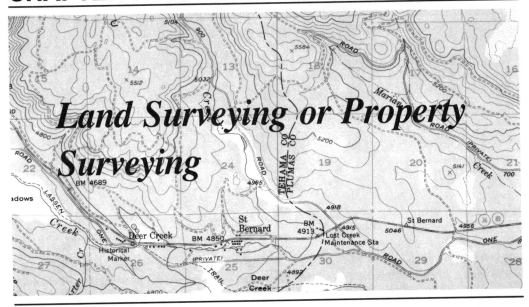

Land Surveying or Property Surveying

19-1 INTRODUCTION

Land surveying or property surveying is concerned with the location of property boundaries and the preparation of drawings (or plats) which show these boundaries. In addition, it involves the writing and interpretation of land descriptions involved in legal documents for land sales or leases. More precisely, land surveys are made for one or more of the following purposes:

1. To reestablish the boundaries of a section of land that has been previously surveyed.
2. To subdivide a tract of land into smaller parts.
3. To obtain the data needed for writing the legal descriptions of a piece of land.

The location of property lines began before recorded history and for all of the subsequent ages it has been necessary for the surveyor to reestablish obliterated land boundaries, establish new boundaries, and prepare boundary descriptions. The need for people to locate, divide, and measure land caused the development of the land surveying profession. Today, our expanding population, the demand for second homes, and the formation of many new industries are creating a demand for more and more property surveys.

In general, land prices have been climbing rapidly and the accurate location of property lines is becoming increasingly important. A century or two ago land was so plentiful that describing a distance such as ''as far as a fast horse can run in 15 minutes'' might have been adequate, but it is certainly no longer true. A few decades ago the compass and chain were satisfactory for making property surveys, but this is no longer the case. Today, the transit and steel tape are still in common use, but the situation is rapidly changing. Recent technical advancements, such as electronic distance-measuring devices, are commonly applied in the field all over the United States.

19-2 TITLE TRANSFER AND LAND RECORDS

Title to land may be transferred by deeds, wills, inheritance without a will, or by adverse possession (discussed in Section 19-8). A deed for a piece of land contains a description of the boundaries, the monuments, and important information concerning adjoining property.

For a deed to be legally effective, it must be recorded in the office of the proper public official. (Though recording is voluntary it provides public notice that someone claims an interest in the property.) Usually, this office is located in the county courthouse and called the Registry of Land Deeds, Recorder of Deeds, or similar name. At this office the information is copied into the records and is available for anyone to see. Legally, anyone can go to the office and see the records that show the description of anyone's land in the county and, in addition, can determine how much he or she paid for it.

These offices keep indexes so that it is reasonably easy to find the records desired. The indexes are kept chronologically and indexed in the name of both the grantor (the seller) and the grantee (the buyer). When the surveyor goes to this office, he or she is usually looking up the legal description of certain pieces of land. There the surveyor will find copies of the land description and generally copies of plats of the land which have been filed at the office.

Anyone who becomes involved in land purchases will want to be sure to obtain full or clear title to the land. Various attorneys and private title searching companies are by far better prepared to search these indexes for information relating to a title than is the ordinary individual. They will, for a fee, search the records and provide information regarding the title. The reader will readily understand the importance of such careful checks if simply by thinking for a moment about the legal problems that could arise if a person were to build a house on a lot believing that he or she had title to the lot but actually not having it. For many important property sales the purchaser may want the seller to provide title insurance guaranteeing clear title to the property before he or she will buy. Title companies usually combine or can arrange both services (title checks and title insurance).

19-3 COMMON LAW

Most of the law pertaining to property surveying is common law or, as it is frequently called, unwritten law (i.e., it is not statute law passed by political bodies). Common law is that body of rules and principles that has been adopted by usage from time immemorial. It has been formed by transforming custom into rules of law. When disputes between two or more parties arose in early days in England, the courts decided what to do from established custom. If situations occurred that did not seem to be adequately covered by custom, the judges based their decisions on their own ideas of right and wrong.

As an illustration of the kind of problems settled by these courts, consider the case in which a person excavated for a building foundation on his land and a nearby building on another person's land caved in because of the subsidence of the ground. The injured party went to court to ask for compensation for the loss. From such cases as these the principle evolved that each landowner owes lateral support to adjoining landowners.

Beginning with the Year Book of 1272 A.D., the English courts began to keep records of their decisions and have kept them continuously since. Since the beginning of these records, judges have searched through them to determine if the particular case had been decided earlier.

English common law is the basis of jurisprudence in 49 of the states of the United States. Louisiana was originally French and its legal foundation is based on the Roman common law. Even though the Romans occupied most of the British Isles for six centuries, England was the only European country to develop an independent system of common law. (English common law, however, did draw heavily from Roman common law.)

19-4 MONUMENTS

Corners are points that are established by surveys or by agreements between adjacent property owners. The usual practice is to mark these corners with a relatively permanent object called a *monument*. Monuments may be natural features such as rocks, trees, springs, and so on, or they may be artificial objects such as iron pipes driven into the ground, posts of concrete or stone, mounds of stone, wooden stakes perhaps with some more permanent material buried at their bases such as charcoal or glass, or (in areas of low rainfall) mounds of earth or pits. Although wooden stakes alone would seem inadequate as monuments because of their temporary nature, some courts have held that substantial wooden stakes may be so classified.

The surveyor must be reasonable in the method of marking corners being used. He or she may describe a particular corner in a manner that is perfectly clear to everyone at the time of marking, for example, the northeast corner of Joe Smith's barn. A few decades later, however, Mr. Smith's barn may be completely obliterated and there will be no one around who can prove where it was located. There are plats on record that describe a corner as being that spot where so and so shot a bear on such and such a date. Needless to say, it is quite a challenge to locate that point 5 or 50 years later. Clearly, that kind of description is a good start toward future litigation.

Just about anything may be used for monuments, but long iron pipes or concrete monuments are generally more satisfactory and may even be required by law in some areas. Whatever type of monument is used, it should be described carefully in the surveyor's notes on the plats.

The life of the land surveyor and the problems of landowners (small or large) would be much simpler if they both would understand and follow this important rule: *Establish property monuments so carefully that their obliteration is unlikely, but reference them in such a manner that they may be replaced easily and economically if they are obliterated.*

Unfortunately, monuments are frequently destroyed and resurveys are required to replace them. (Deliberate damage or destruction of monuments is prohibited by law.) In those places where monuments may easily be destroyed unintentionally (e.g., near busy streets or in construction areas), it is a good practice to use *witness corners*. These might be iron pipes (or other types of monuments) placed a convenient distance back along property lines in more protected locations. These monuments are also described in the surveyor's notes and are shown on the plats. Another place where witness corners are used is where the actual corners are located in places difficult of access, for example, in streams or lakes.

If the location of a property corner can be fixed beyond any reasonable doubt, it is said to *exist*. If its position cannot be found, it is said to be *lost*. On occasions when the monument used to mark the corner cannot be found, the corner is said to be *obliterated*. This does not necessarily mean that the corner is lost, because it may very well be possible to reestablish its original position.

Corners may often seem lost when they really are not. For instance, if the corners had been marked with wooden stakes, cutting off slices of earth with a shovel might reveal a change in soil color where the stake had decayed. If an iron pipe or concrete monument had been removed, the void space filled in with the surrounding soil might show a slight change in coloration. Among several other methods that may be used to find old corners are the locating of old fence corners, statements by neighboring people, the use of metal detectors to locate iron pipes, and so on.

19-5 BLAZING TREES

Another aid in locating property lines and corner monuments in forest areas is blazing and hacking trees. This practice was very common with surveyors in the past, but it is not as common today because so many people object to having their trees marked. A *blaze* is a flat scar made with an ax (or machete) on a tree at about breast height. The bark and a small amount of the tree tissue are removed and the tree is marked so that it can be identified for several decades. A *hack* or *notch* is referred to as a V-shaped cut in the trunk of a tree.

One common method of marking trees is described in this section, but the reader should realize that surveyors in various parts of the country mark trees in different ways. In the author's locality it was the practice along boundary lines to put two hacks on the sides of trees nearest the boundaries. Two hacks were used to distinguish them from accidental marks from other causes. The trees hacked were those which the surveyor could reach while walking along the line. If a tree was exactly on the boundary (sometimes called a *line tree*), it was marked with a blaze with two hacks underneath. These marks were placed on both sides of the tree in line with the boundary.

It was also the custom to place three hacks on trees near property corners. These hacks were placed on the trees facing the corner and at a distance above the ground equal to the distance from the particular tree to the corner. If a tree was the corner monument, it was marked with an × and three hack marks underneath it. Blazing trees may often aid the surveyor in locating property lines and corners. It is true, however, that in many cases lumber operations, forest fires, and other causes may have destroyed the marked trees.

19-6 THE LAND SURVEYOR—A SPECIALIST

Disputes over land boundaries are an everyday occurrence, as may be seen by checking through court records. Although the land surveyor may on occasion be involved in original surveys or in subdividing tracts of land into smaller tracts, a very large proportion of his or her work deals with resurveys in which the surveyor tries to reestablish old boundaries that have previously been surveyed.

The surveyor is faced with many problems in relocating old land boundaries. Incomplete data on plats and deeds, missing monuments on the ground, conflicting claims by adjoining property owners, and inaccurate original measurements greatly magnify a surveyor's problems. These are only a few of the problems besetting the land surveyor, but they show that for the person to do the work, he or she must be something of a specialist. As a result, land surveying is a profession that is to a large extent learned only after considerable experience in a particular locality.

Land surveying may be learned by a combination of formal study and field experience and by a thorough study of the laws pertaining to the subject. For a surveyor to become familiar with the conditions in a particular area, he or she must have a great deal of experience in that locality with regard to the methods of surveying used by the previous surveyors, court interpretation of land problems, and so on. As an example, the weight given to different items varies from state to state, for instance, the relative importance of monuments in place when they do not agree with the recorded values.

One frequent purpose of resurveys is to attempt to settle disputes between adjacent property owners as to where the lines should actually be. *In resurveying, the land surveyor's goal is to reestablish lines and corners in their original positions on the ground, whether or not those locations are in exact agreement with the old land descriptions. The surveyor's duty is not to correct old surveys but to put the corners back in their original positions.*

Too often the inexperienced surveyor thinks that land surveying merely involves the careful measurement of angles and distances. These measurements are actually only one means of reestablishing old boundaries. This comment is not meant to imply that careful measurements are not important but rather to remind the surveyor that his or her duty is to find where the original corners and boundaries were regardless of the precision with which the original survey was conducted.

The location of corners and property lines is determined from the *intent* of the parties to the original establishment of the boundary. The law is not concerned with their secret intentions but with their intent as expressed by the action of their surveyors as evidenced in plats, deeds, existing monuments, and so on. As time goes by it becomes more and more difficult to determine the original intent of the parties.

The reader should understand that the surveyor has no legal authority to establish boundary locations. On many occasions, however, he or she is called upon to relocate lines which are so difficult to find that the results are doubtful. In such cases as these, an agreement between the parties involved is probably the most sensible and economical solution. Property owners are always free to do this as long as their agreement does not affect some other party.

In most lawsuits over land boundaries, even the winner loses unless the land is extremely valuable, because of the cost and ill-will connected with such actions. If the surveyor is able to bring about an amicable agreement between the two parties, he or she will probably have served them both from an economic viewpoint and perhaps even more in preserving friendship. An experienced land surveyor is probably in a better position to suggest a just settlement of this type of problem than is a court of law. If the surveyor is able to persuade the landowners to compromise in a boundary dispute, he or she must survey the new line and the landowners should then go through a formal and recorded acceptance of the line.[1]

19-7 MONUMENTS, BEARINGS, DISTANCES, AND AREAS

Among the factors involved in relocating boundaries are existing monuments, adjacent boundaries, bearings, distances, and areas. In considering the relative importance of these items in reestablishing property boundaries, natural monu-

[1] A. H. Holt, "The Surveyor and His Legal Equipment," *Transactions of the ASCE*, 1934, vol. 99, pp. 1155–1169.

ments are given the greatest preference. Artificial monuments are given the next-highest preference. It is thought that natural objects (springs, streams, ridge lines, lakes, and beaches) offer a greater degree of permanence (they are less likely to be destroyed or relocated) than do artificial monuments such as set iron pipes, concrete markers, or stones. Of course, some people will deliberately move one of the latter monuments in order to better their land position. Other people, not realizing the importance of these monuments, will pick them up and take them home for their own use. Occasionally, concrete monuments are found being used as doorsteps at nearby houses. A person who deliberately moves a property monument is breaking the law and may be fined and/or imprisoned.

The courts feel that because the original owners could *see* the monuments, they more nearly express the original intent of the parties to land agreements than do measurements of directions and distances, which are subject to so many errors and mistakes. After natural and artificial monuments come adjacent boundaries and then bearings and distances, generally in that order. It should clearly be noted that bearings and distances cannot control the reestablishment of land boundaries if the monuments that are in place actually show the original boundaries. Bearings and distances can control the outcome, however, if there have been mistakes in the placing of the monuments or if the monuments are lost.

Land area is considered the least important of the factors listed, unless the area is the very essence of the deed as, for example, when an exact amount of land is clearly conveyed. A layperson may be puzzled by the following description commonly used in deeds regarding acreage: "26 acres more or less." The purpose of the words "more or less" is to indicate that all of the land within the specified boundaries is being conveyed, even though the area stated may vary greatly from the actual area. *This does not mean that the surveyor can do a sloppy job and get by with it by using such a term as "more or less." The surveyor is responsible for the quality of his or her work, and if it is not up to the standard expected of a member of the profession, he or she is liable for any damages that result.*

19-8 MISCELLANEOUS TERMS RELATING TO LAND SURVEYING

In this section are defined several terms that frequently occur in dealing with the transfer of land title.

Adverse Possession or "Squatters Rights"

If a person occupies and openly uses land not belonging to him or her for a specific length of time and under the conditions described by the laws of the state, he or she may acquire title to the land under the U.S. doctrine of adverse possession. The possession must be open and hostile and is usually for a period of 20 years but perhaps less under certain conditions, as, for example, when the original title to the land was not clear.

Even though a surveyor is able to reestablish without question the original boundaries of a tract of land, it is possible that because of adverse possession the original lines are no longer applicable. If the owner fails to act during the designated time, he or she loses the right to act. It is generally held that a private citizen does not have the right to acquire government land by adverse possession. If the owner of the land gives permission to another person to occupy the land, that person can never acquire title to the land by the adverse possession doctrine no matter how long he or she uses the land after the permission is given.

Riparian Rights

The rights of persons who own property along bodies of water are referred to as riparian rights. Following are a few comments concerning these rights as they apply to property lines. Creeks, rivers, and lakes are very natural and convenient boundaries between various tracts of land because they are easy to describe and are easy to locate. For small streams that are not navigable, property lines usually go to the center of the stream or to the "thread of the stream." The thread of the stream is generally thought of as being the center line of the main channel.

The U.S. Supreme Court has decided that it is up to the individual states to decide where private property lines run along navigable streams. Some states have decided on the center lines of streams, some on the threads of the streams, and others have selected the high water marks (those marks to which the water has risen so commonly that the soil is marked with a definite character change as to vegetation, and so on).

Erosion and Avulsion

Although a stream or river is an excellent boundary, it cannot always be depended upon to stay in the same place. Its position may shift so very slowly and imperceptibly as a result of erosion, current, or the force of waves that the property owners cannot recognize the short-term change in position. When bodies of water change by this method, property lines are held to move with the change. However, if the change is very sudden and perceptible, as, for example, when a large amount of soil is suddenly moved from one landowner to another or when a stream changes its bed completely, it is called *avulsion*. When avulsion occurs, property lines are held not to change from their original positions. This description seems simple enough, but unfortunately, many years after the fact it is difficult to tell whether a change was caused slowly or suddenly.

On some occasions we cannot even find the stream. Such a case might occur where beavers have built a continuous series of dams, making the entire area a swamp.

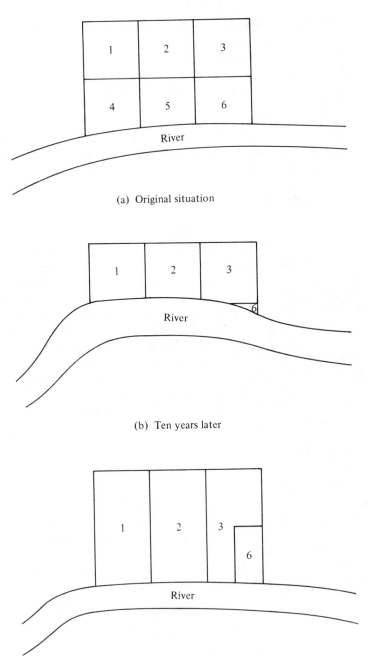

(a) Original situation

(b) Ten years later

(c) Twenty years later

Figure 19-1   Possible ownership changes as water boundaries change (depending on state).

Accretion and Reliction

When water slowly and imperceptibly deposits materials on the bank of a stream or other body of water, it is called *accretion*. Should a body of water recede, as when it dries up partly or wholly, the land area adjoining it is increased and this is called *reliction*.

When land areas change by accretion or reliction the courts will attempt to establish the new boundaries of the adjoining tracts of land in an equitable manner. If water frontage is involved, it may be of the greatest importance in the decisions made. They may divide the water frontage in some proportion to the frontage of the old boundaries.

Further Comments on Riparian Rights

There are so many complicated problems that can arise regarding riparian rights that a special body of law has evolved on the subject. Some of the numerous problems that may be encountered are related to property lines, navigation, docks, water supply, fishing rights, oyster beds, artificial fills or cuts, erosion, accretion, and many others. The laws pertaining to these situations vary quite a bit from state to state, and court decisions in different states have been entirely different for similar cases.

One illustration is presented here to show how involved the legal situation can become concerning riparian rights. For this discussion reference is made to Fig. 19-1, where both erosion and accretion are involved. The discussion presented here can apply to the legal situation in some states, whereas it may not apply to others.

The purpose of this discussion is to show how complicated the legal problem may be. In part (a) of Fig. 19-1, six lots are shown. The lots numbered 4 through 6 bordered the river and had riparian rights. During the next 10 years lots 4, 5, and most of 6 eroded away, leaving the situation shown in part (b) of the figure. At this time lots 1, 2, and the parts of 3 and 6 bordering the river have water rights. In the next 10 years it is assumed that the river gradually moved back to its original position and the replaced land now could belong to lots 1, 2, 3, and 6, as shown in part (c) of the figure. A similar discussion applies to beachfront property in some states.[2]

19-9 RESURVEYS

The average client of a surveyor has no idea what a time-consuming and hair-pulling job a resurvey can be. Ideally, the original surveyor painstakingly measured distances and directions and carefully set monuments and witnessed them

[2] P. Kissam, *Surveying for Civil Engineers*, 2nd ed. (New York: McGraw-Hill Book Company, 1981), pp. 327–328.

with equal care. Sadly, the truth is that often the monuments were wooden stakes placed at points that were located by haphazardly determined distances and bearings. The result is that there is no problem in surveying so difficult and requring so much patience, skill, and persistence as that of relocating old property lines.

If, as stated in Section 19-6, the old lines and monuments can be reestablished, they still govern the boundaries no matter how poorly the old survey was performed. For a surveyor to reestablish old property lines, he or she must follow in the footsteps of the original surveyor. To be able to do this, he or she must have a good idea of how the original survey was performed. Was a compass or transit used to measure the angles? Were distances measured on slopes or horizontally?

When the surveyor cannot definitely relocate boundaries, he or she has no power to establish them. If there are disputes between adjacent landowners over boundaries and if the surveyor cannot persuade the owners to compromise, the courts will have to settle the matter. In court the surveyor can only serve as an expert witness and present the evidence that he or she has found. Once a dispute is settled, whether by mutual agreement or by court action, the surveyor should do all he or she can to see that a precise survey is made of the settled boundaries and good monuments established.

To begin a resurvey the surveyor carefully studies available plats and deeds of the property as well as those of adjacent tracts. As a part of this study he or she typically calculates by latitudes and departures the precision obtained in the original survey of the tract in question. Surprisingly enough, the surveyor may frequently spend more than half of his or her time studying and planning, while the actual field work may not take very long at all.

Clients often do not understand the amount of research that will be needed to conduct a property survey properly. In this regard the surveyor is well advised to inform his or her clients of the potential time and resulting costs that may be needed for such research so that an accurate survey can be made. It may be necessary to study county records, to interview neighboring property owners, to contact other surveyors, and so on. In this regard innumerable land surveys have been made in the United States—monumented, platted, and described but never properly recorded. These surveys can provide important information as to the accurate location of land tracts, and they can be a real source of trouble if they are neglected or not discovered.

Next, the surveyor begins a careful examination of the tract of land in the field. If the original survey was done precisely and if one or more of the original corners can be found, the survey can be handled just as was the five-sided traverse used for an example in Chapters 4 and 10 in which the distance and angle measurements were made.

The trouble, however, is that so many original surveys contain major errors and mistakes, and often several or all of the monuments are gone. One frequent cause of poor measurements was the equipment used, often the compass and link chain. If the original survey was made with a surveyor's compass, the surveyor may attempt to rerun the lines by using a compass and by making proper allowance for the estimated change in magnetic declination since the time of the original survey.

If a surveyor can definitely establish the corners at the ends of at least one line, he or she is off to a good start in the entire relocation job. He or she can measure that line and compare its distance with the original measurement and thus obtain an idea as to how the original value was obtained. Was it measured on the slope or horizontally? By writing the proper proportions he or she can compute proportional lengths of the other sides. The surveyor can determine the astronomic bearing of the known side (as described in Chapter 16) and from that value compute the estimated magnetic declination at the time of the original survey. With this magnetic declination he or she can compute the estimated astronomic bearings of the other sides.

With these computed distances and bearings the surveyor can start running the sides. At each estimated corner location he or she will carefully search for evidence of the original monuments. If the surveyor finds one of the old monuments, he or she sets a new one (if necessary) and carefully references it. If unable to find a monument, he or she sets a temporary point and moves on to the next line. If the surveyor finds monuments farther along, he or she tries to work back to those that he or she couldn't find by using new proportions based on the found monuments. The surveyor continues in this fashion until he or she locates all the old monuments or sets temporary monuments at all of the various corners. Several ideas will be helpful to the surveyor when looking for old corners: hacked trees, old fence lines, roadbeds, places where vegetation is different, and so on. After studying the gathered information carefully, he or she may very well return to the field for more measurements.

In a resurvey the surveyor may initially be able to find only one corner or perhaps, as is often the case, none at all. If only one corner is evident and the original survey was run with a compass, he or she can estimate the magnetic declination at the time of the survey, convert the bearings to estimated astronomic bearings, and try to run the lines as described in the preceding paragraph.

If no corners are evident, the surveyor will probably begin his or her work by studying the records of the adjoining land tracts and try to run their lines in order to see if he or she can locate some of their monuments.

When the surveyor has finished working on any of the cases mentioned, he or she will give the client his or her best judgment as to the location of the lines and corners based on his or her opinion as to the true intent of the parties to the original survey.

19-10 METES AND BOUNDS

The oldest method of land surveying is the metes and bounds system. (The term *metes* means to measure or to assign a measure, and the term *bounds* refers to boundaries.) Thus a metes and bounds description gives the measurements and limits of the outside boundaries of the tract of land in question. The length and direction of each of the sides of a tract of land are determined and monuments are established at each of the property corners. Creeks, lakes, or other natural landmarks are occasionally used to define property corners. Most land surveys in the original 13 states of the United States, in Kentucky and Tennessee, and in some surveys elsewhere were made by the metes and bounds method.

A *plat* is a dimension drawing that shows the data pertaining to a land survey and is primarily a legal instrument becoming a matter of public record. It provides the information necessary for finding, describing, and preparing a description of the land.

In Fig. 19-2 the author has presented a portion of a plat. In addition to the information shown, there will a title box, a scale, and a legend or any necessary special notes. Any nonstandard symbols used in the drawing and other special information will be included. The surveyor will provide a statement as to the precision of the field survey and the method used for determining the land area. Furthermore, in areas where houses or other buildings are located or are to be constructed, a note should be given as to whether or not the land is above or below the floodplain. Finally, deedbook numbers and page numbers in the court-

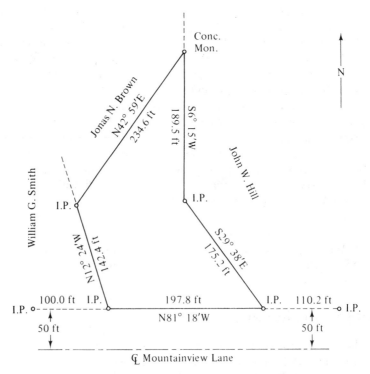

Figure 19-2

house are often given wherein previous owners of the property are identified and previous descriptions of the property presented.

Property descriptions are usually written by lawyers, real estate persons, or surveyors. The accuracy of their descriptions is of the utmost importance because one simple mistake in their writing may lead to property disputes between adjoining landowners for several generations. A person who writes the descriptions for land should try to put himself or herself in the shoes of someone else 5, 10, or 100 years later who will try to interpret the description as to ground location. Following is a typical metes and bounds deed description (see also Fig. 19-2).

"All of that certain piece, parcel or tract of land situated, lying and being in state and country aforesaid about ½ mile west of the town of Tamassee, and lying on the north side of Mountainview Lane. Beginning at an iron pin at the southwesterly corner of the property of the grantor located 50 ft from the center line of Mountainview Lane and running thence along the easterly line of the William G. Smith property N12°24'W 142.4 ft to an iron pin; thence with the southerly line of the Jonas N. Brown property N42°59'E 234.6 ft to a concrete monument; thence _____to the point of beginning. All bearings are referred to the astronomic meridian, the tract contains 0.83 acres more or less, and is shown on a plat dated October 16, 1981, drawn by Arthur B. McCleod, registered land surveyor #1635, state of _____.

19-11 THE U.S. PUBLIC LAND SYSTEM

In colonial America, the manner of obtaining land from the governments involved (Dutch, Spanish, English, French, etc.) and the methods of describing the land grants varied widely. The boundaries of these tracts of land consisted of streams, roads, fences, trees, stones, and other natural features. The tracts were generally irregular in shape except for some subdivision in towns and cities. Furthermore, no overall control method for the surveys was available. The legal descriptions of the land were often vague and the measurements involved contained many mistakes. As a result of all these factors, when boundaries were obliterated it was and is difficult, if not impossible, to restore their original positions.

Because of the problems involved in the early colonies the Continental Congress in 1785 established the U.S. Public Lands Survey System with the objective that the same mistakes would not be repeated for the remaining land which the federal government possessed. This land, called the *public domain*, is the land held in trust for the people by the federal government. About 75% of the land in the United States was once part of the public domain. Our congress apparently thought that the sale of this land would yield sufficient funds to pay off the public debt.

The first survey was begun on September 30, 1785, near Liverpool, Ohio, under the direction of Thomas Hutchins, who was the official geographer of the

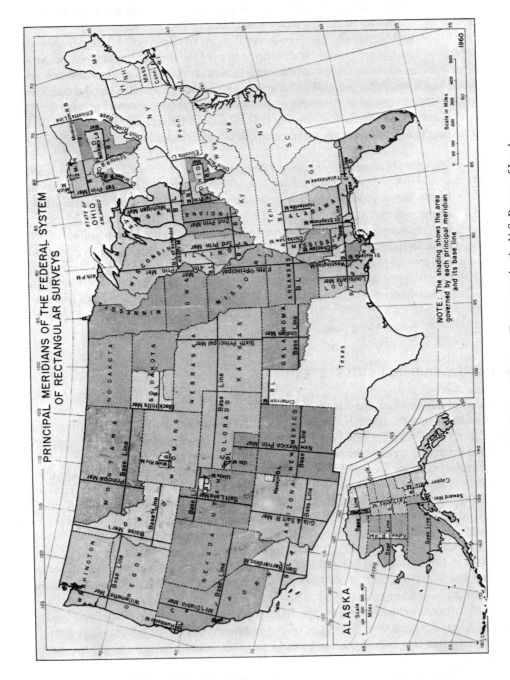

Figure 19-3 U.S. Public Land System. (From a map by the U.S. Bureau of Land Management.)

United States. This point of beginning is now marked with a national historic monument.

Today, approximately 30% of the country's land is still in the public domain. Of the approximately 2.3 billion acres of land in the 50 states, over 1.35 billion have been surveyed by the public land survey system. There are about 350 million acres in Alaska which are unsurveyed. Hawaii is not included in the public land survey system.

The U.S. Public Land System, which is a rectangular system, has been used to subdivide the states of Alabama, Alaska, Florida, and Mississippi as well as all the states to the west and north of the Mississippi and Ohio Rivers except for Texas. Actually, Texas has a system similar to that of the U.S. Public Land Survey, but it was affected by the Spanish settlers before its annexation into the United States. Figure 19-3 shows the parts of the country covered by the U.S. Public Land System.

It would certainly have been ideal if all the land of the United States could have been obtained at one time and set up under one survey system. Although this was not possible, the country is indeed fortunate to have so much of its land under the public land survey system. Although the system does have its problems, it has nevertheless been a tremendous asset to the country. The success of the system can be verified by the fact that the original procedures have not been greatly changed up to the present day.

With the public land survey system, each parcel of land, whether it be $2\frac{1}{2}$, 5, 10, 40, or 160 acres, is described in such a manner that the description will not apply to any other parcel of land in the entire system. This simplicity of describing land tracts has made the system one of the most practical methods ever devised for land identification and description. When the rectangular system was introduced it was not applied to the original 13 states because of the enormous problems involved in changing the existing descriptions of the countless thousands of land tracts involved.

Most of the property surveys in the original 13 states, as well as Kentucky and Tennessee, were made by separate closed traverses, that is, by the metes and bounds system described in Section 19-10. There is no definite overall control system in those areas, and very often some boundaries, such as the edges of lakes or streams, were not even measured but were merely stated as the boundaries.

19-12 EARLY DAYS OF THE SYSTEM

The purpose of the public land survey was to devise a rectangular system and establish it on the ground so that it would provide a permanent basis for describing land parcels. In 1796 the post of Surveyor General was established at an annual salary of $2000; the first appointment was given to General Rufus Putman, an experienced surveyor and aide to George Washington during the Revolutionary War. In 1812 the General Land Office (now the Bureau of Land Management) was established to manage, lease, and sell the vacant public lands of the United States. To identify and describe the land involved, it was of course first necessary to survey it.

At that time there was a great demand for land, yet its price was very low (about $1.25 per acre for "homesteads" of 160 acres each). For such low prices it was impossible to justify very accurate surveys. The surveying instruments available were rather crude in today's terms and the survey points were marked with wooden stakes, mounds of earth placed over buried pieces of charcoal, or pieces of stone. Many of these monuments have been destroyed through the years.

The surveys were handled on a contract basis and sometimes different rates were paid depending on the relative importance of the lines and on their characteristics (as wooded or swampy or hilly). The surveyors were paid about $2 per mile until 1796 and about $3 per mile thereafter. The amount of money surveyors could make depended entirely on how fast they could complete a survey.

Although the maximum permissible errors were specified from the early days of the system, the standards were vague and little supervision or checking was done to see if the work met the requirements until the 1880s. Another reason for large errors and mistakes in the system was the speed with which some surveys were handled in Indian territory (where often little field work was actually carried out).

Despite all of these problems a large part of the public land survey was handled very well. Generally, inaccuracies in the old surveys cannot be corrected if there is evidence of the location of the original land corners. In other words, once corners are established and used to mark property boundaries, they cannot be changed regardless of the magnitude of the mistakes made.

19-13 OUTLINE OF THE SYSTEM

This section presents a quick summary of the procedure involved in the U.S. public land survey system, and subsequent sections present a more detailed description of the system. The area is divided into smaller and smaller approximate squares as follows:

1. The land is divided into *quadrangles* approximately 24 miles on each side (see Fig. 19-7).
2. The quadrangles are each divided into 16 *townships*, the sides of which are approximately 6 miles (see Fig. 19-8).
3. The townships are divided into 36 sections, each approximately 1 mile square (see Fig. 19-9).
4. The sections are further subdivided into quarter sections and smaller sections, as will later be described (see Fig. 19-10).

Since meridians converge to the north, it is impossible for all townships, sections, and so on to be in the shape of exact squares. In the subdivision of townships it is, however, desirable to lay out as many sections as possible in 1-mile squares. To do this, convergence errors are thrown as far west as possible by running the lines parallel to the eastern boundary. In a similar manner the errors in distance measurement are thrown as far to the north as possible by locating the monuments at 40-chain (or ½-mile) intervals along the lines parallel to the eastern boundary so that any accumulated error will occur in the most northerly half mile.

19-14 MEANDER LINES

Navigable waterways or streams of three or more chains in width, as well as lakes covering 25 acres or more (except those formed after the state in question was admitted to the union) are not part of the public domain and thus were not surveyed nor disposed of by the federal government. The individual states have sovereignty over such bodies of water.

The traverses of the margins of these bodies of water are called *meander lines*. These traverses consist of straight lines that conform as closely as possible to the mean high water marks along the banks or shore lines involved. Meander lines were not run as boundary lines and when the bed of the lake or stream changes the high water marks and the property lines change also.

19-15 DETAILED SUBDIVISION

From 1815 to 1855 the subdivision of the public lands were handled in accordance with instructions from the office of the Surveyor General. Since 1855, however, the detailed directions for subdivision have been given in the *Manual for Instructions for the Survey of Public Lands of the United States* published by the U.S. Government Printing Office. This manual has been revised at various intervals since its original publication. When surveying is being done in a particular area it is necessary to be familiar with the instructions in effect when that land was originally surveyed. The following paragraphs contain a detailed description of the procedure used in subdividing the public land.

Initial Points

Beginning points called initial points were established in each area. They were selected with the intention of controlling large agricultural areas within reasonable geographical limitations. Their locations were determined on the basis of astronomical observations. There are 37 initial points, 5 of which are in Alaska.

Principal Meridians

Astronomic meridians called principal meridians were run through each initial point and extended as far as necessary to cover the area involved. They were identified by number or by name as, say, the sixth principal meridian, which is in Nebraska and Kansas or the Willamette Meridian, which is in Washington and Oregon. These are shown in Fig. 19-3.

As the principal meridians were run, monuments were set at intervals of 40 chains and at the intersection with meander lines. The directions of the meridians were usually determined from solar attachments with which true directions could be mechanically obtained. Supposedly, the directions obtained for the meridians were to be within 3' of astronomic directions. The solar compass was developed by William A. Burt and introduced in 1836. In 1850 the Surveyor General specified that surveyors were to use Burt's improved solar compass or its equivalent for public land surveys. This was a much better compass for determining astronomical north than were the usual magnetic compasses.

With the Manual of 1890 the use of the magnetic compass was forbidden except for subdivision and meandering, and in those situations it could supposedly be used only in areas that were free of local magnetic attractions. With the Manual of 1894 the use of the magnetic compass was discontinued for all new public land surveys. In attempting to rerun old lines or "follow in the footsteps of the original surveyors," the regular magnetic compass is still frequently used.

Base Lines

Through each initial point, lines are run out at 90° to the principal meridian and extended as far to the east and/or west as required to cover the area involved. These lines, referred to as base lines, are run as astronomic parallels of latitude. They are, therefore, curved lines and are run by laying off 40-chain chords along the curves. Three different methods may be used for laying out the base lines: the solar, tangent, and secant methods.

In the *solar method* astronomic north is measured at each 40-chain interval, usually with a solar attachment; a 90° angle is measured from astronomic north and another 40 chains is laid off and the process is repeated. This procedure is illustrated in Fig. 19-4.

With the *tangent method* a 90° angle is laid off to the east and/or west as required from the principal meridian and a straight tangent line is laid off for 480 chains or 6 miles, as shown in Fig. 19-5. Corners are set at 40-chain intervals. Since this is a straight line, corrections will have to be made to set the corners correctly on the curved base lines. The corrections are made by offsets from the originally set straight line. The magnitudes of the offsets are available in the *Standard Field Tables* of the U.S. Bureau of Land Management. They are given in links (1 link = $\frac{1}{100}$ × 66 ft = 0.66 ft) and run as high as 37 links at the 480-chain corner. The offset corrections, a few of which are shown in Fig. 19-5, are

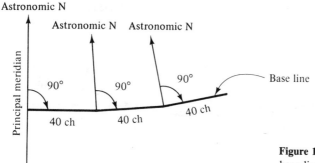

Figure 19-4   Solar method of laying off base lines.

measured in astronomic north directions to the base line. The errors resulting for this correction procedure are thought to be negligible.

Finally, the *secant method* of laying out base lines is illustrated in Fig. 19-6. This method is quite similar to the tangent method. However an angle of less than 90° is measured from the meridian and a straight line is run as shown in Fig. 19-6. Since the rate of curvature of base lines depends on the latitude, the sizes of the angles laid off from the meridian vary. Their values are available in the *Standard Field Tables*. It is necessary to set a point south of the beginning corner at a distance specified in the tables. Then the correct angle is measured from the meridian east and/or west and projected for 480 chains or 6 miles. The offset corrections are then measured to the north or south of the projected lines as required (see Fig. 19-6). These offsets are relatively small compared to some of the tangent offsets. Therefore, this method, with its smaller offsets, is very desirable in areas with heavy timber and/or underbrush.

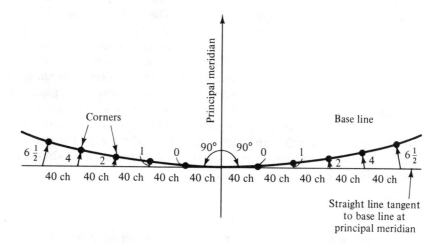

Figure 19-5   Tangent method of laying off base lines.

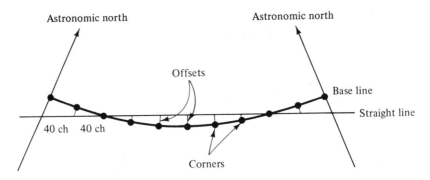

Figure 19-6 Secant method of laying off base lines.

Standard Parallels

After the principal meridians and base lines have been established, the standard parallels (often called correction lines) are run in astronomic directions at 24-mile intervals along the principal meridian by the same method as that used for laying out the base line. (In some of the earlier public land surveys the intervals between the parallels were 30 or 36 miles.) Corners are also set at 40-chain intervals along the parallels. As shown in Fig. 19-7, the parallels are numbered north and south as the 1st Standard Parallel North, 2nd Standard Parallel North, and so on.

Guide Meridians

The so-called guide meridians are run as astronomic directions at 24-mile intervals east and west of the principal meridians. These also are shown in Fig. 19-7. The guide meridians and standard parallels form approximately square quadrangles. From the 24-mile corners on the most southern standard parallel, the guide meridians are run in astronomic northerly directions until they intersect the next standard parallel. At these points corners are established as shown in Fig. 19-7 and the distances to the standard township corners already in place are measured. Then the surveyor moves to those township corners and runs astronomic north meridians to the next standard parallel. The surveyor continues this procedure until he or she reaches the upper boundary of the area being surveyed. These guide meridians are designated at 1st Guide Meridian East, 2nd Guide Meridian East, and so on, as shown in the figure.

Townships

Once the quadrangles are established they are divided into townships (approximately 6 miles by 6 miles). Range lines are established at 6-mile intervals along

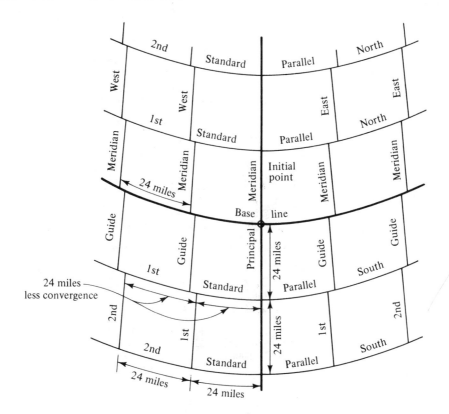

Figure 19-7 Subdivision of land into 24-mile quadrangles.

the parallels between the guide lines. These lines are run continuously between standard parallels as shown in Fig. 19-8.

The surveyor starts at the southeast corner of the southwest township (point *A* in Fig. 19-8) and runs a line in an astronomic northerly direction for 480 chains, setting corners at 40-chain intervals along the way. He or she then sets township corner *B* and runs a random line due west for 480 chains toward the 1st guide Meridian West, setting temporary corners at 40-chain intervals. If the random line hits within 3 chains in the east–west direction and within 3 chains in the north–south direction of the previously set corner *F*, the surveyor's work can be considered acceptable. Then the line and the temporary corners are adjusted proportionately in their positions to agree with the correct position of corner *F*. Any error in length from *B* to *F* is put in the most westerly half mile.

The same procedure is followed in setting points *C* and *D* in the figure. Then the range line is continued north until it intersects the standard parallel at *E*. If there is an error in the 24 miles from points *A* to *E*, it is put in the last or most northerly half mile of range line *DE*. In the same fashion the range lines *GH* and

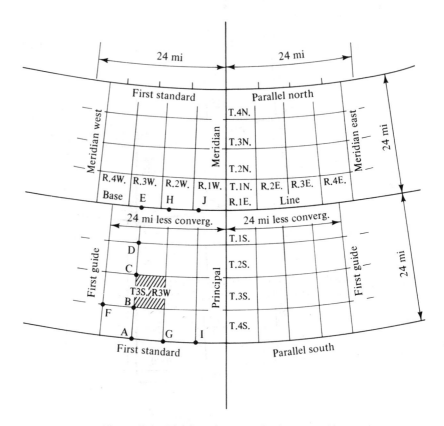

Figure 19-8　Division of quadrangles into townships.

IJ are run. As range line *IJ* is established, the random lines are run to the east and corrected backward so that any error is placed in the most westerly half mile.

A row of townships extending north and south is called a *tier*, and a row extending east and west is called a *range*. Figure 19-8 shows this arrangement in detail. Notice the numbering system for the crosshatched township (Tier 3 south, Range 3 west).

Sections

The final work done in the rectangular survey is that of laying out the sections as shown in Fig. 19-9. The townships are divided into 36 sections each approximately 1 mile square and containing 640 acres. The sections are numbered as shown in the figure. In addition, the government surveyors mark the quarter section corners at 40-chain intervals and the local surveyors (state, county, or private) perform any further subdivision.

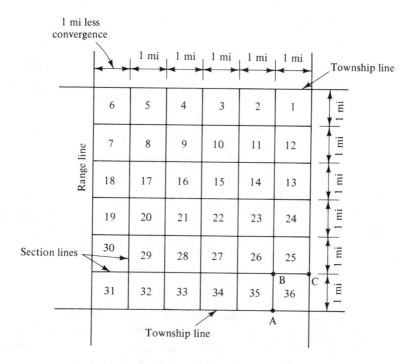

Figure 19-9 Subdivision of township into sections.

To lay out the sections in a township the surveyor starts at the southeast corner of the township (Section 36 in Fig. 19-9) and reruns the south and east boundaries for 1 mile to check the earlier work and to check instruments against the ones used previously. The surveyor then starts at point A in the figure and runs a line for 80 chains or 1 mile to the north, parallel to the eastern side of the township, and sets point B as well as a one-quarter-section corner at the 40-chain point in the middle. He or she runs a random line from B to the east toward the previously set section corner at C setting a temporary half-mile corner in between. If the surveyor comes within 50 links of C, he or she adjusts proportionately the position of the quarter section corner at the midpoint of line BC.

A similar procedure is followed for setting the remaining section and quarter section corners for the township. Space is not taken up here to describe this rather repetitious process. The laying out of the U.S. public land system throughout its existence is covered in great detail in an excellent book by McEntyre.[3]

[3] J. G. McEntyre, *Land Survey Systems* (New York: John Wiley & Sons, Inc., 1978), pp. 32–199.

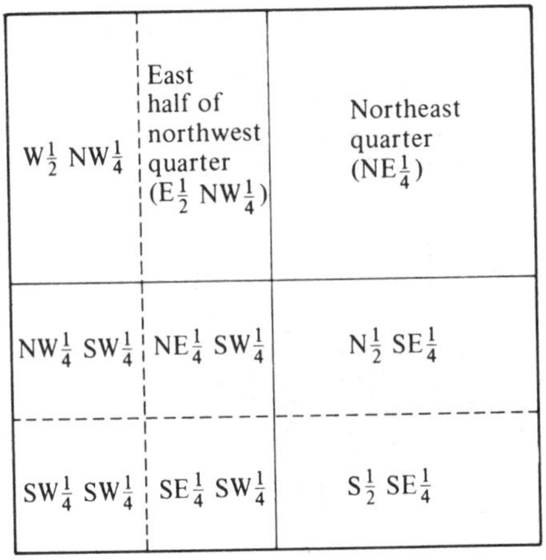

W$\frac{1}{2}$ NW$\frac{1}{4}$	East half of northwest quarter (E$\frac{1}{2}$ NW$\frac{1}{4}$)	Northeast quarter (NE$\frac{1}{4}$)
NW$\frac{1}{4}$ SW$\frac{1}{4}$	NE$\frac{1}{4}$ SW$\frac{1}{4}$	N$\frac{1}{2}$ SE$\frac{1}{4}$
SW$\frac{1}{4}$ SW$\frac{1}{4}$	SE$\frac{1}{4}$ SW$\frac{1}{4}$	S$\frac{1}{2}$ SE$\frac{1}{4}$

Figure 19-10   Subdivision of Section 22 of Fig. 19-9.

The public system called for the disposal of the land in units equal to quarter-quarter sections of 40 acres each. This division is seen in Fig. 19-10. It will be noted that this procedure can be continued because the quarter-quarter sections can be divided into quarters, containing 10 acres each.

In Fig. 19-10 a particular quarter or half of a quarter section is readily identified. In the full description of one of these 40-acre pieces, the quarter of the quarter section is listed, then the quarter section, then the section number, then the township, range, and principal meridian. In this manner a tract of land could be described as the NE 1/4 SW 1/4, Section 22, T2S, R2E of the third principal meridian.

19-16 CORNER MARKERS

Quite a few different materials have been used for marking the township, section, and quarter-section corners. Markers have included wood posts, stones, broken bottles, pits filled with charcoal, and other types. When stones or posts were used, it was common to mark them with from one to six notches to identify the particular section or township corners. Quarter-section corners usually had $\frac{1}{4}$ marked on them.

Since 1910 iron pipes with brass caps on which the designation of the particular corner is given have been in common use. It is thought that they will last and remain legible for perhaps 100 years or more. Stones are usually piled around the pipes to protect them and to make them easier to find.

19-17 WITNESS CORNERS

Should the location of a corner fall in an unmeandered lake or stream or be on a steep cliff or in a marsh or other inaccessible spot, a witness corner is established. A witness corner is usually placed on one of the regular survey lines of the property. However, if a satisfactory point for such a corner cannot be occupied within 10 chains along one of the survey lines, it is permissible to locate a witness corner in any direction within 5 chains of the corner position.

19-18 RESTORING LOST OR OBLITERATED CORNERS

When resurveys are made in the public land area it is sometimes necessary to restore lost corners. Restoration is made only as a last resort and then is done by a procedure that is carefully spelled out in *Restoration of Lost or Obliterated Corners and Subdivision of Sections*, 1974 edition, published by the Bureau of Land Management, U.S. Department of the Interior. To make the restoration it is first necessary for the surveyor to understand fully the manner in which the original survey was conducted.

The corners are restored by either single or double proportionate measurements from existing corners in the area. *Single proportionate measurement* involves the new measurement of an old line to determine one or more positions on that line. As the term implies, the direction of a line is established from two existing corners, and intermediate positions on that line are established by proportionate measurements.

Double proportionate measurement involves the new measurements made between four known corners (two of which establish a north–south or meridional direction and two of which establish an east–west or latitudinal direction). The lost corner is reestablished by double proportionate measurements between the four corners.

The above-mentioned publication of the Bureau of Land Management presents rules of precedence as to whether single or double proportionate measurement should be used. For instance, single proportionate measurement may be used where one line on which the lost corner is located holds precedence over the line passing through the corner in the other direction. One such case would be a section corner on a township line where the east–west township line takes precedence over the north–south section line.

19-19 DEED DESCRIPTIONS OF LAND

To describe regular tracts of land within the U.S. public lands system for legal purposes is quite simple. An acceptable description of a 40-acre quarter section was given in Section 19-15 with reference to Fig. 19-10.

When an irregular tract is involved, however, or one that is not a regular part of the public land system, it is first necessary to tie the description carefully into the rectangular system. Then a length and bearing or metes and bounds description of each side of the tract is given. A description of such a tract might be as follows: "Beginning at a point, marked by an iron pin 300.00 ft North of the NE corner of the SE1/4 of the SW1/4 of Section 28, T3S, R1E, 3rd P.M.; thence North 998.00 ft to an iron pin; thence East 864.00 ft to an iron pin; and so on back to the point of beginning.

CHAPTER TWENTY

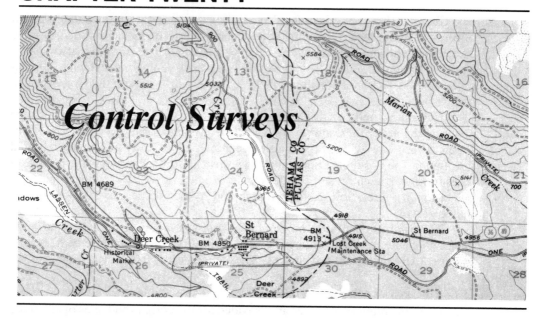

Control Surveys

20-1 INTRODUCTION

The determination of the precise position of a number of stations, usually spread over a large area, is referred to as *control surveying*. Control surveys can be divided into two general types: *horizontal* and *vertical*.

The objective of horizontal control surveys is the establishment of a network of triangulation stations. In 1927 the NGS performed a least squares adjustment of all their previous geodetic surveys. The result was referred to as the North American Datum of 1927 (NAD 27). For this datum an initial point was selected at Meades Ranch in Kansas. This point was held fixed in the adjustment as was an azimuth from this point to a nearby station called "Waldo". Adjusted latitudes and longitudes were established for about 25,000 stations around the continent. Until the past few years horizontal positions were established with reference to these monuments.

In 1973 the NGS began another adjustment this time incorporating over 250,000 stations and all geodetic surveys on record. Originally the work was to be completed in 1983 and thus is called the North American Datum of 1983 (NAD 83). It, however, was not completed until 1986. This adjustment is not based on a single station of reference as was NAD 27. The references used are the mass center of the earth and quite a few points on the earth's surface whose latitudes and longitudes were very precisely established by satellite observations and radio astronomy.

The horizontal position of a station is given in latitude and longitude referenced to the North American 1983 Datum (NAD 83) and state plane coordinates (see Chapter 21). The objective of vertical control surveys is the establishment of a network of reference bench marks. The elevations of these bench marks were determined with respect to mean sea level as defined by the National Geodetic Vertical Datum of 1988 (NAVD-88). This vertical control net which is almost complete (1990) consists of almost 600,000 bench marks in both the United States and Canada.

Horizontal and vertical control systems are laid out in the form of nets covering the areas to be surveyed. The control points for the systems are established at locations where other surveys can be conveniently and accurately tied into them. The results of horizontal and vertical control surveys are used as a basis from which surveys of smaller extent can be originated. Boundary surveys, construction route surveys, topographic and hydrographic surveys, and others may be involved.

The establishment of a horizontal control system for a state or other large geographical area (which is the general topic of this chapter and the next) results in quite a few important advantages. These include the following:

1. All types of surveys are referred to a common datum.
2. Common numerical values are provided for the corners and lines of adjoining tracts of land.
3. A common basis is provided for restoring lost corners.
4. The system reduces error accumulation.
5. Blunders can easily be discovered.
6. A common basis is provided for tying together public works projects.
7. It is very helpful for computer programming, in that all land tracts are placed on a consistent system.

20-2 HORIZONTAL CONTROL

Horizontal control can be carried out by precise traversing, by triangulation, by trilateration, and perhaps by some combination of these methods. The exact method or methods used depends on the terrain, equipment available, information needed, and economic factors.

With *traversing* (which has been described in previous chapters), a series of horizontal distances and angles are measured. With respect to the other two methods it is generally cheaper and can be extended in any direction. It is not limited by the lack of intervisibility between widely spaced stations as are the other methods. As a result it can be accomplished under somewhat less favorable weather conditions than triangulation and trilateration. Traversing, however, has the disadvantages that there are fewer checks available for locating mistakes in the work and the whole system can rather easily sway or bend. To check a traverse it is necessary to form a loop returning to the starting point or to tie it into previously established control points.

With *triangulation* two interior angles of a triangle and the length of one side must be measured. If all three angles are measured the results can be checked (i.e., 180° total for the three) and used to improve the accuracy of the computed distances of the missing side lengths. Triangulation is most often encountered in old surveys of government land by federal agencies. Today EDMIs have fairly well made the method of triangulation obsolete.

With *triangulation* the angles in a series of connected triangles are measured as is the length of at least one side of one triangle, after which the missing lengths are computed. Triangulation has several particular advantages. There are more redundancies or checks in the measurements. In other words, more than one route can be used in moving through the system to calculate desired lengths. There is little tendency for the system to bend or sway—that is, azimuths can easily and accurately be carried or established throughout the system. Another special advantage of triangulation is that outstanding landmarks such as steeples and water tanks can be located by establishing directions to them from different stations. A particular disadvantage of triangulation is that long-range intervisibility is needed between stations, and that usually means that special towers are necessary. Intervisibility also means that reasonably good weather is required. Triangulation is probably the most expensive of the control methods.

With *trilateration* the lengths of the sides of a series of triangles are measured (usually with EDMIs) and the angles are computed from the lengths. Trilateration is the most accurate of the three methods because it is possible today to measure distances more accurately than angles. Moreover, it is generally less expensive than is triangulation. Like triangulation, it has the advantage that checks can be made in the calculations while moving through the system by more than one route. On the other hand, specially constructed towers are generally needed for trilateration, and relatively good weather is required to permit the needed intervisibility of stations. Another disadvantage of this method is that it is not easy to position distant transmission towers, steeples, water tanks, and so on, by electronic distance measurements, because to do so usually requires the placing of reflectors on those landmarks. They can, of course, be located if angle measurements are made from two or more points.

The majority of the first-order control monuments in North America were established by triangulation. In those years the equipment available enabled surveyors to measure angles more quickly and more accurately than they could measure distances. Today, however, this is no longer true and with EDMIs and total station instruments distances can be measured as well or better than angles. As a result control surveys today are commonly handled by (1) triangulation, (2) combined trilateration and triangulation, or (3) very precise traversing. Perhaps these methods will themselves be replaced in the near future by the Global Positioning System.

The remainder of this chapter is devoted to a discussion of triangulation and trilateration and the presentation of simple numerical examples for those control methods. A brief study of triangulation will provide the reader with a good background for the study of trilateration.

20-3 TRIANGULATION

In the past, triangulation was the most important method used for establishing horizontal control for large areas. Distances and directions are determined by using triangles and making a maximum number of angle measurements and a minimum number of distance measurements. The triangulation method was originally adopted because it eliminated the need for a great deal of difficult taping of long distances perhaps over difficult terrain. The sides whose lengths are measured are referred to as *bases* or *base lines*.

Although triangulation is not ordinarily used for surveys of small areas, it may be needed for construction projects where a high degree of precision is required, say 1/25,000 or better such as for bridges, tunnels, and similar projects.

A simple form of triangulation was mentioned in Chapter 6, where the determination of distances across rivers, ravines, or other relatively inaccessible areas was discussed. Such a situation is shown in Fig. 20-1, where it is desired to determine the distance *AB* across a river. A base line *AC* is established and carefully measured. Its distance should be at least half as long as that of *AB*. The angles at *A* and *C* are desirably measured by repetition with a repeating instrument or from several positions with a direction instrument. It is also well to measure the angle at *B* as a check. After these measurements are taken, the length *AB* can be computed by applying the sine law as follows, with reference being made to Fig. 20-1:

$$\frac{l_{AC}}{\sin \sphericalangle B} = \frac{l_{AB}}{\sin \sphericalangle C} = \frac{l_{BC}}{\sin \sphericalangle A}$$

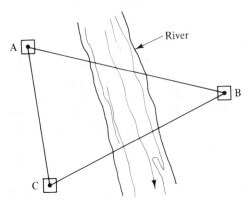

Figure 20-1   Triangulation across a river.

20-4 ACCURACY STANDARDS AND SPECIFICATIONS FOR CONTROL SURVEYS

In this section we consider survey or accuracy standards and specifications. *Survey* or *accuracy standards* are generally defined as being the minimum accuracies deemed necessary to obtain certain specific objectives. *Survey specifications* can be defined as the field operations or "recipes" needed to achieve the particular standards desired. For instance, to achieve a position closure with an accuracy of 1/15,000 it may be specified that distances should be accurate to ± 0.002 ft per 100 ft, that angles be accurate to $\pm 5''$, that angular closure be equal to $\pm 8''\sqrt{n}$ or better where n is the number of angles measured, and so on.

If the standard desired is to be achieved, there can be few deviations from the specifications. One of the great weaknesses in past and present land surveys in the United States is the lack of surveying specifications and the failure to meet specifications that are provided. If a survey accuracy standard for a particular closed traverse is required to be 1/10,000, the average surveyor will attempt to obtain a precision closure of 1/10,000 or better. Yet the closing precision may not have a great deal to do with accuracy. For instance, a surveyor may take a 101-ft steel tape that he or she thinks is 100 ft long and obtain a 1/10,000 closure precision. The probable result is an accuracy of less than 1/100.

The accuracy standard selected for a particular survey depends on its anticipated use. For most control surveys the cost of using a standard a little higher than what is thought to be necessary is very small. Over the years control systems are generally used for all sorts of purposes often far in excess of what was generally anticipated. As a result, it is often thought desirable to select a standard a little higher than what is thought to be needed.

The Federal Geodetic Control Committee has established a set of accuracy standards for horizontal and vertical control surveys. There are in descending order three major classifications: first-order, second-order, and third-order. For horizontal control the latter two classifications are subdivided into two categories labeled classes I and II, while for vertical control the first two orders are subdivided into classes I and II.

In Table 20-1 the minimum accuracies required for horizontal control are briefly summarized. This information is taken from the publication "Classification, Standards of Accuracy, and General Specifications of Geodetic Control Surveys," published in February 1974 by the U.S. Department of Commerce, Rockville, Maryland. The minimum relative accuracy between two directly connected points is specified. For vertical control the accuracy standards are specified in terms of K, the distance between bench marks in kilometers. The minimum accuracies required for vertical control were given in Table 7-1. Also included in the tables is information concerning the recommended uses for each classification.

There is an extensive network of triangulation stations across the United States that were established with different orders of accuracy. Most of the higher-order work has been done by the National Geodetic Survey, while other federal

TABLE 20-1. STANDARDS OF CLASSIFICATION—HORIZONTAL CONTROL

	First-order	Second-order		Third-order	
		Class I	Class II	Class I	Class II
Recommended uses	Primary national network. Metropolitan area surveys. Scientific studies.	Area control which strengthens the national network. Subsidiary metropolitan control.	Area control which contributes to, but is supplemental to, the national network.	General control surveys referred to the national network. Local control. Surveys.	
Base measurement standard error not to exceed	1 part in 1,000,000	1 part in 900,000	1 part in 800,000	1 part in 500,000	1 part in 250,000
Relative accuracy between directly connected adjacent points (at least)	1 part in 100,000	1 part in 50,000	1 part in 20,000	1 part in 10,000	1 part in 5000
Triangle closure. Average not to exceed	1.0″	1.2″	2.0″	3.0″	5.0″
Maximum—seldom to exceed	3.0″	3.0″	5.0″	5.0″	10.0″

Source: Federal Geodetic Control Committee.

agencies such as the U.S. Geological Survey, the Army Corps of Engineers, and others have extended the system, particularly with third-order work.

The primary control network consists of first-order east–west triangulation arcs spaced at about 60 or 70 miles, and north–south arcs with approximately the same spacing. There is next a secondary control system in areas surrounded by the primary control network. The secondary system, which is conducted to second-order class I standards, increases the network density and is used particularly in areas of high land values.

There is also supplemental control executed to second-order class II standards which enhances control in lightly developed areas. It is used particularly

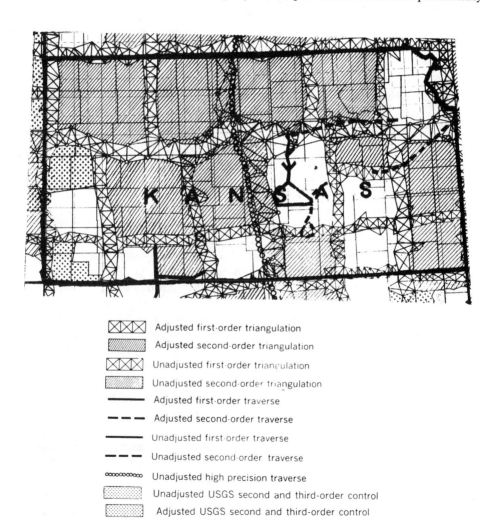

⊠⊠⊠	Adjusted first-order triangulation
⫽⫽⫽	Adjusted second-order triangulation
⊠⊠⊠	Unadjusted first-order triangulation
⫽⫽⫽	Unadjusted second-order triangulation
——	Adjusted first-order traverse
– – –	Adjusted second-order traverse
——	Unadjusted first-order traverse
– – –	Unadjusted second-order traverse
∞∞∞∞∞	Unadjusted high precision traverse
░░░	Unadjusted USGS second and third-order control
▓▓▓	Adjusted USGS second and third-order control

Figure 20-2 Status of horizontal control in Kansas, 1979.

along coastlines and for extensive mapping on construction projects. Finally, local control is used for local construction projects and small-scale topographic maps done to third-order classes I or II. Figure 20-2 shows the status of the control system in the state of Kansas as of October 1, 1979. This information is taken from a U.S. Department of Commerce map of that date entitled "Status of Horizontal Control, United States."

20-5 TRIANGULATION STATIONS

Over the past 200 years the Coast and Geodetic Survey and, more recently, the National Geodetic Survey have established a set of triangulation stations across the United States. This control net has also been tied into the Canadian and Mexican control nets. The private surveyor might very well become involved in a triangulation survey as for a large construction project or for a metropolitan area. He or she will probably use the national system as a starting point for the project. In most areas of the country there will be NGS triangulation stations in the vicinity. Information concerning nearby stations can be obtained from the Director, National Geodetic Information Center, National Ocean Survey, NOAA, Rockville, MD 20852.

A large percentage of NGS surveys have been conducted by the triangulation method in which an accurately measured base line serves as one side of a group of triangles formed by a number of widely spaced points. The points are selected on the basis of their visibility as for example on the tops of hills or church steeples or radio towers or water tanks. As a result, the points are not uniformly spaced. Some of the points are obviously inaccessible and it is necessary to establish eccentric stations from them and to determine the distances and directions from the main points to the eccentric ones.

Sometimes it is necessary to build special towers for making the observations. These towers in effect consist of one tower built inside another because the structural members that support the observers are not the same members that support the instrument (Fig. 20-3). Lines of sight should be kept at least 10 ft above the ground because of possible refraction effects.

For first-order triangulation the stations are spaced at large distances apart— perhaps 20, 40, or even 100 miles. Second-order triangulation stations are placed between the first-order stations. It is the ultimate goal of the National Geodetic Survey to have stations of second-order no farther apart than 7 miles.

Referring back to the earth's curvature and atmospheric refraction expression ($C = 0.574M^2$) in Chapter 7, it can be proved that if towers are placed at both ends of a line each tower will only have to be one-fourth as high as would one tower placed at one end of the line only. Proof of this statement is given in Fig. 20-4.

Figure 20-3 Bilby steel tower that consists of two entirely separate structures. The outer tower supports the observers, while the inner tower supports the instrument only and is thus free from vibrations caused by movements of the observers. (Courtesy of National Geodetic Survey.)

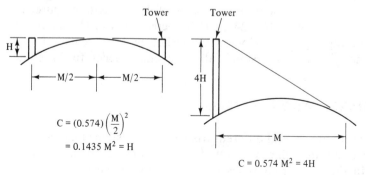

$$C = (0.574)\left(\frac{M}{2}\right)^2$$
$$= 0.1435\,M^2 = H$$

$$C = 0.574\,M^2 = 4H$$

Figure 20-4

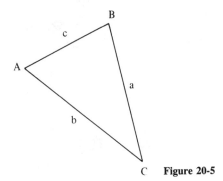

C   **Figure 20-5**

20-6 STRENGTH OF FIGURES

The strength of figures is an important subject in triangulation. This term refers to the effect of the proportions of a triangle on the accuracy with which the lengths of the sides can be computed. When small errors in angle measurement affect the computed distances very little, the figure is said to be *strong*.

In triangulation the lengths of triangle sides are computed with the law of sines. When triangles are used that contain small angles, the best results may not be obtained because of the fact that the rate of change of the sines of angles near 0 or 180° is quite large compared to the rate of change for angles near 90°. Thus angles near 90° are the optimum ones to use, with those from 30 to 150° being acceptable.

It is not correct to say that small angles should always be avoided—rather, they should not be used where they will weaken the system. For instance, in calculating the length of a given side of a triangle there are two angles that are used: the angle opposite the known side and the angle opposite the side whose length is to be determined. In Fig. 20-5 the length *b* is assumed to be known and it is desired to determine the length *c*. With the sine law only the angles *B* and *C* are used, and as a result the angle *A* has no direct effect on the calculations and thus can be quite small without affecting the strength of the figure.

A comprehensive discussion of the strength of figures is presented in "Special Publication No. 247," U.S. Department of Commerce, Coast and Geodetic Survey, by F. R. Gossett, published in 1950, revised in 1959. Included in the material is a discussion of the frequency needed for base lines.

20-7 TRIANGULATION SYSTEMS

There are several different triangulation systems that can be used for a particular survey. In each case, a set of triangles that adjoin or overlap each other are used. In Fig. 20-6, four types of systems that have been used are presented. These

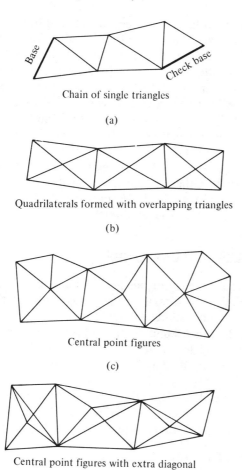

Chain of single triangles

(a)

Quadrilaterals formed with overlapping triangles

(b)

Central point figures

(c)

Central point figures with extra diagonal

(d)

Figure 20-6   Triangulation systems.

systems include a chain of single independent triangles, a chain of quadrilaterals formed with overlapping triangles, a chain of central point figures, and a chain of central point figures each with an extra diagonal.

The use of a single chain of triangles such as the one shown in part (a) of Fig. 20-6 does not provide the most accurate results. Such a system is usually employed in rather long and narrow surveys of low precision such as for a valley or a narrow body of water. There is only one route through this type of system, while the other systems provide at least two routes. The single triangle system is not as satisfactory as the other triangulation systems, and it is frequently necessary to measure base lines as checks for the work. The only other checks on the work are the summations of the angles in each triangle.

Figure 20-6(b) shows the most common triangulation system—a chain of quadrilaterals. Systems of quadrilaterals are best adapted to rather long and narrow surveys where a high degree of precision is required. With a system such as this, the lengths of the sides can be computed with different routes as well as different triangles and angles, offering excellent checks on the computations. Most of the major triangulation arcs in the United States consist of quadrilaterals.

When horizontal control is to be extended over a rather wide area involving a rather large number of points, as might be the case in metropolitan areas, a chain of polygons or central points figures such as the one shown in Fig. 20-6(c) may be used. These figures are very strong and are often quite easy to arrange. The central point figure can itself be further strengthened by using a diagonal of the type shown in Fig. 20-6(d). Other systems are available which are combinations of the types mentioned here. In this book, only the chain of single triangles and the chain of quadrilaterals are considered further.

20-8 MEASUREMENT OF ANGLES AND BASELINES

The instruments to be used for measuring angles for triangulation depend on the accuracy desired in locating the positions of the triangulation stations. If first-order work is desired, directional theodolites should be used with which directions can be read directly to 0.2″. For second-order work, it is necessary to use instruments capable of being read to 1″. For third-order triangulation, engineer's transits that can be read to 20″ or 30″ may be used if the angles are measured by repetition.

When the directional theodolite is used, it is set up over a particular station and pointed to each of the desired stations. For first-order triangulation the set of readings is repeated from 8 to 16 times, while for second-order work they are repeated from 4 to 8 times.

For many decades base lines were measured with precise Invar tapes. The locations of the base lines were selected in relatively smooth open areas to facilitate taping. Today, however, while Invar-taped sides are still acceptable, electronic distance-measuring devices are almost always used. As a result, base lines can be located in much rougher country. The longer the lines selected, the more accurate will be the other lengths determined in the system. Slope distances have to be reduced to horizontal distances, and those distances reduced to sea-level distances.

To correct a distance to sea level, it is possible to write a simple proportion. For this purpose R is considered to be the radius of the earth (usually taken as 20,906,000 ft or 6 372 200 m), L is the measured distance at some elevation H, and C is the correction to be made. With reference to Fig. 20-7, it can be seen that for a certain angle θ the arcs shown are directly proportional to their radii. Thus the following ratio can be written:

$$\frac{L - C}{L} = \frac{R}{R + H}$$

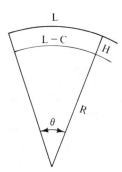

Figure 20-7

from which the sea-level distance is

$$L - C = \frac{RL}{R + H}$$

Example 20-1

A base line was measured at an elevation of 1462 ft and found to be 3692.320 ft. Convert this value to a sea-level distance.

Solution

$$L - C = \frac{RL}{R + H} = \frac{(20,906,000)(3692.320)}{20,906,000 + 1462} = \textbf{3692.062 ft}$$

20-9 WORK INVOLVED IN TRIANGULATION

The work involved in triangulation usually includes the following steps:

1. Selection of the stations.
2. Construction and placement of monuments, erection of the signals, and in many cases construction of towers on which the signals and perhaps the instruments are to be placed.
3. Measurement of the required angles.
4. Astronomical observations to establish the azimuths of the lines.
5. Measurement of the base lines.
6. Necessary office calculations, including adjustment of the angles, calculation of the lengths of triangle sides, and coordinates of the station.

20-10 ADJUSTMENT OF A CHAIN OF SINGLE TRIANGLES

The system of triangles is placed around the exterior of the area being considered and the triangles are established so as to be as nearly equilateral as possible. All of the angles of the triangles are measured, as are the lengths of at least two sides.

As an illustration of the distance measurements, part (a) of Fig. 20-6 shows the base and the check base.

Before the length computation can begin, it is necessary to make the sum of the angles around each point total exactly 360° (called the *station adjustment*), and the sum of the measured angles in each triangle total exactly 180° (called the *figure adjustment*). If precise triangulation is being used, these adjustments are made all in one operation by the method of least squares, but for triangulation of ordinary precision it is possible to use a simpler process as described herein.

For the angles about a point the difference between the sum of the angles and 360° is balanced equally between the number of angles. In the same fashion the difference between the sum of the angles in each triangle and 180° is balanced equally between the angles. This procedure can be adjusted somewhat if it is known that some of the angles were measured with a higher degree of precision than some of the others.

Example 20-2 shows the adjustment of the angles of a chain of two triangles. It is assumed that in each case the angles were measured with equal precision.

Example 20-2

For the two triangles shown in Fig. 20-8 the measured angles are given in the table in the solution part of the problem. Make the station and figure adjustments to the angles.

Solution Make the station adjustment (total of angles around each station adjusted to 360°00′00″).

Station	Angle no.	Measured angle	Adjusted angle
A	1	41°16′10″	41°16′05″
	2	53°36′20″	53°36′15″
	3	265°07′45″	265°07′40″
	Σ	360°00′15″	360°00′00″
B	4	91°16′10″	91°16′20″
	5	268°43′30″	268°43′40″
	Σ	359°59′40″	360°00′00″
C	6	35°07″00″	35°06′50″
	7	78°42′30″	78°42′20″
	8	246°11′00″	246°10′50″
	Σ	360°00′30″	360°00′00″
D	9	60°01′05″	60°00′55″
	10	299°59′15″	299°59′05″
	Σ	360°00′20″	360°00′00″

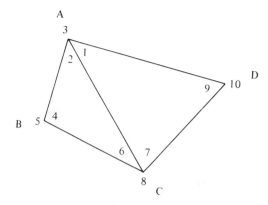

Figure 20-8

Now make the figure adjustment (total of angles in each triangle adjusted to 180°00'00").

Triangle	Angle No.	Angle Value after Station Adjustment	Angle Value from Figure Adjustment
ABC	2	53°36'15"	53°36'27"
	4	91°16'20"	91°16'32"
	6	35°06'50"	35°07'01"
	Σ	179°59'25"	180°00'00"
ACD	1	41°16'05"	41°16'18"
	7	78°42'20"	78°42'34"
	9	60°00'55"	60°01'08"
	Σ	179°59'20"	180°00'00"

20-11 ADJUSTMENT OF A QUADRILATERAL

For a quadrilateral to be adjusted properly, two conditions must be satisfied. The first of these is the *geometric condition* that the sum of the interior angles must equal $(n - 2)(180°)$, where n equals the number of sides of the figure. Second, there is the *trigonometric condition* by which the sine of each angle must be proportional to the length of the opposite side of that triangle.

The adjustment of a quadrilateral is illustrated in Fig. 20-9, where it is as-sumed that the base line distance AD has been determined as have the eight lettered angles. It is desired to determine the length BC.

To make the adjustments and to compute the desired length, the following steps are taken:

1. The angles around each point are balanced to a total of 360° by distributing the error equally (or approximately so) among the angles.

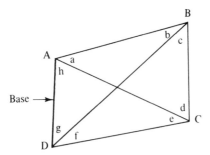

Figure 20-9

2. The sum of all the angles (*a* through *h* in Fig. 20-9) is adjusted to equal $(n - 2)(180°) = 360°$. This is accomplished by adding the angles together and balancing the discrepancy from 360° between the angles.

3. The opposite angles at the intersection of the diagonals should also be equal. In other words, the following condition should apply:

$$a + b = e + f$$

The values of these angles which were adjusted in step 1 are compared and the difference between them is divided by 4 and distributed to the angles, adding to the smaller pair of angles and subtracting from the larger ones. The following relation should also exist:

$$c + d = g + h$$

A similar procedure is used to adjust these angles as used for the previous four.

4. For Fig. 20-9 the length of *BC* is desired. It can be determined by two different routes. Working with triangle *ABD*, the length of side *BD* can be determined with the sine law. Then from triangle *BCD* and the sine law the length of side *BC* can be determined.

In a similar fashion, side *CD* can be determined by using triangle *ACD* and then side *BC* can be determined from triangle *BCD*. With perfect trigonometric adjustment the length of *BC* is the same no matter which route is taken. If the results are close to each other, they may be averaged for work of ordinary precision. For more precise work there are various methods available by which the trigonometric adjustment may be made. Such adjustments are described in advanced surveying textbooks.

If a chain of quadrilaterals is being used, the adjustment in each quadrilateral is made as described for the single quadrilateral.

The angles of a triangle on the earth's curved surface are spherical. As a result, the sum of the spherical angles in such a triangle is slightly larger than 180°. This value is called the *spherical excess* and equals approximately 1″ for a triangle covering about 75 square miles of the earth's surface. This excess is neglected in the example problems of this chapter.

Examples 20-3 and 20-4 illustrate the calculations necessary to adjust the angles of a quadrilateral and to compute the lengths of its sides.

Example 20-3

The angles in the quadrilateral of Fig. 20-9 have been measured with the following results:

$$a = 36°12'24''$$
$$b = 43°58'40''$$
$$c = 47°29'54''$$
$$d = 52°18'46''$$
$$e = 39°08'28''$$
$$f = 41°02'40''$$
$$g = 59°12'16''$$
$$h = 40°36'36''$$
$$\overline{\Sigma = 359°59'44''}$$

Adjust the angles so they satisfy the geometric condition that the sum of the interior angles must equal $(n - 2)(180°)$.

Solution   The total value of the interior angles in the four-sided figure should be $(n - 2)(180°) = 360°$. The sum of angles a through h is 16″ less than 360°. As a first step each of the angles is increased by 2″ to make their total exactly 360°. The corrected angles are

$$a = 36°12'26''$$
$$b = 43°58'42''$$
$$c = 47°29'56''$$
$$d = 52°18'48''$$
$$e = 39°08'30''$$
$$f = 41°02'42''$$
$$g = 59°12'18''$$
$$h = 40°36'38''$$
$$\overline{\Sigma = 360°00'00''}$$

As the next step in the adjustment, the following angle relations should be true:

$$a + b = e + f$$
$$a + b = 80°11'08''$$

$$e + f = 80°11'12''$$

Since the sum of these four angles is 4″ in error, 1″ is added to angles a and b and 1″ is subtracted from angles e and f.

Finally, the following angle relation should be true:

$$c + d = g + h$$

$$c + d = 99°48'44''$$

$$g + h = 98°48'56''$$

Since the total of these angles is in error by 12″, they are adjusted by adding 3″ to each of angles c and d and subtracting 3″ from each of angles g and h. The geometrically balanced angles are as follows:

$$a = \quad 36°12'27''$$

$$b = \quad 43°58'43''$$

$$c = \quad 47°29'59''$$

$$d = \quad 52°18'51''$$

$$e = \quad 39°08'29''$$

$$f = \quad 41°02'41''$$

$$g = \quad 59°12'15''$$

$$\underline{h = \quad 40°36'35''}$$

$$\Sigma = 360°00'00''$$

Example 20-4

It is assumed that the length of side AD of the quadrilateral of Fig. 20-9 has been measured and found to be 864.52 ft. Determine the length of side BC by two different routes through the figure, using the adjusted angles from the solution of Example 20-3.

Solution Using the sine law and triangle ABD, the length of side BD is determined as follows:

$$\frac{AD}{\sin b} = \frac{BD}{\sin (a + h)}$$

$$BD = \frac{(864.52)(\sin a + h)}{\sin b} = \frac{(864.52)(0.9736475)}{0.69438979}$$

$$= 1212.20 \text{ ft}$$

Then from triangle BCD the length of side BC is determined as follows:

$$\frac{BD}{\sin (d + e)} = \frac{BC}{\sin f}$$

$$BC = \frac{BD \sin f}{\sin (d + e)} = \frac{(1212.20)(0.65664792)}{0.99967733}$$

$$= \mathbf{796.25 \ ft}$$

Using a similar procedure and triangles ACD and ABC, the lengths AC and BC are determined.

$$AC = \frac{AD \sin (f + g)}{\sin e} = \frac{(864.52)(0.98404415)}{0.63123624}$$

$$= 1347.71 \ ft$$

$$BC = \frac{AC \sin a}{\sin (b + c)} = \frac{(1347.71)(0.59071129)}{0.99966715}$$

$$= \mathbf{796.37 \ ft}$$

20-12 TRILATERATION

Trilateration is a method of horizontal control in which the lengths of all the lines in geometric figures are measured directly and with the angles of the figures being computed subsequently. The acceptance this method has gained has been due to the tremendous advances made in recent years with EDMIs. Trilateration possesses some advantages over triangulation because measurement of the distances with EDMIs is so quick, precise, and economical, while measurement of the angles needed for triangulation may be more difficult and expensive.

To obtain first- or second-order class I accuracy temperature, humidity, and barometric pressures must be measured along the routes. This is sometimes done by flying the lines, and it is also possible to use some types of laser EDMIs to determine refraction corrections.

Originally, triangulation was adopted because it quickly enabled the National Geodetic Survey to extend their horizontal control for long distances over rough and forbidding terrain without extensive taping. EDMIs enable the surveyor to accomplish very much the same objectives. It may very well be that in the future triangulation and trilateration will be used in combination.

To obtain suitable accuracy with trilateration it is necessary to follow certain guidelines, such as using certain minimum lengths and geometric configurations. The geometric figures used for trilateration are not as standard as those used for triangulation, but in general the figures are similar. It is better in trilateration to use approximately square figures because slender figures are weaker in the short directions transverse to the figures. Should relatively long, narrow figures have to be used because of terrain or other reasons, it is well to strengthen the network by measuring some horizontal angles. The publication "Classification, Standards of Accuracy, and General Specifications of Geodetic Control Surveys" presents standards for trilateration surveys as it does for triangulation and traversing.

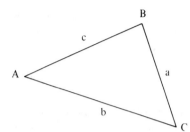

Figure 20-10

If trilateration is used alone, it has the disadvantage that there is no check on the values obtained as the work progresses. If, however, some angles are measured, the disadvantage is eliminated. Notice that with modern equipment the measurement of distances and angles at one setup is easily handled.

Once the distances are obtained for each triangle, the angles may be calculated with the law of cosines. With reference to Fig. 20-10, the law of cosines can be expressed with the following equation, in which a, b, and c are the lengths of the sides and A, B, and C are the opposite interior angles:

$$\cos \angle A = \frac{b^2 + c^2 - a^2}{2bc}$$

The sum of the interior angles should be exactly 180°. In trilateration as in any other form of surveying, it is wise to make checks on the work as it progresses. For instance, the measurement of occasional angles, the calculation of azimuths and the comparison of them with observed azimuths are very helpful. Example 20-5 illustrates the calculation of the angles in a triangle for which the distances are given.

If EDMIs are calibrated properly, it is better to calculate angles (except when distances are short) than to directly measure them. It is wise to use a combination of trilateration and triangulation in which all distances and the largest angles are measured. However, proper weight must be carefully given to the distances and angles in the calculations. This is because of the fact that with properly adjusted EDMIs, it is more accurate to measure distances (other than short ones) and compute angles than to just measure the angles.

Example 20-5

The following sea-level distances in meters were obtained for a triangle by a trilateration survey. Compute the interior angles of the figure with the law of cosines.

$$a = 1226.423 \text{ m}$$

$$b = 1354.677 \text{ m}$$

$$c = 1416.224 \text{ m}$$

Solution

$$\cos \angle A = \frac{(1354.677)^2 + (1416.224)^2 - (1226.423)^2}{(2)(1354.677)(1416.224)}$$

$$= 0.6089901$$

$$\cos \angle B = \frac{(1416.224)^2 + (1226.423)^2 - (1354.677)^2}{(2)(1416.224)(1226.423)}$$

$$= 0.48208416$$

$$\cos \angle C = \frac{(1226.423)^2 + (1354.677)^2 - (1416.224)^2}{(2)(1226.423)(1354.677)}$$

$$= 0.40133838$$

from which

$$\angle A = \mathbf{52°29'0.5''}$$
$$\angle B = \mathbf{61°10'42.2''}$$
$$\angle C = \underline{\mathbf{66°20'17.3''}}$$
$$180°00'00.0''$$

PROBLEMS

20-1. Determine the distance AB across the river shown in the accompanying illustration using the distance AC and the angle values given.

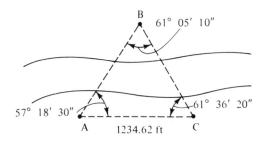

(Ans.: $AB = 1240.75$ ft)

20-2. For the two triangles shown in the accompanying illustration, the measured values of the angles are as follows:

$$1 = \ \ 98°16'32''$$
$$2 = 261°43'38''$$
$$3 = \ \ 48°51'26''$$

$$4 = 43°36'08''$$

$$5 = 267°33'11''$$

$$6 = 85°18'19''$$

$$7 = 274°42'01''$$

$$8 = 51°06'18''$$

$$9 = 32°51'42''$$

$$10 = 276°02'04''$$

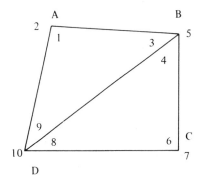

Make the station and figure adjustments for the angles.

20-3. For the quadrilateral shown, the angles have been measured with the following results:

$$a = 49°17'05''$$

$$b = 37°14'24''$$

$$c = 40°07'31''$$

$$d = 53°20'32''$$

$$e = 43°08'12''$$

$$f = 43°23'45''$$

$$g = 56°00'44''$$

$$h = 37°28'11''$$

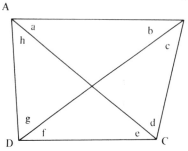

(*Ans.*: $a = 49°17'09''$, $d = 53°20'42''$)

Adjust the angles to their most probable values.

20-4. The following distances in feet were determined for a triangle for a trilateral survey. Compute the interior angles of the triangle.

$$AB = 1642.32 \text{ ft}$$

$$BC = 1583.95 \text{ ft}$$

$$CA = 1296.84 \text{ ft}$$

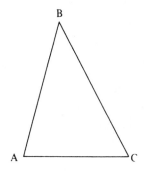

CHAPTER TWENTY-ONE

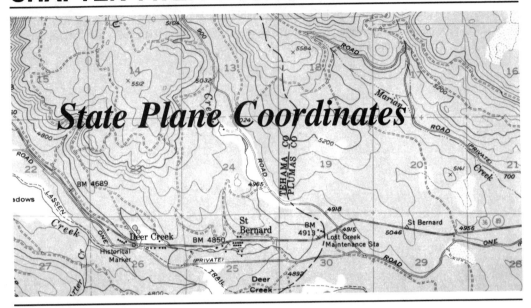

State Plane Coordinates

21-1 INTRODUCTION

In earlier chapters the metes and bounds system and the U.S. public land system for describing property were described. Each of these methods has its shortcomings, and many persons in the surveying profession feel very strongly that a better system is needed for the future. In the United States, a good state plane coordinate system may be the answer. The use of such a system has been recommended by the American Society of Civil Engineers, the American Congress on Surveying and Mapping, and the American Bar Association.

If a coordinate system were adopted in each state, it would make possible the mathematical determination of the location of each property corner. Thus the position of a particular corner would be known in relation to all the other corners in that vicinity whose coordinates were known. If a corner of known coordinates were lost, it would be possible to reestablish it by means of the coordinates of the other points in that area. If a number of corners for a large area were destroyed, they could again be laid out from the coordinates of points outside the area.

The National Geodetic Survey (NGS) has developed a system for each state in which plane or grid coordinates are provided at various control stations. The surveyor can merely run traverse lines (lengths and bearings) from those points to other points whose coordinates are desired. The surveyor will need to adjust the measured distances to sea level and apply a scale correction to them depending on the method used by the NGS in that state. The x and y components of the traverse lines (latitudes and departures) can then be computed and the plane or grid coordinates of the points in question determined. Although the work of the

444

NGS was very complicated, the use of plane coordinates by the practicing surveyor is quite simple.

The reader should clearly understand that a state plane coordinate system cannot replace local monument control. In other words, if a monument exists and no evidence is available to show that it has been disturbed, it represents the corner and any measurements (metes and bounds, coordinates, etc.) are just secondary evidence.

21-2 COMMENTS ON COSTS OF ESTABLISHING AND USING A STATE PLANE COORDINATE SYSTEM

If the land surveyor is required to tie property corners into a coordinate system with control monuments some distance away, there will be appreciable extra costs to the landowners. As a result, surveyors will object and unless laws are passed requiring the use of the system, it will seldom be used. Thus the density or closeness of the control monuments is a very important factor in obtaining voluntary usage of a coordinate system. For instance, if monuments are spaced 10 miles on center and the surveyor has to measure 6 miles in one direction and 4 miles or more in another to tie into the system, the cost will be high and the surveyor probably will not use the system. If the monuments are spaced every half mile, the likelihood that the system will be used will be greatly increased.

Should the monuments be widely spaced the uncertainty of the surveyor's measurements will also be a major problem. If he or she has to measure for 4 miles to tie into a control monument, there may well be an uncertainty in the work of ± 1 ft or more. As a result, a system such as metes and bounds tied into neighboring property monuments may be just as good or better than a coordinate system based on widely spaced control monuments.

The establishment of a closely spaced set of coordinate positions (say, $\frac{1}{2}$ mile on center) would cost a great deal of money—far more than private surveyors could afford to bear. If, however, such a system were established on the ground and paid for by government agencies, it would be quite reasonable to require land surveyors to tie into the system. In urban areas such as Los Angeles a closely spaced state plane coordinate network is in place. For such cases all surveys should definitely be tied into the system.[1]

21-3 TYPES OF COORDINATE SYSTEMS

The purpose of coordinate systems is to take advantage of the very precise work of the National Geodetic Survey and use it in a simple fashion to control ordinary surveying work. In other words, it is desired to use the mathematics of plane surveying, yet take into account the earth's curvature.

[1] C. M. Brown, W. G. Robillard, and D. A. Wilson, *Evidence and Procedures for Boundary Location*, 2nd ed. (New York: John Wiley & Sons, Inc., 1981), pp. 342–343.

The idea is to project points from the earth's spheroid to some imaginary surface that can be rolled out flat without substantially destroying its shape or size. A plane rectangular grid system is then superimposed on to that flat surface and the location of points specified with x and y components. Several of these so-called map projections have been devised through the years, but only three will be discussed here. These are the tangent plane projection, the Lambert projection, and the transverse Mercator projection.

21-4 DISADVANTAGES OF STATE PLANE COORDINATES

State plane coordinates have been available for almost 50 years, yet despite their many advantages, they are not commonly used throughout the country, for the following reasons:

1. The control monuments are too far apart.
2. The surveys are too expensive.
3. Quite a few states (12 to 14) do not have legislation formally permitting their use, although the systems were established more than half a century ago.
4. A large percentage of surveyors and attorneys do not understand the system.
5. So many surveys of the past and the present have such a low degree of precision that nothing would be gained by using the coordinates.

21-5 TANGENT PLANE PROJECTION

For the survey of many urban areas it has been customary to refer points to a rectangular coordinate system. A point within the area, such as one whose geographic coordinates have been established by the National Geodetic Survey, is selected as the origin for the system. The coordinates of all points in the area are calculated as though they are on a plane tangent to the earth at the origin. The astronomic meridian through the origin is taken as the y axis, while the astronomic east–west line at the origin is taken as the x axis. Sometimes the y axis is referred to as grid north. Obviously, the trouble with this method is that the farther a particular point is from the origin or point of tangency, the greater will be the error of the work. It is usually not used much farther than about 20 miles in each direction from the origin.

For average-sized cities, tangent plane coordinates are quite satisfactory, but there has been little need for them since the state plane coordinate systems were introduced. The latter systems are the subject of the remainder of this chapter. For areas that have satisfactorily used tangent plane coordinates there is no need to discard them once state plane coordinates are adopted. They can easily be transferred into the new system. The use of tangent plane coordinates is made relatively simple by Publication No. 71 of the U.S. Coast and Geodetic Survey (now the National Geodetic Survey) entitled "Relation Between Plane Rectangular Coordinates and Geographic Position."

21-6 COMMENTS ON EARTH'S CURVATURE

In the early days of the United States, all surveys were plane surveys; however, the true relative positions of different points on the earth's surface cannot be given unless the earth's curvature and thus spherical coordinates are used. While the average surveyor is acquainted with plane trigonometry, goeodesy is usually unfamiliar to him or her.

Today a large amount of surveying is of such an extensive nature that the true shape of the earth must be considered. As noted previously, this shape is close to an oblate spheriod of revolution in which the axis at the equator is about 27 miles larger than is the polar axis. In proportion to the overall size of the earth, this difference is small, and for all but the most precise geodetic work, an average diameter can be used.

In 1933 the North Carolina State Highway Commission asked the U.S. Coast and Geodetic Survey to design a statewide system of coordinates by which plane surveying coordinates could be used, yet take advantage of the very precise geographic coordinates of that organization. A system was designed for the state of North Carolina as well as for the other states of the country. There are actually two basic systems or projections employed: the Lambert conformal conic projection and the transverse Mercator projection. The first of these methods uses an imaginary cone as its developable surface and the second uses an imaginary cylinder. The mathematics for these systems is described in the publication *Manual of Plane Coordinates Computation* by O. S. Adams and C. N. Claire, Special Publication No. 193 of the National Geodetic Survey.

21-7 LAMBERT CONFORMAL CONIC PROJECTION

As the name implies, this method involves the projection of a section of the earth onto the surface of an imaginary cone. The term *conformal* means that true angular relations (or very nearly so) are retained for all points. There are different forms of the Lambert projection, but the one used for state plane coordinates consists of a cone such as the one shown in Fig. 21-1, whose axis *OE* coincides with the polar axis of the earth. In the figure it will be noted that one element of the cone cuts through the earth's surface at points *B* and *C*. All the elements of the cone cutting through the earth create the two circles called *standard parallels* which are shown in the figure. A section of the cone between the parallels and outside them is rolled out flat and the points from the sphere projected on to it. The scale of the projected plane will clearly be exact along these parallels.

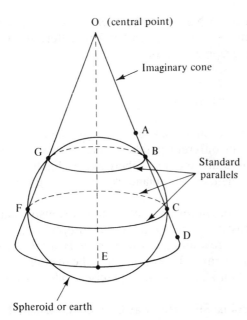

O (central point)

Imaginary cone

A

G   B

Standard
parallels

F   C

D

E

Spheroid or earth

Figure 21-1 Lambert conformal conic
projection.

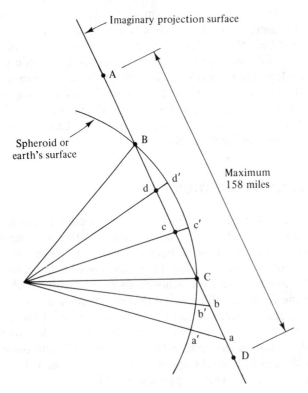

Imaginary projection surface

A

Spheroid or
earth's surface

B

d'
d

Maximum
158 miles

c'
c

C

b
b'

a'   a

D

Figure 21-2

Figure 21-2 shows a little more detail of the intersection picture for the element of the cone that intersects the earth at points *B* and *C* in Fig. 21-1. It can be seen from this second figure that points on the earth's surface are projected along radial lines from the earth's center to the surface of the imaginary cone.

When the cone surface is outside the spheroid, a projected distance on the cone (as *ab* in the figure) is greater than the actual distance (*a'b'*) on the spheroid. When the cone surface is inside the spheroid, a projected distance (as *cd*) is less than the distance (*c'd'*) on the spheroid.

Figure 21-3 shows the Lambert projection developed into a plane surface for the frustrum of the cone *ADHI*. For this figure the longitude of any point as *D* or *H* can be determined by adding or subtracting the angle θ at point O to the central meridian. The longitude at any point on the plane *ADHI* will be exact, but the latitude of any point above or below the standard parallels will be slightly in error. The magnitude of the error will not be very large if the height of the plane *AD* is kept within certain limits. The heights have been limited to a maximum of 158 miles, with the standard parallels located at two-thirds of that distance. Unless these values are exceeded, there will be no point where the discrepancy between a sea-level distance and a grid distance will be greater than 1/10,000.

The Lambert projection can be extended indefinitely in the east–west di-

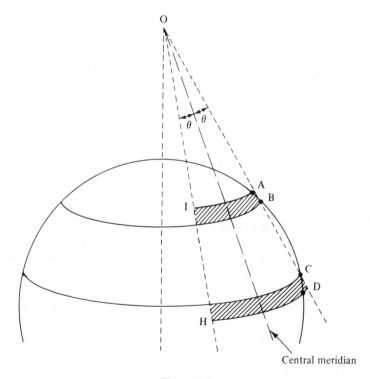

Figure 21-3

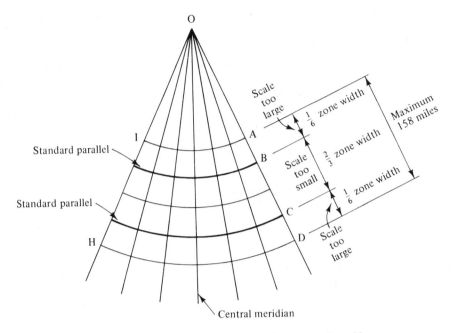

Figure 21-4 Lambert projection showing rectangular grid.

rection without affecting the accuracy of the work, but it must be limited in the north–south direction as described. The method is therefore most suitable for areas that are wide but not deep. Figure 21-4 shows the grid system superimposed onto the plane surface of the cone.

21-8 TRANSVERSE MERCATOR PROJECTION

The transverse Mercator projection uses an imaginary cylinder that has its axis perpendicular to the polar axis of the earth. The diameter of the cylinder is a little smaller than that of the earth, with the result that it intersects the earth with two parallel circles equidistant from the central meridian as shown in Fig. 21-5. In part (a) of the figure the cylinder is shown passing through the earth, and the projected area to be used is diagrammed in part (b).

In the projection the scale is exact on the two lines of intersection, but between the lines a distance on the projection is smaller than the corresponding distance on the sea-level surface. Outside the two lines the projected distance is larger. Notice in the figure that neither lines of latitude nor longitude (except the central meridian) will appear as straight lines on the projection. The projection can be extended indefinitely in the north–south direction without affecting its accuracy but in the east–west direction its width is limited to 158 miles with the

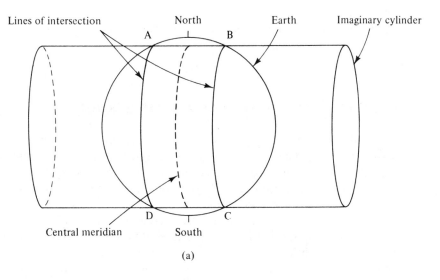

Lines of intersection   North   Earth   Imaginary cylinder

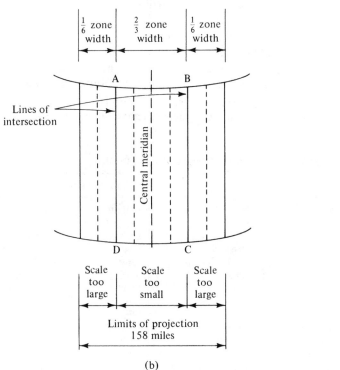

Figure 21-5   Transverse Mercator projection.

exact intersection lines separated by about two-thirds of the distance. Again using these limits the discrepancy between a projected or grid distance and a sea-level distance will not be greater than 1/10,000. The Mercator projection was designed to meet the requirements of states whose largest distances are in the north–south direction.

21-9 ADOPTION OF STATE PLANE COORDINATE SYSTEMS

The need for establishing plane coordinate systems that could make use of the existing geodetic data of the National Geodetic Survey in the various states led to the development of state plane coordinate systems. The systems adopted were based on the Lambert conformal conic projection and the transverse Mercator projection. The Lambert system is used for the states whose greatest dimension is in the east–west direction while the Mercator system is used for those states whose greatest dimension is in the north–south direction. Actually, Alaska, Florida, and New York make use of both projections because of their shapes. For some states one zone of projection is sufficient to cover the entire state, while for some large ones several zones will be needed. California has seven Lambert zones; Texas has five. In Table 21-1 the system or systems used for each of the states of the United States is presented.

The National Geodetic Survey has set up state plane coordinate systems for each of the 50 states. In addition, they have published computed grid coordinates for various control stations throughout the country. Some states, however, have not passed the necessary legislation to officially recognize the applicable coordinate system. Approximately 36 states have laws permitting the use of state plane

TABLE 21-1   STATES WITH THEIR GRID SYSTEMS

Lambert system		Transverse mercator system		Both
Arkansas	North Dakota	Alabama	Mississippi	Alaska
California	Ohio	Arizona	Missouri	Florida
Colorado	Oklahoma	Delaware	Nevada	New York
Connecticut	Oregon	Georgia	New Hampshire	
Iowa	Pennsylvania	Hawaii	New Jersey	
Kansas	South Carolina	Idaho	New Mexico	
Kentucky	South Dakota	Illinois	Rhode Island	
Louisiana	Tennessee	Indiana	Vermont	
Maryland	Texas	Maine	Wyoming	
Massachusetts	Utah			
Michigan	Virginia			
Minnesota	Washington			
Montana	West Virginia			
Nebraska	Wisconsin			
North Carolina				

coordinates, but they do not make their use mandatory. If a point is within 800 m of an NGS control station in North Carolina, the surveyor is supposed to use state plane coordinates to describe its position.

21-10 CALCULATION OF PLANE COORDINATES ON A LAMBERT GRID

Before actually considering a traverse with state plane coordinates, we shall first consider how the geographic coordinates of a point can be transformed into plane coordinates. For a particular Lambert projection zone a central meridian is established near the center of the zone, and is given an x value of 2,000,000 ft. The x axis is placed well below the southern edge of the zone and is given a y value of zero. Thus all x and y positions in the projection area will be positive. This information is shown in Fig. 21-6 together with the other information needed to transform the geographic coordinates of point P into Lambert grid coordinates.

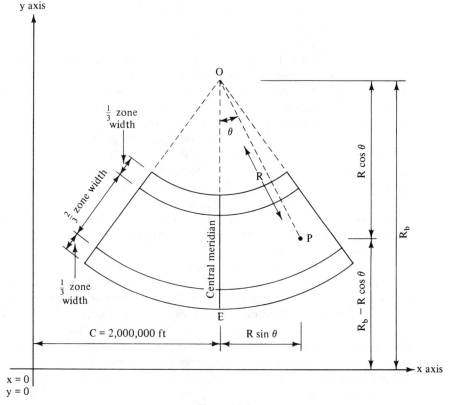

Figure 21-6

In the figure point O is the apex of the cone from which the area is projected, line OE is the central meridian and C is the x coordinate of that meridian equal to 2,000,000 ft. The meridians are straight lines that converge at point O. The parallels of latitude are arcs of concentric circles that have their centers at point O (see Figs. 21-4 and 21-6).

Line OP represents part of a meridian through point O and has a length equal to R. The y coordinate of point O is a constant equal to R_b. The angle θ between lines OE and OP is called the *mapping angle* or *convergence angle*.

The National Geodetic Survey has available for each state using the Lambert projection the necessary information needed to use plane coordinates. Included in this information are values of R for each whole minute of latitude and values of θ for each whole minute of longitude as well as the latitudes and longitudes for each of their control stations. Tables E and F in Appendix A show this type of information. The calculations necessary to compute state plane coordinates for these stations are illustrated in this section even though the National Geodetic Survey provides these coordinates for many of their stations. From Fig. 21-6 the x and y plane coordinates of point P can be seen to equal the following:

$$x = R \sin \theta + C$$

$$y = R_b - R \cos \theta$$

If the angle θ is to the left of the central meridian, it will have a negative value and thus the value of $R \sin \theta$ will be negative for computing the x coordinate. In computing the y coordinate for point R, $\cos \theta$ will always be subtracted. Example 21-1 that follows illustrates the determination of the Lambert plane coordinates for a particular station in South Carolina.

South Carolina is divided into two Lambert zones, north and south, both having a central meridian of 81°00′00″W. The north zone has standard parallels at 33°46′ and 34°58′N latitudes, and the south zone has standard parallels at latitudes 32°20′ and 33°40′N, as shown in Fig. 21-7, where they are labeled *Scale Exact. It should be noted in this figure that the exact division between the north and south zones is not a line of constant latitude. Rather, the dividing line follows the somewhat irregular county lines. This means that all counties fall completely in one zone or the other.*

It is considered necessary to make calculations for plane coordinates to at least 10 place accuracy, and this practice is followed in this chapter. Trigonometric functions can be determined to 10 places with some pocket and desk calculators. Special Publication No. 246 of the U.S. Coast and Geodetic Survey, entitled "Sines, Cosines, and Tangents Ten Decimal Places with Ten Second Interval 0°-6°," can be helpful in this regard. This publication can be obtained from the U.S. Department of Commerce, National Oceanic and Atmospheric Administration, National Geodetic Survey, Rockville, MD 20852.

Tables E and F of Appendix A are taken from Special Publication No. 273 of the U.S. Department of Commerce Coast and Geodetic Survey, entitled "Plane

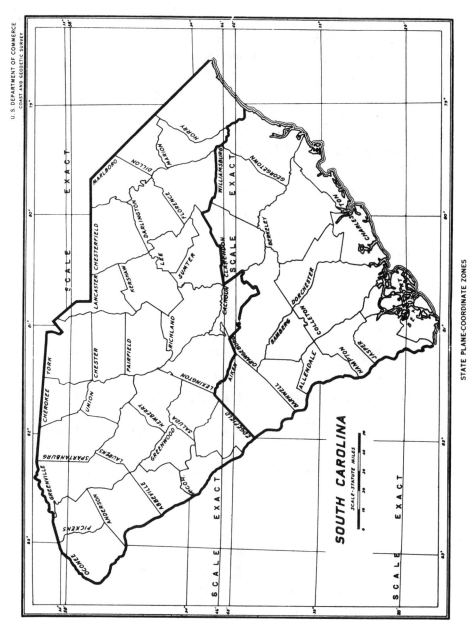

U.S. DEPARTMENT OF COMMERCE
COAST AND GEODETIC SURVEY

SOUTH CAROLINA

SCALE-STATUTE MILES

STATE PLANE-COORDINATE ZONES

Figure 21-7

455

Coordinate Projection Tables South Carolina (Lambert)." The reader can obtain the corresponding publication for his or her state by writing the address given in the preceding paragraph. Table E provides information as to R values for different latitudes, and Table F provides values of θ for different longitudes. These values are needed for the solution of Example 21-1.

Example 21-1

Determine the x and y Lambert coordinates for the following NGS monument in the South Carolina North Zone:

Name and location of station: "Blaney," 12 miles southeast of Camden, SC

Geodetic latitude = 34°10′18.873″N

Geodetic longitude = 80°47′30.989″W

$$C = 2,000,000.00 \text{ ft}$$

R_b for North Zone of SC = 31,127,724.75 ft

Solution

R = 30,703,171.70 at 34°10′N (Table E)

  minus correction for 18.873″ farther north (Table E)

 = 30,703,171.70 − (18.873)(101.08333)

 = 30,701,263.95 ft

Θ = +0°07′20.3080″ for longitude 80°47′W (Table F)

  minus $\left(\dfrac{30.989}{60}\right)$ (change in θ from 80°47′ to 80°48′W longitude)

 = +0°07′20.3080″ minus $\left(\dfrac{30.989}{60}\right)$ (33.8699)

 = +0°07′2.8148″

Then

  $x = R \sin \theta + C$

   = (30,701,263.95)(0.0020498626) + 2,000,000.00

   = **2,062,933.373 ft**

  $y = R_b - R \cos \theta$

   = 31,127,724.75 − (30,701,263.95)(0.9999978990)

   = **426,525.3024 ft**

21-11 GRID AZIMUTHS

It is to be remembered for this discussion that there is a very slight difference between a geodetic azimuth and an astronomic azimuth. A geodetic azimuth is equal to the astronomic azimuth plus the Laplace correction. In the sections that follow a geodetic azimuth is used. Should the surveyor be interested in the value of the Laplace correction for a particular location, he or she may contact the National Geodetic Survey. That organization will provide an approximate value for a given latitude and longitude.

At a distance of from about $\frac{1}{4}$ to 2 miles from each triangulation station the National Geodetic Survey has established a monument or azimuth mark which can be sighted on from the station. The grid azimuth from the station to the mark or monument is usually given together with the other data for the station. The reader should be sure that he or she understands the azimuth definition here. A geodetic azimuth is one that takes into account the convergence of the meridians, while a plane or grid azimuth is constant for the entire zone (i.e., parallel to the central meridian). Thus grid azimuths and geodetic azimuths will be identical only for points located on the central meridian. It will also be remembered that south azimuths were used until recently by the National Geodetic Survey.

The farther a point is located from the central meridian, the greater will be the difference between the grid and geodetic azimuths. The difference between the two is *substantially* equal to the angular convergence between the central meridian and a true meridian passing through the point in question. This difference is denoted as θ in the Lambert projection and $\Delta\alpha$ in the Mercator projection. In the difference for both systems there is a second term (due to meridian curvature) which may be neglected for all third-order traverses except those in which the orientation sight is more than 5 miles. For points to the east of the central meridian, the grid azimuth is less than the geodetic azimuth, while for points to the west of the central meridian, it is greater than the geodetic azimuth. This situation is illustrated in Fig. 21-8.

To make use of the plane coordinate system in a particular state, the local surveyor sets up over one of the triangulation stations, sights on the line of known grid azimuth and then runs the survey by traversing (or perhaps by triangulation or trilateration) to the area under consideration. He or she then by the usual method of plane coordinates determines the coordinates of any point in the survey. Of course, if the plane coordinates of two stations are known, the grid azimuth between these points can be determined by the following expression:

$$\tan \text{azimuth} = \frac{\Delta x}{\Delta y}$$

where Δx and Δy are the differences between the x coordinates and the y coordinates of the two stations, respectively.

If the azimuth mark is not available or if the grid azimuth to that point is

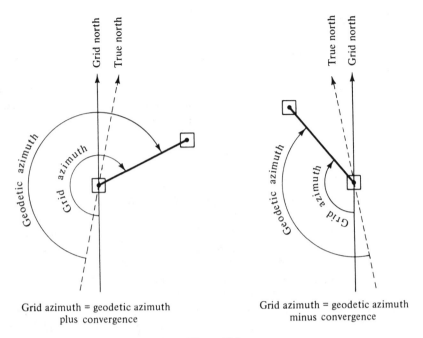

Grid azimuth = geodetic azimuth
plus convergence

Grid azimuth = geodetic azimuth
minus convergence

Figure 21-8

not known at a particular station, it will be necessary for the surveyor to establish a geodetic azimuth by means of an astronomic observation. The azimuth obtained can then be converted to a grid azimuth by one of the following expressions:

For Lambert:

$$\text{Grid azimuth} = \text{geodetic azimuth} - \theta + \text{second term}$$

For Mercator:

$$\text{Grid azimuth} = \text{geodetic azimuth} - \Delta\alpha - \text{second term}$$

21-12 OBTAINING GRID DISTANCES FROM MEASURED GROUND DISTANCES

Before the measured ground distances between traverse points and the grid azimuths or bearings can be used to calculate grid coordinates, it is necessary to reduce the measured distances to sea level or that is to their equivalent geodetic distances. Then these distances must have scale corrections applied for the particular state plane system being used. These two conversions are considered in the paragraphs to follow, where the grid distance to be used equals the measured ground distance times the sea-level factor times the scale factor.

Reduction to Sea Level

It is necessary to make grid projections on a sea-level basis. For measurements taken at elevations other than sea level, the ground distances can be converted to sea-level distances by similar triangles. For this discussion reference is made to Fig. 21-9, in which R, the mean radius of the earth, is assumed to equal 20,906,000 ft (for Alaska and Hawaii different values are used). Also in the figure, H is the elevation of the line above sea level and C is the correction to be made to the measured ground distance L to obtain the geodetic distance.

$$\frac{C}{L - C} = \frac{H}{R}$$

$$C = \frac{H}{R}(L - C)$$

$$CR = HL - CH$$

$$CR + CH = HL$$

$$C(R + H) = HL$$

$$C = \frac{HL}{(R + H)}$$

But the value of H is so small compared to R that it can be neglected and the expression becomes

$$C = \frac{HL}{R} = \frac{HL}{20,906,000}$$

It is common to convert the measured ground distance to sea level by mul-

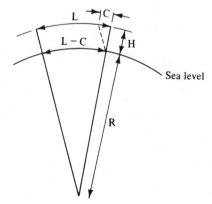

Figure 21-9

tiplying it by a factor called the *sea-level factor* (SLF). This factor equals the following:

$$\text{SLF} = \frac{L - C}{L} = \underbrace{1 - \frac{HL}{R}}_{L} = 1 - \frac{H}{R}$$

$$= 1 - \frac{H}{20,906,000}$$

Scale Correction

As indicated previously (see Figs. 21-2 and 21-5), there will be errors in scale unless the points in question are exactly at the intersection lines. The scale correction can be taken from prepared tables of the National Geodetic Survey. With the Lambert projection, the scale factor tables are entered with the latitude. When working with a particular traverse it is necessary to determine the mean latitude for the line. This can be done by plotting the traverse lines to scale in the appropriate direction and scaling off the distance parallel to grid north from a station of known latitude. The distance can also be computed by multiplying the length of each line by the cosine of its bearing. The distance from the reference station can be converted to seconds of latitude with sufficient accuracy by dividing the distance by 100. The scale factors for each state are published by the National Geodetic Survey. In Table E the Appendix A, scale factors are given for South Carolina for the Lambert projection. Scale factors for all other states and the projection used are available from the National Geodetic Survey. A similar discussion can be made for the Mercator scale factor where the tables are entered with the longitude.

21-13 CALCULATION OF STATE PLANE COORDINATES

Example 21-2 illustrates the calculations of state plane coordinates for a traverse point in South Carolina. The problem is worked with respect to a National Geodetic station for which the following detailed description applies.

> Blaney (Kershaw County, C. L. Garner, 1918; 1934).—About 20 miles northeast of Columbia, 12 miles southeast of Camden, at village of Blaney, on top of high bank, 300 meters (984 feet) east-northeast of railway station, about 100 meters (328 feet) north-northwest of center line of U.S. Highway 1, directly opposite Blaney High School, and 68.4 feet south-southeast of south-southeast rail of main track of Seaboard Air Line Railway. Surface mark, standard disk in concrete, note 1a, was reported in 1934 as having been removed. Underground mark, recovered in 1934, is bottle in concrete, note 7d, 2 feet below surface of ground. Reference mark, standard

disk in concrete, note 11e, is on top of bank. 4.21 meters (13.8 feet) east of east rail of Seaboard Air Line Railway, about 4 meters (13 feet) above roadbed, 3 feet east of edge of cut and 20.78 meters (68.2 feet) from station in azimuth 80°05′. Station *Blaney 2* (see description thereof) is 574.03 feet from station in azimuth 304°03′53″. Plane coordinates:** (N), x = 2,062,933.37 feet; y = 426,525.30 feet.

For this example the reader should carefully note the manner in which the grid bearings were obtained. The geodetic bearings of the sides of the traverse were measured in the field. Each of these bearings was converted to grid bearings by adding or subtracting the angle θ. To decide whether to add or subtract θ, a sketch was made for each side showing astronomic north, grid north, and the geodetic bearing. From these sketches the correct grid bearings are obvious. One of the sketches is shown in Fig. 21-10, where θ is + or east 0°07′03″ and the geodetic bearing of the line is S85°22′36″W. (This is line *BC* in Example 21-2.) The grid bearing of the line equals its geodetic bearing minus θ in this case.

Using the grid bearings and distances, the east and west projections of each side of the traverse are computed and the state plane or grid coordinates of any point of the traverse can be computed. For the solution of the problem, use is made of the information contained in Tables E and F of Appendix A.

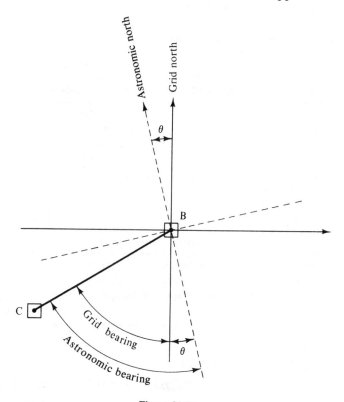

Figure 21-10

Example 21-2

A traverse is run from NGS monument "Blaney" (a description of which was presented earlier in this section) to a property corner D with the following results:

Line	Geodetic Bearing	Distance Measured (ft)	Average Elevation (ft)
AB	N71°16′20″W	3622.20	400
BC	S85°22′36″W	4736.24	360
CD	S19°46′11″W	2942.41	320

"Blaneys" x and y grid coordinates are 2,062,933.37 ft and 426,525.30 ft, respectively, as given in the description of the station and as calculated in Example 21-1. Its geodetic latitude is 34°10′18.873″N and its geodetic longitude is 80°47′30.989″W. The value of θ is 0°07′03″, as determined from Table F for Example 21-1. Compute the grid bearing of the traverse lines and the grid coordinates of property corner D.

Solution

Side	Sea-level factor $= 1 - \dfrac{\text{elevation}}{20{,}906{,}000}$	Scale correction factor from table E in Appendix A (computed at mean latitude of each line)	Grid distance = distance measured × sea-level factor × scale-correction factor
AB	0.9999808667	0.9999511	3621.953573 ft
BC	0.9999827801	0.9999511	4735.926844 ft
CD	0.9999846934	0.9999513	2942.221668 ft

Grid bearing = geodetic bearing ± θ as illustrated in Fig. 21-10 for line BC	Cosine	Sine	Latitude N	Latitude S	Departure E	Departure W
N71°23′23″W	0.3191293159	0.9477111795	1155.872	X	X	3432.566
S85°15′33″W	0.0826487674	0.9965787381	X	391.419	X	4719.724
S19°39′08″W	0.9417513143	0.3363100683	X	2770.841	X	989.499
				$\Sigma = 2006.388$S		$\Sigma = 9141.789$W

Coordinates for point D:

$$x = 2{,}062{,}933.37 - 9141.79 = \mathbf{2{,}053{,}791.58 \text{ ft}}$$

$$y = 426{,}525.30 - 2006.39 = \mathbf{424{,}518.91 \text{ ft}}$$

21-14 DETERMINATION OF THE ANGLE BETWEEN GEODETIC AND GRID AZIMUTHS

Example 21-3 shows the calculation of the angle between geodetic north and grid north at point D of the traverse considered in Example 21-2. In Fig. 21-11, point O is the apex of the cone for the Lambert system for the north zone of South Carolina. It has an x coordinate of 2,000,000.00 ft and a y coordinate of $R_b = 31,127,724.75$ ft. The grid or state plane coordinates of point D were determined in Example 21-2 and from them it is possible to determine the x distance (x') and the y distance ($R_b - y$) from point O to point D as shown in the figure.

With the x' and $R_b - y$ values known it is possible to determine the angle θ, which is the angle between geodetic north and grid north at point D:

$$\theta = \tan^{-1} \frac{x'}{R_b - y}$$

Example 21-3

Determine the angle between geodetic north and grid north at point D of the traverse of Example 21-2.

Solution

$$x' = 2,053,791.58 - 2,000,000.00$$

$$= 53,791.58 \text{ ft}$$

$$R_b - y = 31,127,724.75 - 424,518.91$$

$$= 30,703,205.84 \text{ ft}$$

$$\theta = \tan^{-1} \frac{x'}{R_b - y} = \tan^{-1} \frac{53,791.58}{30,703,205.84}$$

$$= 0.0017519857$$

$$= \tan^{-1} (0.1003812882)$$

$$= 0°06'1.37''$$

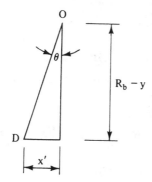

Figure 21-11

21-15 CALCULATION OF GEODETIC POSITION FROM STATE PLANE COORDINATES

The geodetic position, that is, the geodetic latitude and longitude of a point, can be determined from the grid coordinates in approximately the reverse order to that used to determine the grid coordinates from the geodetic position illustrated in Example 21-1. Such a procedure is described here.

1. *Geodetic latitude.* The grid coordinate y of a particular point has previously been written as follows for the Lambert system:

$$y = R_b - R \cos \theta$$

Thus R is

$$R = \frac{R_b - y}{\cos \theta}$$

With the value of R known, the latitude ϕ can be determined from Table F in Appendix A.

2. *Geodetic longitude.* The convergence angle θ can be determined as illustrated in Example 21-3. Then $\Delta y = \theta/l$, where l is a constant for the particular Lambert zone. The longitude λ is determined as follows:

$$\lambda = \text{longitude} = \text{central meridian} - \Delta y$$

Example 21-4 presents the calculations necessary to determine the geodetic position of point D of Example 21-2.

Example 21-4

Determine the latitude and longitude of point D of Example 21-2. The given values are:

$$C = 2,000,000.00 \text{ ft}$$

$$R_b = 31,127,724.75 \text{ ft for north zone}$$

$$\left. \begin{array}{l} x = 2,053,791.58 \text{ ft} \\ \\ y = 424,518.91 \text{ ft} \end{array} \right\} \text{from Example 21-2}$$

$$\theta = 0.1003812882°\} \text{ from Example 21-3}$$

Solution

$$R = \frac{R_b - y}{\cos \theta} = \frac{31,127,724.75 - 424,518.91}{0.99999847}$$

$$= 30,703,252.82 \text{ ft}$$

Latitude = **34°09′59.198″** from Table E in Appendix

$$\Delta y = \frac{\theta}{l} \quad \text{where } l \text{ is a constant for the north zone}$$

$$l = 0.56449738$$

$$\Delta y = \frac{0.1003812882}{0.56449738} = 0.1778234188$$

$$= 0°10′\ 40.167″$$

$$\lambda = \text{longitude} = 81°00′00.000″ - 0°10′40.167″$$

$$= \textbf{80°49′19.833″}$$

21-16 CALCULATION OF PLANE COORDINATES FOR THE MERCATOR GRID

For the Mercator projection the central meridian is set at an x distance usually equal to 500,000 ft for most states, while the x axis, as in the Lambert projection, is placed well below the southern edge of the zone. Again the x and y coordinates of all points in the projected area will be positive.

The x and y Mercator grid coordinates of some point P can be determined with the equation given at the end of this paragraph, in which $\Delta\lambda''$ is the difference in seconds between the longitude of the central meridian and point P. The value of $\Delta\lambda''$ will be positive if P is to the east of the central meridian and negative if it is to the west. In the expressions that follow x' is the distance to point P either east or west of the central meridian. In the expression for x and y the terms y_0, H, V, and a are values based on the geographic latitude, while b and c are based on $\Delta\lambda''$. The magnitudes of these values are given in the tables available from the NGS for individual states. If the sign of ab is positive, it increases the value of x'; a negative value decreases it.

$$x' = H \cdot \Delta\lambda'' \pm ab$$

$$x = x' + 500,000$$

$$y = y_0 + V\left(\frac{\Delta\lambda''}{100}\right)^2 \pm c$$

Space is not taken here to present a transverse Mercator projection example, as it is felt that if the surveyor can apply one system, he or she will easily be able to use the other one.

21-17 CONCLUDING REMARKS ON STATE PLANE COORDINATE SYSTEMS

Sometimes surveys are made in border areas in between different zones or may even extend over into an adjoining state. Such a situation really does not cause a problem as the surveyor can easily change state plane coordinates into those of the other zone or state. This is done for a point by converting the state plane coordinates of the point to its geodetic latitude and longitude. Then with these values known the state plane coordinates of the other zone or state can be computed.

The Lambert and Mercator projections that have been described briefly in this chapter are not the only coordinate systems available. These two systems were developed to be used primarily for areas running long distances in the east–west or north–south directions, respectively. They are not conveniently applied to long areas that run in other directions or for circular areas. Other systems have been proposed for such areas, such as the *oblique Mercator projection* (now used for parts of Alaska and by the U.S. Lake Survey for the area of the Great Lakes Erie and Ontario and the St. Lawrence River) and the *horizon stereographic projection* (used in Canada and some other countries).[2] Among other proposals being considered is one which suggests that a national Transverse Mercator system be used using 2° bands of longitude.

The use of state plane coordinates is gradually increasing not only by governmental agencies but by private organizations as well. State plane coordinate systems can be used to show the location of important government or private structures, such as dams, pipelines, bridges, oil wells, industrial plants, and many more. For such projects good vertical control is needed as well as the horizontal control provided by the coordinates.

[2] R. C. Brinker and P. R. Wolf, *Elementary Surveying*, 8th ed. (New York: Harper & Row, Publishers, 1989), pp 458–459.

CHAPTER TWENTY-TWO

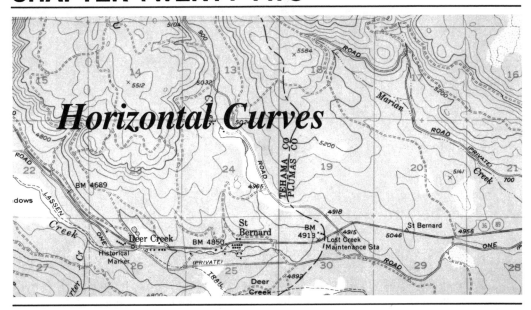

Horizontal Curves

22-1 INTRODUCTION

The centerlines of highways and railroads consist of a series of straight lines connected by curves. The curves for fast traffic are normally circular, although spiral curves may be used to provide gradual transitions to or from the circular curves. Three circular curves are shown in Fig. 22-1. The *simple curve* consists of a single arc. The *compound curve* consists of two or more arcs with different radii. The *reverse curve* consists of two arcs that curve in different directions. Only the simple curves are discussed here.

Several definitions relating to curves are presented in the next few paragraphs and are illustrated in Fig. 22-2. A curve is initially laid out with two straight lines or tangents. These lines are extended until they intersect and that *point of intersection* is called the P.I. The first tangent encountered is called the *back tangent* and the second one is called the *forward tangent*.

The curve is laid out so that it joins these tangents. The *points on tangents* (P.O.T.s) are the points where the curves run into the tangents. The first of these points is on the back tangent at the beginning of the curve and is called the *point of curvature* (P.C.). The second one is at the end of the curve on the forward tangent and is called the *point of tangency* (P.T.). In another notation the point of curvature may be written as T.C., indicating that the route changes from a tangent to a curve; the point of tangency may be written as C.T., indicating that the route goes from curve back to tangent.

The angle between the tangents is called the *intersection angle* and is labeled I. The *radius of the curve* is R, while T is the *tangent distance* and equal to the

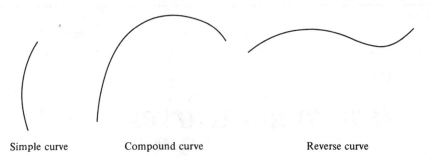

Simple curve   Compound curve   Reverse curve

Figure 22-1

length of the back or forward tangents. The distance from the P.I. to the middle point of the curve is called the *external distance* and is denoted by *E*. Finally, the chord of the arc from the P.C. to the P.T. is called the *long chord* (L.C.) and the distance from the middle of the curve to the middle of the long chord is labeled *M*, the *middle ordinate*, and *L* is the actual curve length.

The discussion of horizontal curves presented in this chapter makes use of the foot unit. Although this is the case, it should be realized that all of the equations used herein are perfectly valid for metric distances as long as a full station is understood to be 100 m.

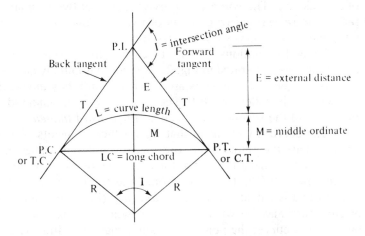

Figure 22-2 Curve notation.

22-2 DEGREE OF CURVATURE AND RADIUS OF CURVATURE

The sharpness of a curve may be described in any of the following ways.

1. *Radius of curvature.* This method is often used in highway work where the radius for the curve is frequently selected as a multiple of 100 ft. The smaller the radius, the sharper the curve. Should the degree of curvature (defined

Figure 22-3   Horizontal curves on I-285 in Atlanta, Georgia. (Courtesy of Georgia Department of Transportation.)

in the next two paragraphs) be specified rather than the radius of the curve, the radius can be computed. It will be an odd number of feet.

2. *Degree of curvature, chord basis.* In this method, the degree of curvature is defined as the central angle subtended by a chord of 100 ft, as illustrated in Fig. 22-4. The radius of such a curve may be computed with the following expression:

$$R = \frac{50}{\sin \frac{1}{2}D}$$

3. *Degree of curvature, arc basis.* As shown in Fig. 22-5, the degree of curvature on the arc basis is the central angle of a circle that will subtend an arc of 100 ft. It will be noted that a sharp curve has a large degree of curvature and a flat curve a small degree of curvature. For a particular curve with a degree of curvature D, the radius of the curve, R, can be computed as follows:

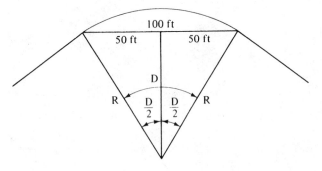

Figure 22-4 Degree of curvature, chord basis.

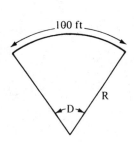

Sharp curve (large D, small R)

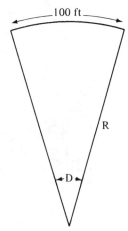

Flat curve (small D, large R)

Figure 22-5 Degree of curvature, arc basis.

$$\text{circumference of circle} = \left(\frac{360°}{D}\right)(100) = 2\pi R$$

$$R = \left(\frac{360°}{D}\right)\left(\frac{100}{2\pi}\right) = \frac{5729.58}{D}$$

The arc basis is used for the calculations presented in this chapter. Actually, both the chord and arc methods are used extensively in the United States. The method selected often depends on the experience of the surveyor involved. For long gradual curves, which are common in railroad practice, the chord basis (where the lengths along the arcs are considered to be the same as along the chords) is normally used. For highway curves and curved property boundaries, the arc basis is more common.

It will be noted that the difference between the chord and arc bases is normally not large. For instance, for a 1° curve, R on the arc basis is 5729.58 ft, and 5729.65 ft on the chord basis. Corresponding values for a 4° curve are 1432.39 ft and 1432.69 ft, respectively.

From a practical standpoint a horizontal curve defined by the arc basis would appear to be the same as a horizontal curve defined by the chord basis. However, the parameters for the two curves would be significantly different when computed to the nearest $\frac{1}{100}$th of a foot.

22-3 CURVE EQUATIONS

The formulas needed for circular curve computation are presented in this section, with reference to Fig. 22-2. The radius of curvature was given previously as

$$R = \frac{5729.58}{D}$$

The tangent distance T is the distance from the P.I. to the P.C. or P.T. and can be computed from

$$T = R \tan \tfrac{1}{2}I$$

The length of the long chord is

$$\text{L.C.} = 2R \sin \tfrac{1}{2}I$$

The external distance E is

$$E = R\left(\sec \frac{I}{2} - 1\right) = R \text{ ex. sec } \frac{I}{2}$$

where

$$\text{ex. sec} = 1 - \sec$$

A more convenient form with today's calculators is

$$E = R \left[\frac{1}{\cos (I/2)} - 1 \right]$$

The middle ordinate M equals

$$M = R - R \cos \frac{1}{2} I = R \text{ vers } \frac{I}{2}$$

where

$$\text{vers} = 1 - \cos$$

or, more conveniently,

$$M = R \left(1 - \cos \frac{I}{2} \right)$$

The length of the curve is

$$L = \frac{100I}{D}$$

For the chord definition of D,

$$L = \frac{RI\pi}{180}$$

where I is measured in degrees.

For surveyors who frequently work with curves, the use of versines and external secants simplifies the calculations. Various books provide tables of their values.[1] Use of these terms, is, however, becoming obsolete, because the equations using the secant and cosine values can be so easily solved with modern hand-held calculators (or with computer programs) without reference to tables.

Curves are staked out using straight-line chord distances. If the degree of curvature for a particular curve is 3° or less, the curve can be staked out using 100-ft chords as shown in Fig. 22-6, yet keep the arc length and chord length values sufficiently close to each other so as to be within the precision of tape measurements. As the usual curve will be of an odd length (i.e., not a whole number of hundreds of feet), there will certainly be subchord lengths at the ends of the curve as indicated in the figure. For curves from 3 to 7°, it is necessary to go to 50-ft chords, and for those from 7 to 14°, to 25-ft chords to maintain satisfactory precision.

For flat curves, say 2 or 3°, the chord distances and arc distances are almost the same to two places beyond the decimal, whereas for sharper curves the dif-

---

[1] For example, T. F. Hickerson, *Route Location and Design*, 5th ed. (New York: McGraw-Hill Book Company, 1967), pp. 559–588.

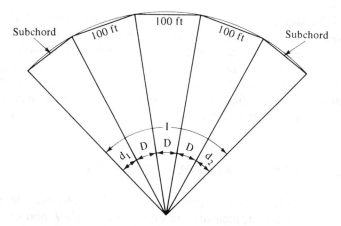

Figure 22-6   Length of curve, chord basis.

ference is more pronounced. For a 100-ft long 2° curve, the chord is 99.995 ft, and for a 100-ft 3° curve, it is 99.989 ft. For a 6° curve, the chord is 99.954 ft, and for a 10° curve, it is 99.873 ft.

22-4 DEFLECTION ANGLES

The angle between the back tangent and a line drawn from the P.C. to a particular point on a curve is called the *deflection angle* to that point. Circular curves are laid out almost entirely by using these angles. From the geometry of a circle, the

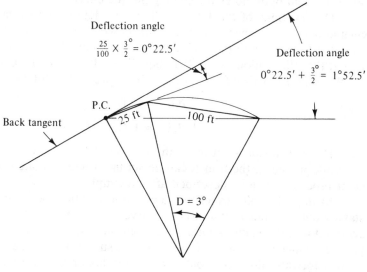

Figure 22-7

angle between a tangent to a circular curve and a chord drawn from that point of tangency to some other point on the curve equals one-half of the angle subtended by that chord. Thus for a 100-ft chord, the deflection angle is $D/2$, and for a 50-ft chord is $50/100 \times D/2$. These values are illustrated in Fig. 22-7, where D is $3°$. It will be noted that the deflection angle from the P.C. to each succeeding 100-ft station can be calculated by adding $D/2$ to the last deflection angle, or to each succeeding 50-ft station by adding $D/4$ to the last deflection angle.

22-5 SELECTION AND STAKING OUT OF CURVES

Before a horizontal curve can be selected, it is necessary to extend the tangents until they intersect, measure the intersection angle I, and select D, the degree of curvature. For any pair of intersecting tangents, an infinite number of curves can be selected, but the field conditions will narrow down the choices considerably. For high-speed highways, the degree of curvature is generally kept below certain maximums, while for twisting mountain roads, the lengths of tangents may be severely limited by the topography. Should a road be running along the bank of a river, the external distance E may be restricted.

Actually, the curve to be used can be selected by assuming a value for D, R, T, E, or L. This permits the calculation of the other four since they are dependent on each other. Usually, the radius of the curve or the degree of curvature is the value assumed. Then the necessary computations can be made as illustrated in Example 22-1. In addition to the determination of R, T, L.C., E, M, and L, these computations include the calculations of the deflection angles and the setting up of the field-book notes for staking out the curve.

The positions of the P.C. and P.T. are determined by measuring the calculated tangent distance T from the P.I. down both tangents. The reader should note that the station of the P.C. equals the station of the P.I. minus the distance T, and that the station of the P.T. equals the P.C. plus the curve length L (not the P.I. plus T). This information can be expressed as follows:

$$\text{P.C.} = \text{P.I.} - T$$

$$\text{P.T.} = \text{P.C.} + L$$

The instrument is set up at the P.C. or P.T. and the curve staked out. As the work proceeds, the transit can be set up at intermediate points as well. The latter procedure will be described after Example 22-1.

Example 22-1 illustrates the calculations and field-book notes needed for the staking out of a horizontal circular curve. It will be noted that these notes are arranged and the stations numbered from the bottom of the page upward. This practice, which is common for route surveys such as this one, enables the surveyor to look forward along the route and follow his or her notes while going in the same direction. In the same manner, as the surveyor looks at the sketch page and forward along the route, items to the right of the project center line on the ground are to the right on the sketch page, and vice versa.

As regards stationing the reader will remember that if we say T is 308.82 ft that's the same as 3 + 08.82 stations or if L is 982.36 ft that's 9 + 0.8236 stations.

Example 22-1

For a horizontal circular curve, the P.I. is at station 64 + 32.2, I is 24°20′, and a D of 4°00′ has been selected. Compute the necessary data and set up the field notes for 50-ft stations.

Solution   The values of R, T, L.C., E, M, and L are computed by the formulas presented previously, although tables are given in many books for simplifying the calculations.[2] For horizontal curves it is common to carry calculations to the nearest ±0.01 ft.

$$R = \frac{5729.58}{4} = 1432.39 \text{ ft}$$

$$T = (1432.39)(0.21560) = 308.82 \text{ ft}$$

$$\text{L.C.} = (2)(1432.39)(0.21076) = 603.77 \text{ ft}$$

$$E = (1432.39)\left(\frac{1}{\cos 12°10'} - 1\right) = 32.91 \text{ ft}$$

$$M = 1432.39 - (1432.39)(\cos 12°10') = 32.17 \text{ ft}$$

$$L = \frac{(100)(24.333333)}{4} = 608.33 \text{ ft}$$

From these data, the stations of the P.C. and P.T. can be calculated as follows:

P.I. =	64 + 32.2
− T =	−(3 + 08.8)
P.C. =	61 + 23.4
+ L =	6 + 08.2
P.T. =	67 + 31.6

As the degree of curvature is between 3 and 7°, the curve will be staked out with 50-ft chords. The distance from the P.C. (61 + 23.4) to the first 50-ft station (61 + 50) is 26.6 ft, and the deflection angle to be used for that point is (26.6/100) × (D/2) = (26.6 /100) × (4°/2) = 0°32′. For each of the subsequent 50-ft stations from 61 + 50 to 67 + 00, the deflection angles will increase by D/4 or 1°00′. Finally, for the P.T., the deflection angle will be 11°32′ + (31.6/100)(4°/2) = 12°10′. This value equals $\frac{1}{2}I$, as it should.

The field notes are set up as shown in Fig. 22-8.

[2] T. F. Hickerson, *Route Location and Design*, 5th ed. (New York; McGraw-Hill Book Company, 1967), pp. 396–458.

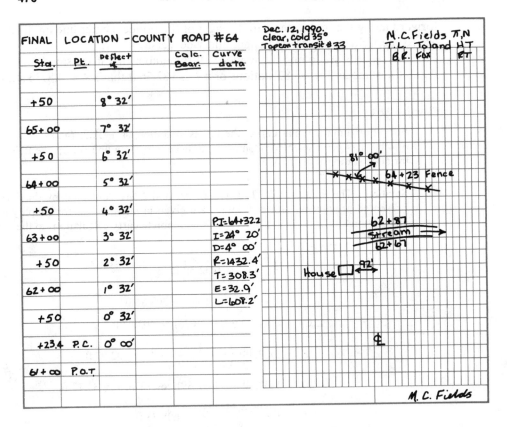

Figure 22-8　Field notes for a horizontal curve.

22-6 FIELD PROCEDURE FOR STAKING THE CURVE

The tangent distances are taped from the P.I. down both tangents to locate the P.C. and P.T. For this discussion the instrument is assumed to be set up at the P.C. and the curve and numbers of Example 22-1 used. It should be noticed, however, that if the entire curve can be seen from the P.T., it is possible to avoid one instrument setup by setting up there, because after the curve is completed, the survey can proceed down the forward tangent from the same setup.

The instrument is set up at the P.C. with the scale reading zero while sighting on the P.I. The first deflection angle (0°32′) is turned and the chord distance of 26.6 ft is taped to locate station 61 + 50. The next deflection angle (1°32′) is turned and 50 ft is taped beyond station 61 + 50 to stake station 62 + 00. This procedure is continued until the P.T. is reached. These points should check very closely. For long curves it is considered better to run in the first half of the curve from the P.C. and the second half back from the P.T. so that small errors that occur can be adjusted at the middle of the curve, where a little variation is not as important as it would be near the points of tangency to the curve.

Very often it may not be possible to set all the points on the curve from one position. The instrument may have to be moved up to one of the intermediate stations on the curve and the work continued. To do this, the instrument is set up on the intermediate station and is backsighted on the P.C. with the telescope inverted and the scale reading 0°00'. The telescope is reinverted, the deflection angle that would have been used for that next station turned, and the process continued.

As an illustration for Example 22-1, it is assumed that the instrument is moved up to station 64 + 00 and backsighted on the P.C. with the telescope inverted. The telescope is reinverted and turned to an angle of 6°32', the distance of 50 ft taped, and station 64 + 50 set.

Another horizontal curve situation is considered in Example 22-2. A surveyor is assumed to be working along the boundary of a piece of private property which adjoins a curved section of a state highway. He would like to determine the information necessary to describe correctly the curved boundary of the land.

Example 22-2

As shown in Fig. 22-9, a surveyor has determined that the intersection angle for a particular highway curve is 28°00'. In addition, he or she has measured the external distance E from the P.I. to the highway center line for this curve and found it to be 73.6 ft.

(a) Determine the values of D, R, and T for the highway center line.

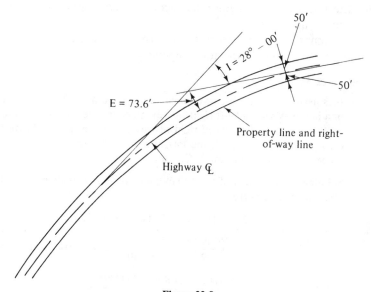

Figure 22-9

(b) If the highway right-of-way is 50 ft from the highway center line as shown in the figure, determine for the property on the inside of the curve the values of R and D.

Solution

(a) $R = \dfrac{73.6}{0.03061} = \textbf{2404.4 ft}$

$D = \dfrac{5729.58}{2404.4} = 2.383° = \textbf{2°23}'$

$T = (2404.4)(0.24933) = \textbf{599.5 ft}$

(b) R of property line or right-of-way line $= 2404.4 - 50 = \textbf{2354.4 ft}$

D of property line curve $= \dfrac{5729.58}{2354.4} = 2.434° = \textbf{2°26}'$

The surveyor who is staking out new lots or resurveying old ones along a highway will often find that parts of the lot boundaries adjacent to the right-of-way (ROW) of the highway will be curved. His or her work should reflect these curved property lines.

From the equations given earlier in this chapter it is evident that only two of the elements (R, T, L.C., E, M, and L) of a curve are necessary to define the curve. It is often desirable, however, for the surveyor to provide three or even more of these elements. If more than two elements are provided, they must be compatible with each other. Among the curve elements most commonly provided are the radius, intersection angle, curve length, and sometimes the length and bearing of the chord.

Example 22-3 illustrates the information necessary to stake out 100-ft lots along a right-of-way curve for a highway.

Example 22-3

It is assumed that the curve considered in Example 22-1 represents the center line of a highway with a 50-ft right-of-way on each side. It is desired to lay off 100-ft lots on the inside of the curved right-of-way as shown in Fig. 22-10. Compute the necessary data for laying off the lots assuming that the last corner on the ROW back tangent is 53.7 ft from the P.C.

Solution   The radius of the ROW curve is 50 ft less than the radius of the center-line curve. Therefore,

$$R = 1432.39 - 50.00 = 1382.39 \text{ ft}$$

$$D = \frac{5729.58}{R} = \frac{5729.58}{1382.39} = 4.1446914° = 4°08.7'$$

$$L = \frac{100I}{D} = \frac{(100)(24.333333)}{4.1446914} = 587.10 \text{ ft}$$

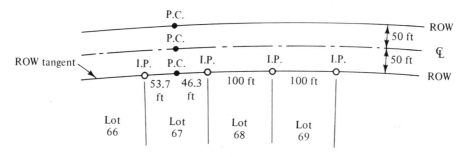

Figure 22-10

Setting up the instrument at the P.C. on the ROW curve the deflection angle to the first lot corner on the curve (a distance equal to $100.0 - 53.7 = 46.3$ ft) is equal to

$$\left(\frac{46.3}{100}\right)\left(\frac{4.1446914}{2}\right) = 0.95949606° = 0°57.6'$$

From the same position of the instrument, the deflection angle to the next lot corner is

$$\left(\frac{146.3}{100}\right)\left(\frac{4.1446914}{2}\right) = 3.0318418° = 3°01.9'$$

or it can be calculated as

$$0°57.6' + \frac{D}{2} = 0.95946606° + \frac{4.1446914°}{2} = 3.0318118° = 3°01.9'$$

The deflection angle for the P.T. can be determined by taking the initial deflection angle $(0°57.6')$ and adding $D/2$ to it for each 100 ft plus the angle value for the partial lot width before the P.T.; or it equals

$$\left(\frac{587.19}{100}\right)\left(\frac{4.1446914}{2}\right) = 12.166742° = 12°10'$$

which is half of I of $24°20'$.

Note: Since the degree of curvature is between 3 and 7°, the curve should be laid out with 50-ft chords with the iron pin corner monuments placed at the 100-ft points.

22-7 HORIZONTAL CURVES PASSING THROUGH CERTAIN POINTS

Sometimes it is necessary to lay out a horizontal curve that passes through a certain point. For instance, it may be desired to establish a curve that passes no closer than a certain distance to some feature such as a building or stream. The

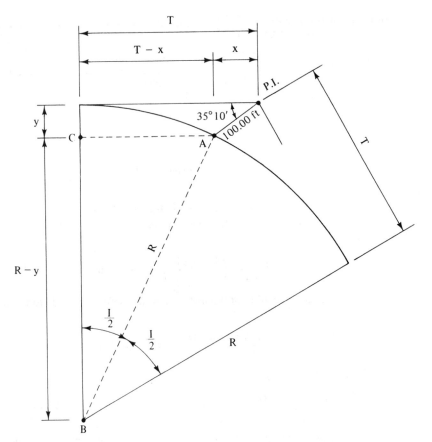

Figure 22-11

distance from the P.I. to the point in question and the angle from one of the tangents can be measured. Such a situation is considered in Example 22-4 and illustrated in Fig. 22-11.

To solve the problem, the distances x and y shown in Fig. 22-11 can be computed by trigonometry. With reference to the figure it can be seen that the radius of the curve R can be determined by considering the right triangle ABC. For this triangle the following quadratic equation applies:

$$R^2 = (R - y)^2 + (T - x)^2$$

The value of T can be expressed in terms of R as $R \tan \frac{1}{2}I$ and substituted into the preceding equation leaving only R as an unknown. Then R can be determined with the quadratic equation, by trial and error with a calculator or by completing the squares. If the quadratic equation is used, the solution will yield two answers, but one of them will be seen to be unreasonable. Hickerson[3] provides a solution to this type of problem which does not involve the use of a quadratic equation.

Example 22-4

A horizontal curve is to be run through point A as shown in Fig. 22-11. From the P.I. the distance to point A is 100.00 ft and the angle from the back tangent to a line from the P.I. to point A is $35°10'$. If I is $65°00'$, determine the values of $R, D, T,$ and L.

Solution

$$y = (100.00)(\sin 35°10') = 57.596 \text{ ft}$$

$$x = (100.00)(\cos 35°10') = 81.748 \text{ ft}$$

$$T = (R)\frac{\tan 65°00'}{2} = 0.63707R$$

$$R^2 = (R - y)^2 + (T - x)^2$$

$$= (R - 57.596)^2 + (0.63707R - 81.748)^2$$

Squaring the terms and simplifying yields

$$R^2 - 540.46R + 24{,}639.137 = 0$$

Using the quadratic equation with $a = 1.00$, $b = -540.46$, and $c = 24{,}639.137$:

$$R = \frac{+540.56 \pm \sqrt{(-540.46)^2 - (4)(1.00)(24{,}639.137)}}{(2)(1.00)}$$

$$= \textbf{490.20 ft} \quad \text{or} \quad 50.26 \text{ ft (not feasible)}$$

$$D = \frac{5729.58}{490.20} = \textbf{11.688°}$$

$$T = (490.20)\left(\tan \frac{65°00'}{2}\right) = \textbf{312.29ft}$$

$$L = \frac{(100)(65.00)}{11.688} = \textbf{556.13 ft}$$

PROBLEMS

All the problems listed here are to be solved on the arc basis.

22-1. For a particular horizontal curve the degree of curvature is $4°00'$. Compute its radius of curvature.

(Ans.: 1432.40 ft)

[3] T. F. Hickerson, *Route Location and Design*, 5th ed. (New York: McGraw-Hill Book Company, 1967), pp. 90–91.

22-2. Repeat Problem 22-1 if the degree of curvature is 12°00′.

In Problems 22-3 to 22-5, a series of horizontal curves are to be selected for the data given. Determine the stations for the P.C.s and P.T.s for each of these curves.

	P.I.	Intersection angle, I	Degree of curvature, D
22-3.	10 + 16.56	9°32′	2°30′
22-4.	32 + 54.92	18°20′	3°30′
22-5.	87 + 09.20	12°42′	4°00′

(Ans.: 8 + 25.45, 12 + 06.78)

(Ans.: 85 + 49.80, 88 + 67.30)

22-6. If I is 33°20′30″ and the maximum value of E is 85.0 ft, determine the degree of curvature D to the nearest full minute that will provide this E.

22-7. Two highway tangents intersect with a right intersection angle of 31°10′. If a 3°30′ horizontal circular curve is to be used to connect the tangents, compute $R, T, L,$ and E for the curve.

(Ans.: 1637.02 ft, 456.55 ft, 890.48 ft, 62.47 ft)

22-8. A horizontal circular curve having a radius of 600.00 ft is to connect two highway tangents. If the chord length L.C. is 700.00 ft, compute $E, L, M,$ and the intersection angle I.

In Problems 22-9 to 22-12, prepare the field-book notes to the nearest full minute for laying out the curves with stakes at full stations for the data given.

22-9. P.I. at 91 + 64.30, $I = 20°00′$, and $D = 3°00′$.

(Ans.: $R = 1909.86$ ft, Deflection angle @ 92 + 00 = 5°35′)

22-10. P.I. at 76 + 32.10, $I = 28°22′$, and $D = 4°00′$.

22-11. P.I. at 117 + 16.60, $I = 18°32′$, and $D = 2°30′$.

(Ans.: $L = 741.33$ ft, Deflection angle @ 117 + 00 = 4°28′)

22-12. It is necessary to have a horizontal curve pass through a specified point. The point was located by distance and angle from the P.I. of the curve (150.00 ft and 32°00′ to the left of the back tangent with the instrument located at the P.I.). If the intersection angle between the back and forward tangents is 50°00′ to the right, determine the curve radius required.

CHAPTER TWENTY-THREE

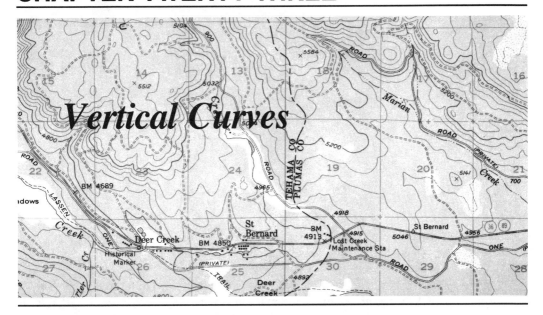

Vertical Curves

23-1 INTRODUCTION

The curves used in a vertical plane to provide a smooth transition between the grade lines of highways and railroads are called *vertical curves*. These curves are parabolic rather than circular. Figure 23-1 shows the nomenclature used for vertical curves. When moving along the road, the first of the grade lines that will be encountered is called the *back tangent*. The other one is called the *forward tangent*. To distinguish the points of tangency and their intersection from the similar terms used for horizontal curves, the letter *V* (for vertical) is added to their abbreviations. For instance, the point of intersection for the tangents is called the P.V.I. (point of vertical intersection) and the points of tangency are referred to as the P.V.C. (point of vertical curvature at start of curve) and P.V.T. (point of vertical tangency at end of curve). The *tangent offsets* are distances measured from the tangents in a vertical direction to the curve.

The center-line stakes for the project are set, the profile level made, and the profile drawn with the exaggerated vertical scale as described in Chapter 8. Then trial tangents or grade lines are selected with the idea of establishing a design with optimum cut and fill and hauling costs, reasonable percent grades, and with satisfactory sight distances at the top of hills. Adjustments to the curve are made by changing the curve length *L*.

Several different methods are available for making the necessary calculations. One very satisfactory method involves the use of tangent offsets from the grade lines and it is the method that is stressed in this chapter. An alternative method involving the equation of a parabola is presented in Section 23-6. The

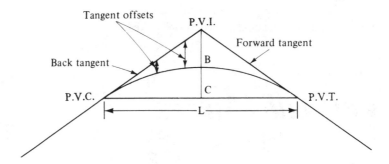

Figure 23-1 Vertical curve nomenclature.

parabola has three mathematical properties that make its application very convenient. These are as follows:

1. The curve elevation at its midpoint will be halfway from the elevation at the P.V.I. to the elevation at the midpoint of a straight line from the P.V.C. to the P.V.T.
2. The tangent offsets vary as the square of the distance from the point of tangency.
3. For points spaced at equal horizontal distances, the second differences are equal. This is a useful property for checking vertical curve calculations as is illustrated in Example 23-1. The differences between the elevations at equally spaced stations are called the *first differences*. The differences between the first differences are called the *second differences*.

23-2 VERTICAL CURVE CALCULATIONS

The station of the vertex or P.V.I. can be determined from the grade lines. Then the length of the curve is selected—usually as some whole number of hundreds of feet or stations. The length of a vertical curve is defined as the horizontal distance from the P.V.C. to the P.V.T. This length is normally controlled by the specification being used. For instance, the length required for a highway vertical curve must be sufficient to provide a sufficiently flat curve over the top of a hill so that certain minimum sight distances of oncoming vehicles or objects in the road are provided. The AASHTO (American Association of State Highway and Transportation Officials) provides such information depending on the design speed of the highway.[1] This topic is continued in Section 23-3.

[1] *A Policy on Geometric Design of Highways and Streets* (Washington, D.C.: AASHTO, 1984), pp. 800–803.

Example 23-1 illustrates the calculations required for determining the elevations along a vertical curve. For this example the given grade lines intersect at station 65 + 00 at an elevation of 264.20 ft, and the curve is assumed to be 800 ft long. From this information the stations of the P.V.C. and P.V.T. can be determined as follows:

$$P.V.C. = P.V.I. - \frac{L}{2}$$

$$P.V.T. = P.V.I. + \frac{L}{2} \text{ or } P.V.C. + L$$

Then the elevations of the intermediate stations along the grade lines or tangents are calculated. The elevation at the midpoint of the long chord is equal to the average elevation of the P.V.C. and the P.V.T.

Next, the elevation of the curve midpoint (point B in Figs. 23-1 and 23-3) is determined by averaging the elevations of point C and the P.V.I. Finally, the tangent offset distances from the grade lines are calculated, thus giving the elevations of the intermediate stations on the curve.

Example 23-1

Figure 23-2 shows the known data for a vertical curve. Assume a curve length of 800 ft and compute the elevation of each full station by using the tangent offset method.

Solution   From the length of the curve the stations of the P.V.C. and P.V.T. are determined (61 + 00 and 69 + 00, respectively) and from the tangent grades the station elevations along the tangents are obtained. Then the elevation of point B, the midpoint of the curve, is calculated as follows:

$$\text{Elevation of P.V.I.} = 264.20 \text{ ft}$$

$$\text{Elevation of } C = \frac{248.20 + 252.20}{2} = 250.20 \text{ ft}$$

$$\text{Elevation of } B = \frac{264.20 + 250.20}{2} = 257.20 \text{ ft}$$

The tangent offset values are calculated as follows:

$$\text{Difference in elevation from P.V.I. to } B = 264.20 - 257.20 = 7.00 \text{ ft}$$

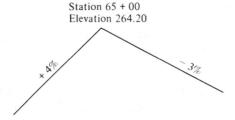

Station 65 + 00
Elevation 264.20

+ 4%   − 3%

Figure 23-2

$$\text{Tangent offset at station } 62 + 00 = \frac{(1)^2}{(4)^2}(7.00) = 0.44 \text{ ft}$$

$$\text{Tangent offset at station } 63 + 00 = \frac{(2)^2}{(4)^2}(7.00) = 1.75 \text{ ft}$$

The elevations of the stations are computed by subtracting the tangent offsets from the grade elevations. These values are shown in Fig. 23-3.

In Table 23-1 all of the preceding information is recorded, and in addition the first and second differences are computed as a math check.

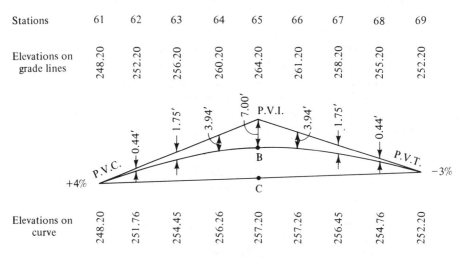

Figure 23-3

TABLE 23-1

Station	Point	Elevation on grade lines	Tangent offset	Curve elevation	First difference	Second difference
61	P.V.C.	248.20	0.00	248.20		
					+3.56	
62		252.20	0.44	251.76		0.87
					+2.69	
63		256.20	1.75	254.45		0.88
					+1.81	
64		260.20	3.94	256.26		0.87
					+0.94	
65	P.V.I.	264.20	7.00	257.20		0.88
					+0.06	
66		261.20	3.94	257.26		0.87
					−0.81	
67		258.20	1.75	256.45		0.88
					−1.69	
68		255.20	0.44	254.76		0.87
					−2.56	
69	P.V.T.	252.20	0.00	252.20		

Example 23-2 provides another vertical curve example. The calculations are made exactly as they were for the curve of Example 23-1. For all vertical curve problems it is necessary for the person making the calculations to be very careful to use the correct signs for determining the elevations along the tangents. He or she must be equally careful with the signs of the offset distances from these tangents, which are used to determine the curve elevations. In this particular case the offsets are measured up from the tangent or grade lines. Space will not be taken here to show the second difference check, which is nevertheless always desirable.

Example 23-2

A $+3\%$ grade line intersects a $+5\%$ grade line at station $62 + 00$, where the elevation is 862.30 ft as shown in Fig. 23-4. Determine the elevations at full 100-ft stations along the curve if a 600-ft long curve is to be used.

Solution

1. The elevations along the grade lines are computed and shown in the figure.

2. Elevation of point C $= \dfrac{853.30 + 877.30}{2} = 865.30$

3. Elevation at midpoint of curve $= \dfrac{862.30 + 865.30}{2} = 863.80$

4. The tangent offsets are computed and subtracted from the grade-line elevations, giving the final curve elevations as shown in the figure.

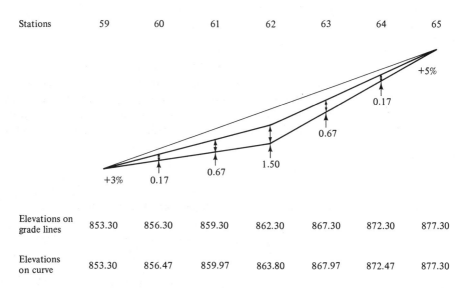

Stations	59	60	61	62	63	64	65
Elevations on grade lines	853.30	856.30	859.30	862.30	867.30	872.30	877.30
Elevations on curve	853.30	856.47	859.97	863.80	867.97	872.47	877.30

Figure 23-4

23-3 MISCELLANEOUS ITEMS RELATING TO VERTICAL CURVES

Elevation of Intermediate Points on Curves

It is often necessary to compute the elevation of points on vertical curves at closer intervals than full stations, for example, at 50 ft or 25 ft, as well as the values for some occasional intermediate point. For such cases as these, the tangent offset method will work as well as it did at full stations, although the numbers are not quite so convenient. As an illustration, the elevation of station 63 + 75.2 for the curve of Example 23-1 equals

$$248.20 + (.04)(275.2) - \frac{(2.752)^2}{(4)^2}(7.00) = 255.90 \text{ ft}$$

Highest or Lowest Point on Curve

When the two grade lines for a vertical curve have opposite algebraic signs, there will be either a low point or a high point between the P.V.C. and the P.V.T. This point will probably not fall on a full station, and yet its position will often be of major significance, as, for example, where drainage facilities are being considered.

For a particular vertical curve the total change in grade (A) between the tangents is determined from the following expression in which g_1 and g_2 are the percent grades of the back and forward tangents, respectively:

$$A = g_2 - g_1$$

The rate of change of grade r is determined by the following expression, in which L is the curve length in stations:

$$r = \frac{A}{L}$$

For this discussion it is assumed that a vehicle moves on to a curve which has an initial grade g_1. As the vehicle moves along the curve it will rotate vertically until finally at the far end it will be sloped at g_2. During its rotation it will eventually become level. If it rotates at a rate of r percent per station, the number of stations to the level point (which is the high or low point) is calculated by dividing r into g_1:

$$X = \frac{-g_1}{r}$$

For the curve of Example 23-1, the high point is as follows:

$$r = -\frac{g_2 - g_1}{L} = -\frac{-3 - 4}{8} = -\frac{7}{8}$$

$$X = -\frac{g_1}{r} = -\frac{4}{-7/8} = +4.57 \text{ stations} = 65 + 57$$

The elevation of the high point equals

$$264.20 - (0.03)(57) - \frac{(3.43)^2}{(4)^2}(7.00) = 257.34 \text{ ft}$$

Sight Distance and Curve Length

As mentioned previously, it is necessary for vertical curves to be so constructed that drivers will have certain minimum sight distances for seeing cars or other objects on the road. The driver should be able to see an object of a given height at no less than the estimated distance that he or she would travel while reacting to put his or her foot on the brake pedal plus the distance required for the car to stop.

The AASHTO Specification goes into great detail in describing minimum lengths needed for sag and crest vertical curves and minimum sight distances. The length of vertical curves is generally predicated on minimum sight distances measured from an assumed eye height of 3.50 ft above the road looking at an object with a 6 in. height in the road.

At least four different criteria are used in establishing the lengths of sag vertical curves. These are (1) headlight sight distance (2) rider comfort (3) drainage control and (4) general appearance. The lengths selected for crest vertical curves are based upon safety, comfort and appearance with some other special considerations. The reader can refer to the AASHTO for this detailed information.[2]

23-4 UNEQUAL-TANGENT VERTICAL CURVES

Almost all vertical curves have equal-length tangents. All those considered so far in this chapter fall into this class. Sometimes, however, it is desirable to use unequal-tangent curves to make them fit unusual topographical situations. An unequal-tangent vertical curve consists of two equal tangent vertical curves each with a different r value (rate of change of grade). The end of the first curve (its P.V.T.) coincides with the start of the second curve (its P.V.C.). This point is referred to as the *compound vertical curvature point* or C.V.C.). This point is shown in Fig. 23-5.

The C.V.C. is located by drawing a straight line from the midpoint of the back tangent to the midpoint of the forward tangent (represented by the dashed line *DE* in Fig. 23-5). The C.V.C. is located on this dashed line directly above or below the P.V.I. The two equal-tangent curves will be selected so as to be tangent to line *DE* at the C.V.C. As a result, there will be a smooth transition from the first curve to the second curve.

The elevation of the C.V.C. can be determined by proportions or it can be

[2] ibid pp. 303–315

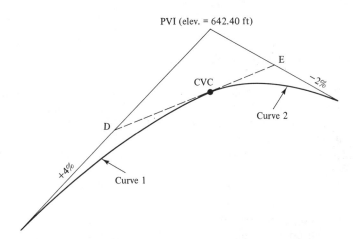

Stations   62   63   64   65   66   67   68   69   70   71   72

Figure 23-5   Compound vertical curvature (CVC) point for an unequal-tangent curve.

obtained from the grade of line DE as follows, noting the elevation of the P.V.I. and the percentages of the grade lines given in Fig. 23-5:

$$\text{Elevation of } D = 642.40 - (0.04)(300) = 630.40 \text{ ft}$$

$$\text{Elevation of } E = 642.40 - (0.02)(200) = 638.40 \text{ ft}$$

$$\text{Grade of line } DE = \frac{638.40 - 630.40}{500} = +0.016 = +1.60\%$$

$$\text{Elevation of C.V.C.} = 630.40 + (0.016)(300) = 635.20 \text{ ft}$$

Now it is possible to proceed with the calculations for the two curves noting that for the first curve the length is 600 ft, $g_1 = +4\%$, and $g_2 = +1.6\%$. For the second curve of 400-ft length, $g_1 = +1.60\%$ and $g_2 = +1.6\%$. In Example 23-3 that follows, elevations are determined along this unequal-tangent curve.

Example 23-3

A $+4\%$ grade line intersects a -2% grade line at station 68 + 00, where the elevation is 642.40 ft, as shown in Fig. 23-5. Determine elevations of 100-ft stations for this 1000-ft unequal-tangent curve.

Solution   The elevation of the C.V.C. and the percent grade of line DE determined before this example are actually the first part of this solution. The elevations along the grade lines are computed and shown in Table 23-2. In computing these elevations, tangent elevations from stations 62 to 65 are changing at $+4\%$, while from stations 65 to 70 they are changing at $+1.60\%$, and from stations 70 to 72 the change is -2%.

TABLE 23-2

	Station	Point	Elevation on grade lines	Tangent offset	Curve elevation	First difference	Second difference
Curve 1	62	P.V.C.$_1$	618.40	0.00	618.40	+3.80	0.40
	63		622.40	0.20	622.20	+3.40	0.40
	64		626.40	0.80	625.60	+3.00	0.40
	65	P.V.I.$_1$ (D)	630.40	1.80	628.60	+2.60	0.40
	66		632.00	0.80	631.20	+2.20	0.40
	67		633.60	0.20	633.40	+1.80	0.40
	68	P.V.T.$_1$ and P.V.C.$_2$ (C.V.C)	635.20	0.00	635.20	+1.15	
Curve 2	69		636.80	0.45	636.35	+0.25	0.90
	70	P.V.I.$_2$(E)	638.40	1.80	636.60	−0.65	0.90
	71		636.40	0.45	635.95	−1.55	0.90
	72	P.V.T.$_2$	634.40	0.00	634.40		

Curve 1:

Elevation of P.V.I. = elevation of D = 630.40 ft

Elevation of midpoint of straight line from station 62 to C.V.C.
$$= \frac{618.40 + 635.20}{2} = 626.80 \text{ ft}$$

Elevation of midpoint of curve $= \frac{630.40 + 626.80}{2} = 628.60 \text{ ft}$

Difference in elevation from P.V.I. to midpoint of curve $= 630.40 - 628.60$
$= 1.80$ ft

Tangent offset at station 63 $= \frac{(1)^2}{(3)^2} (1.80) = 0.20 \text{ ft}$

Curve 2:

Elevation of P.V.I. = elevation of E = 638.40 ft

Elevation of midpoint of straight line from station 68 to station 72
$$= \frac{635.20 + 634.40}{2} = 634.80 \text{ ft}$$

Elevation of midpoint of curve $= \frac{638.40 + 634.80}{2} = 636.60 \text{ ft}$

Difference in elevation from P.V.I. to midpoint of curve $= 638.40 - 636.60$
$= 1.80$ ft

Tangent offset at station 69 $= \frac{(1)^2}{(2)^2} (1.80) = 0.45 \text{ ft}$

23-5 VERTICAL CURVE PASSING THROUGH A CERTAIN POINT

A problem commonly faced in working with vertical curves is that of passing a curve through a definite point. For instance, as shown in part (a) of Fig. 23-6, it may be desired to have a vertical curve pass a certain distance over the top of a

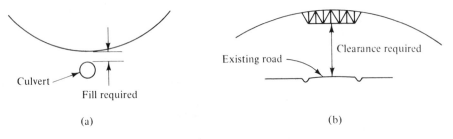

(a)   (b)

Figure 23-6

culvert. The problem is to determine the correct length of a curve which will pass through the point in question. In a similar fashion a vertical curve on a bridge may need to pass a certain distance for clearance over an existing roadway, railroad, or navigable stream as shown in part (b) of the figure. To solve the problem, the tangent offset expression is written from both sides of the curve using the length L as one unknown and the tangent offset at the P.V.I. as the other. The two equations may be solved simultaneously for the unknown values, as illustrated in Example 23-4.

Example 23-4

A -6% grade line intersects a $+5\%$ grade line at station $11 + 00$ (elevation 43.00 ft). The top of a culvert oriented at $90°$ to the road center line at station $10 + 00$ is to be located at elevation 53.00 ft. Determine the vertical curve length in whole stations such that there will be from 1.0 to 3.0 ft of cover material over the top of the culvert.

Solution Assume the material covering the top of the culvert is 2.0 ft thick and determine the theoretical curve length required. Thus the desired elevation on the curve at that station $(10 + 00)$ is $53.00 + 2.00 = 55.00$ ft.

With reference to Fig. 23-7, expressions for the offsets at station $10 + 00$ are written with respect to both tangents and labeled y_{10}.

1. The distance from the back tangent to the desired curve elevation = $55.00 - 49.00 = 6.00$ ft = y_{10}.

2. The distance from the forward tangent (extended to station $10 + 00$) to the desired curve elevation = $55.00 - 38.00 = 17.00$ ft = y_{10}.

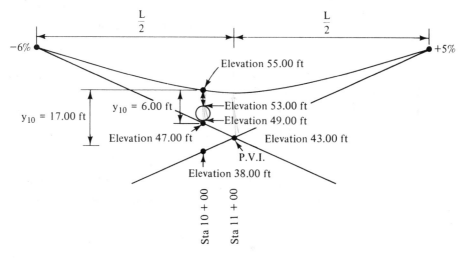

Figure 23-7

In the following two offset expressions, P is the tangent offset at the P.V.I. and L the desired curve length in stations.

For the back tangent:

$$\frac{6.00}{P} = \frac{\left(\frac{L}{2} - 1.00\right)^2}{\left(\frac{L}{2}\right)^2} \qquad P = \frac{6.00\left(\frac{L}{2}\right)^2}{\left(\frac{L}{2} - 1.00\right)^2} \tag{1}$$

For the forward tangent:

$$\frac{17.00}{P} = \frac{\left(\frac{L}{2} + 1.00\right)^2}{\left(\frac{L}{2}\right)^2} \qquad P = \frac{17.00\left(\frac{L}{2}\right)^2}{\left(\frac{L}{2} + 1.00\right)^2} \tag{2}$$

Equating the values for P [Eqs. (1) and (2)] and solving for L yields

$$L = 7.88 \text{ stations}$$

Use $L = 8$ stations and calculate elevations along the curve, making sure that the culvert cover is between 1 and 3 ft.

23-6 PARABOLIC EQUATION

The elevations of points along a vertical curve can be determined using the equation of a parabola instead of the tangent offset method. To use the parabolic equation that follows, a coordinate system is set up with x = the horizontal distance in feet from the P.V.C. to a particular point, y = the elevation of the point in question in feet, and r = the rate of grade change = $(g_2 - g_1)/L$:

$$y = \tfrac{1}{2} rx^2 + g_1 x + \text{elevation of P.V.C.}$$

In Example 23-5 the elevations along the curve of Example 23-1 are recomputed using the preceding equation.

Example 23-5

Repeat Example 23-1 using the parabolic equation.

Solution From Example 25-1, $g_1 = +4\%$, $g_2 = -3\%$, and $L = 8$ stations.

Elevation of P.V.C. $= 264.20 - (0.04)(400) = 248.20$ ft

Being very careful with signs, the value of r is determined.

$$r = \frac{-3 - (+4)}{8} = -\frac{7}{8}$$

The elevations at full stations are computed and shown in Table 23-3.

TABLE 23-3

Station	× (stations)	$\frac{1}{2}rx^2$	$g_1 x$	$y = \frac{1}{2}rx^2 + g_1 x$ + elevation P.V.C.
61	0	0.0000	0.0000	248.20
62	1	−0.4375	4.0000	251.76
63	2	−1.7500	8.0000	254.45
64	3	−3.9375	12.0000	256.26
65	4	−7.0000	16.0000	257.20
66	5	−10.9370	20.0000	257.26
67	6	−15.7500	24.0000	256.45
68	7	−21.4380	28.0000	254.76
69	8	−28.0000	32.0000	252.20

23-7 CROWNS

In establishing the elevations for the construction of highways and streets, two factors in addition to the vertical curve values can effect the elevations at particular points. These factors, which are discussed in this section and the next, are *crowns* and *superelevations*.

To provide adequate drainage for the pavement surface, it is necessary to raise the center of the pavement with respect to the edges. Crowns or cross slopes usually run from about 0.015 to 0.020 ft/ft for heavy-duty pavements. For lower-duty pavements the slopes may be slightly higher.

The crown can be formed by two planes, but more commonly it has a parabolic or circular shape. Since the width of the pavement is large compared to the height of the crown, there is almost no difference between a circle and a parabola. The parabolic vertical curve formula can be used to calculate the amount of crown at a particular point with respect to the pavement edge, as shown in Fig. 23-8.

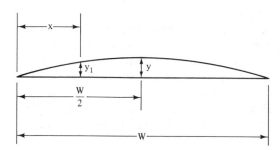

Figure 23-8 Pavement cross section showing pavement crown.

23-8 SUPERELEVATION

For horizontal curves, highway and street pavements are built with a transverse slope called superelevation, which is usually given as a percent. A fairly common value used is 8%. This means that a 24-ft-wide pavement would have its outside edge = (0.08)(24) = 1.92 ft higher than its inside edge.

The amount of superelevation used is controlled by the design speed of the highway and the side friction factor between tires and pavement. The maximum amount that can be used practically depends on the climate and the classification of the area as urban or rural. The maximum amount usually encountered is about 10%. Sometimes it will be a little higher—perhaps up to 12% for highways not subject to appreciable snow and ice and also for infrequently traveled gravel roads to help with cross drainage. These values are normally suitable for the southern states. Situations where combinations of design speeds and degrees of curvature require greater superelevations than the values given here should be avoided.

In parts of the country where snow and ice conditions occur frequently over a period of several months (usually the northern states), maximum rates of 7 to 8% are used to prevent vehicles from slipping sideways across the pavement when they are stopped or moving slowly. For urban streets, where there may be considerable traffic and where speeds often have to be reduced, maximum superelevations of 4 to 6% are common. Sometimes where there are long radius or flat horizontal curves with considerable turning and crossing of traffic, superelevation may be omitted.

Some highways have both fast and slow traffic, and varying weather situations occur during the different seasons. Because of these facts it is impossible to select superelevations that are appropriate for all traffic situations and weather conditions.

There is a transitional area between a normal crown cross section and full superelevation where traffic is on a tangent section of the highway approaching a horizontal curve. On this straight section we will start gradually working into the superelevation and work our way up to the maximum superelevation value, as shown in Fig. 23-9. In a similar fashion we will work our way from full superelevation to the normal cross section on the other end. Notice in this figure how the pavement cross section is revolved around the pavement center line. (Sometimes for divided highways with narrow medians the pavement may be revolved around one of its edges.)

The transition from the normal crown cross section on a tangent to the fully superelevated cross section is referred to as *runoff*. The purpose of runoff is to spread the transition between the two different pavement cross sections over a sufficient distance so as to provide a safe design. Runoff lengths probably range up to as much as 500 or 600 ft. Minimum runoff values vary from about 100 to 250 ft, regardless of superelevation values or pavement widths. These minimum values are used to improve the appearance of the highway and to smooth the edge profile of the pavement.

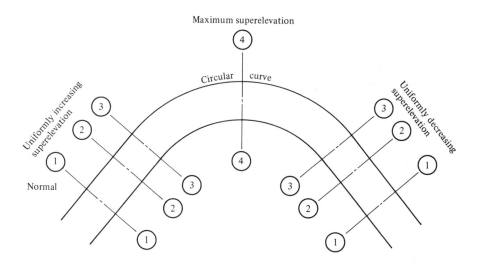

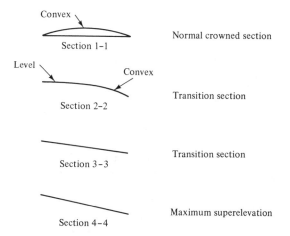

Figure 23-9

PROBLEMS

23-1. A 1000-ft vertical curve is to be used for joining a +4.0% grade line and a −2.0% grade line. The vertex or intersection of the grade lines is at station 62 + 00 and the elevation there is 533.30 ft. Compute the elevations at 100-ft stations throughout the curve.

(*Ans.:* Curve elevation at stations 58, 60, and 64 = 517.00 ft, 522.60 ft, 526.60 ft)

23.2. Repeat Problem 23-1 if the curve is 800 ft long and the first grade is 3.6% instead of +5.0%.

23-3. A +5.0 % grade meets a −2.1% grade at station 46 + 50, where the elevation is 682.24 ft. Compute the elevation of all full stations if the curve is to be 800 ft long. (*Ans.:* Curve elevations at stations 44, 46 + 50, 49 = 668.74 ft, 675.14 ft and 675.99 ft)

23-4. Repeat Problem 23-3 if the second grade is +1.4% instead of −2.1%.

23-5. A −2.6% grade line meets a +3.8% grade line at station 103 + 00. Elevation = 1626.88 ft. Determine the elevation of all full and half stations if a 650-ft curve is to be used.
(*Ans.:* Curve elevations at stations 102 and 104 + 50 = 1631.97 and 1634.09)

23-6. Compute the location and elevation of the high point of the curve of Problem 23-1.

23-7. At what stations is the elevation of the curve of Problem 23-1 equal to 523.05 ft?
(*Ans.:* 60 + 21.06)

23-8. A grade of +4.0% passes station 36 + 00 at an elevation of 622.80 ft. A grade of −5.0% passes station 52 + 00 at elevation 623.40 ft. Compute the station and elevation of the vertex or P.V.I. of these grades.

In problems 23-9 to 23-14 compute the elevation at 100-ft stations throughout the given curves.

	Initial grade (%)	Final grade (%)	Station of P.V.I.	Elevation of P.V.I. (ft)	Length of curve (ft)
23-9.	−3	+4	62 + 00	642.70	800
23-10.	−4	+6	103 + 00	310.20	1000
23-11.	−4	−2	31+ 00	1422.80	1200
23-12.	−3	−6	43 + 00	831.90	600
23-13.	+4	+5	91 + 00	1641.60	600
23-14.	+4	+3	52 + 50	291.70	800

(*Ans.:* 23-9. Curve elevations at stations 59 and 64 = 652.14 ft and 652.45 ft)
(*Ans.:* 23-11. Curve elevations at stations 27 and 36 = 1439.13 ft and 1412.88 ft)
(*Ans.:* 23-13. Curve elevations at stations 90 and 92 = 1637.93 ft and 1646.93 ft)

23-15. A −5% slope meets a +3% grade at elevation 542.80 ft at station 61 + 00. Assuming a 1200-ft vertical curve, determine the elevation on the curve at each 100-ft station and find the station and elevation at the low point on the curve.
(*Ans.:* Low point at 62 + 50, elevation 554.05 ft)

23-16. A −3.2% grade intersects a +2.1% grade at station 103 + 00 at an elevation of 153.20 ft. If a 1200-ft-long vertical curve is to be used to connect the tangent grades, determine the elevation at stations 99 + 00 and 104 + 00.

23-17. A +3% grade line (600 ft length) intersects a −1% grade line (400 ft length) at station 52 + 00, where the elevation is 462.90 ft. Determine elevations at 100-ft stations for this 1000-ft unequal tangent curve.

(*Ans.:* Elevation at station 50 = 454.77 ft, at station 53 = 459.20 ft)

23-18. A −4% grade line (400 ft length) intersects a −2% grade line (200 ft length) at station 63 + 00, where the elevation is 949.50 ft. Determine elevations at 100-ft stations for this 600-ft unequal-tangent curve.

23-19. A −4% downgrade intersects a +3% upgrade at elevation 120.00 ft at station 67 + 00. It is desired to pass a vertical curve through a point of elevation 128.00 ft at station 68 + 00. Determine the curve length required.

(*Ans.: L* = 928.3 ft)

23-20. Repeat Problem 23-1 using the parabolic equation.

23-21. A −3% grade intersects a +4% grade at station 62 + 00 (elevation = 642.70 ft). Determine the elevations at 100-ft stations for an 800-ft long curve using the parabolic equation.

(*Ans.:* Elevation at station 61 + 00 = 649.64 ft).

23-33. Repeat Problem 23-11 using the parabolic equation.

CHAPTER TWENTY-FOUR

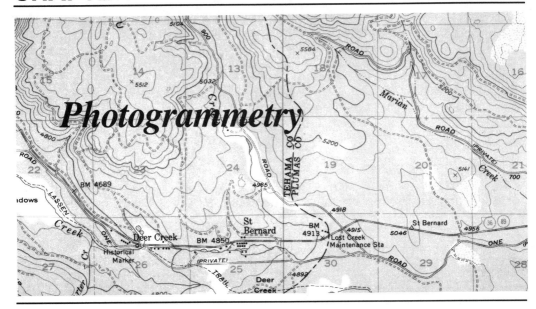

Photogrammetry

24-1 INTRODUCTION

Photogrammetry is a field of study that is very closely related to surveying. Photogrammetry is a tool that can be used effectively as an economical substitute for certain field surveying operations. Also, a certain amount of field surveying is necessary to establish the system of horizontal and vertical control points that is required for many photogrammetric operations. Photogrammetry has been shown to be a faster and less costly method of producing topographic base mapping for many large engineering projects. Photogrammetry has almost completely replaced field methods of topographic mapping for large areas. However, topographic maps for small areas (less than 100 acres) can be produced at lower cost by the field surveying techniques previously described in Chapter 14. Photogrammetry is a tool that is not only useful to surveyors and engineers but also to many other disciplines, such as forestry, geology, archaeology, urban planning, geography, and agriculture. The purpose of this chapter is to provide surveying students with a basic introduction to the field of photogrammetry and its relationship to surveying.

24-2 DEFINITIONS

There is some degree of confusion regarding the definition of the term *photogrammetry*. This confusion occurs because photogrammetry has been used to define a very broad area of study that involves the extraction of quantitative

This chapter was written by Donald B. Stafford, Associate Professor, Department of Civil Engineering, Clemson University, Clemson, S.C.

measurement data and qualitative interpretation data from aerial photographs, as well as a more restricted area of study that applies only to the quantitative use of aerial photographs. This confusion can be overcome by using terms to identify specific types of photogrammetry. However, this practice has not been generally accepted or used widely by the engineering profession and other disciplines that use photogrammetry.

In the broad context, photogrammetry is defined by the American Society for Photogrammetry and Remote Sensing as the art, science, and technology of obtaining reliable information about physical objects and the environment through processes of recording, measuring, and interpreting photographic images and patterns of radiant energy and other phenomena. This broad definition can be subdivided into two components: *metric photogrammetry* and *interpretative photogrammetry*. Metric photogrammetry is generally understood to refer to the use of measurements made on aerial photographs to obtain quantitative data about the earth's surface. In many instances, the end product that results from metric photogrammetry operations is a map of a project area.

The term *interpretative photogrammetry* is used to describe the situation in which aerial photographs or images produced by electronic sensors are carefully studied to produce an interpretation of the conditions existing in the area covered by the photographs or images. The product of interpretative photogrammetry operations is usually a map showing the distribution of soil types, rock types, land use, vegetation species, or other environmental factors. For many years this field of study was known as aerial photographic interpretation, and this term is still appropriate when only aerial photographs are being interpreted. Since the mid-1960s, the term *aerial remote sensing* or simply *remote sensing* has been used to identify a broader area of application in which the product being interpreted may be an image produced by an electronic sensing device or an aerial photograph. Thus aerial remote sensing includes aerial photographic interpretation and also the interpretation of a variety of other sensor products, such as infrared imagery, radar imagery, multispectral imagery, satellite photography, satellite imagery, and other products.

A much simpler definition of photogrammetry is widely used by those involved in the practical applications of photogrammetry and by many users of photogrammetric products. This definition states that photogrammetry is the art and science of measuring elevations, distances, and directions from aerial photographs and a minimum amount of field work or known ground control. In this context, the ultimate objective of most photogrammetric operations is the production of a base map. In the remainder of this chapter, the term *photogrammetry* is used in the context of this simple definition or as a shorter phrase for the term *metric photogrammetry*.

The use of aerial photographs is a fundamental element in photogrammetric operations. An aerial photograph is defined as a photograph made from an aerial platform that shows a portion of the earth's surface to some scale. The aerial platform is usually an airplane, although it may be a balloon or a spacecraft. The scale of an aerial photograph is a very important characteristic because it influences the accuracy that can be obtained from many photogrammetric operations. An understanding of the characteristics of aerial photographs is a prerequisite to the study of photogrammetry.

24-3 TYPES OF AERIAL PHOTOGRAPHS

Aerial photographs can be divided into two groups: vertical aerial photographs and oblique aerial photographs. Vertical aerial photographs are taken with the optical axis of the aerial camera lens pointed along a line perpendicular to the earth's surface. The geometry of vertical aerial photographs is relatively simple

Figure 24-1   Vertical aerial photograph of the Washington, D.C., area. (Courtesy of the U.S. National Ocean Survey.)

and therefore vertical aerial photographs are usually preferred for photogram-metric applications. Although few aerial photographs strictly fulfill the definition of vertical aerial photographs, aerial photographs that have less than 3° of tilt are considered to be vertical for many applications. Figure 24-1 shows a vertical aerial photograph of the Washington, D.C., area.

Oblique aerial photographs are photographs taken with the optical axis of the camera lens purposely tilted at an angle other than 90° to the earth's surface. The angle of tilt usually ranges between 30 and 60° for oblique aerial photographs. It is common practice to designate oblique aerial photographs as either low or high obliques on the basis of the nature of the image of the earth's surface depicted in the photograph. Low oblique aerial photographs do not have an image of the horizon in the photograph because small values of the tilt angle are used. High oblique aerial photographs show the earth's horizon because the angle of tilt is relatively large.

Although oblique aerial photographs can be used for certain photogram-metric applications, these photographs are not used extensively because of the complex geometry of the photograph. The primary problem with oblique aerial photographs is that the scale of the photograph varies throughout the photograph. This scale variation means that measurements on the oblique photograph cannot easily be converted to distances on the ground. Oblique aerial photographs can be used effectively to portray visual information about terrain features because the oblique view is more easily understood by the public than is the plan view contained in vertical aerial photographs.

Other classifications of aerial photographs can be made based on a variety of factors, such as film type, type of aerial camera, altitude of the aerial platform, and other parameters. However, these classifications are not sufficiently impor-tant to warrant detailed coverage in a basic introduction to photogrammetry.

24-4 SCALE OF AERIAL PHOTOGRAPHS

As mentioned earlier, the scale of an aerial photograph is a basic parameter. The scale of the photograph is the property that relates distance measurements made on the aerial photograph to ground distances. One approach to expressing the scale of an aerial photograph is to use a dimensionless ratio of distance on the photograph to a corresponding distance on the ground. The scale expressed in this manner is usually referred to as a *representative fraction*. This form of the scale has the advantage that any system of units can be used as long as the same units are used for both the distance on the photograph and the distance on the ground. Examples of the scale expressed in this manner are typical aerial pho-tograph scales of 1/6000 (or 1:6000) and 1/20,000 (or 1:20,000). Note that the numerator of the representative fraction is always 1.

Another approach to expressing the scale of an aerial photograph is the method commonly used by engineers in which the number of feet on the ground that corresponds to a 1-in. distance on the photograph is given. This form of the scale is commonly referred to as an *engineer's scale*. Although the engineer's scale is usually computed in the form of x feet per inch, the scale can be expressed as 1 inch equals x feet.

The simplest way to look at the scale of an aerial photograph is to assume that the ground surface depicted in the photograph is a plane, as shown in Fig. 24-2. This condition would be representative of the situation in which the ground surface being photographed consists of level terrain. By using similar triangles, the following equations for aerial photograph scales can be developed:

$$S_{RF} = \frac{\text{photo distance}}{\text{ground distance}} = \frac{f}{H'}$$

$$S_{ENGR} = \frac{\text{ground distance}}{\text{photo distance}} = \frac{H'}{f}$$

where S_{RF} is the scale expressed as a representative fraction, S_{ENGR} is the scale expressed as an engineer's scale, f is the camera focal length, and H' is the height of the camera above the ground surface.

In computing the scale as a representative fraction, the same units must be used for both numerator and denominator so that a dimensionless fraction will

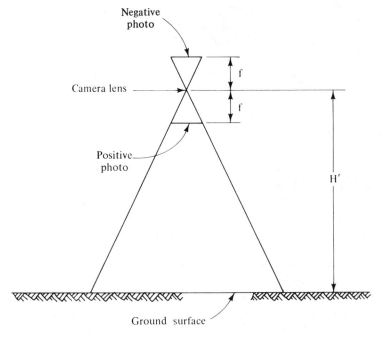

Figure 24-2 Simplified cross section of the geometry of a vertical aerial photograph in flat terrain.

be obtained. In computing the scale as an engineer's scale, the numerator must be in feet and the denominator must be in inches. Conversion from one form of the scale to the other form can readily be made by converting the units and inverting the values. Special care must be exercised in computing aerial photograph scales to ensure that errors in units are not made.

    There are instances where the user of an aerial photograph does not have information on camera focal length or aircraft altitude above the ground. In this situation, the aerial photograph scale can be computed by using known or measured ground distances between points that can be identified on the aerial photograph. For example, the distance between highway intersections or public land survey lines can be measured on the aerial photograph and related to the corresponding ground distance. The ground distance can be obtained by ground measurement, maps, or known public land survey distances. Example 24-1 illustrates this procedure.

Example 24-1

    The distance between two adjacent street intersections on an aerial photograph of an urban area has been measured as 2.25 in. The block lengths are known to be 900 ft long in the area shown on the photograph. Compute the scale of the aerial photograph as an engineer's scale and as a representative fraction.

Solution

$$S_{\text{ENGR}} = \frac{\text{ground distance}}{\text{photo distance}} = \frac{900 \text{ ft}}{2.25 \text{ in.}} = \textbf{400 ft/in.}$$

$$S_{\text{RF}} = \frac{\text{photo distance}}{\text{ground distance}} = \frac{2.25 \text{ in.}}{900 \text{ ft } (12 \text{ in./ft})} = \frac{1}{\textbf{4800}}$$

    Aerial photograph scales are influenced by topography or changes in elevation. Special care must be taken in computing the scale of aerial photographs in areas of rolling terrain because of this effect. Figure 24-3 is a schematic diagram that can be used to show the effect of elevation on scale. The equations for photograph scale now contain an elevation term as follows:

$$S_{\text{RF}} = \frac{f}{H - h}$$

$$S_{\text{ENGR}} = \frac{H - h}{f}$$

where h is the elevation of the point for which the scale is being computed, and H is the height of the aircraft above a vertical datum such as mean sea level. S_{RF}, S_{ENGR}, and f are as defined previously.

    These equations show that for all points, the aerial photograph scale is dependent on the elevation. In other words, a particular scale is applicable only for points at the same elevation. However, for many applications, the average scale

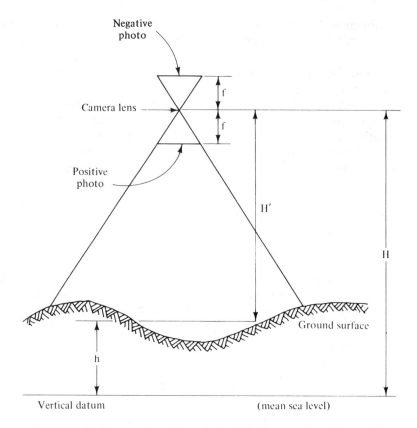

Figure 24-3   Cross section of a vertical aerial photograph in rolling terrain.

can be computed for an aerial photograph by using the average elevation of the terrain. The user of aerial photographs should recognize that an error is introduced when the average scale is used for points having an elevation significantly different from the average elevation of the terrain.

24-5 RELIEF DISPLACEMENT

The geometry of aerial photographs is different from the geometry of a topographic or planimetric map. This is related to the fact that aerial photographs are central perspectives, while maps are orthographic perspectives. One important effect of this difference is that aerial photographs exhibit *relief displacement*. Relief displacement is the condition in which the image of a point is displaced from its true map position. The magnitude of the relief displacement of a point is related directly to the relief or elevation difference of the point and some elevation reference.

Relief displacement is best illustrated by observing the photograph of a tall thin object such as a flagpole, as shown in Fig. 24-4.

On an aerial photograph, the image of the top of the flagpole appears at a different location than the image of the bottom of the flagpole. The difference in the location of the two images represents the amount of relief displacement. It can be shown that the relief displacement is always along a radial line from the center of vertical photographs. Points above the average terrain elevation are displaced outward from the center of the photograph and points below the average elevation are displaced inward toward the center. The amount of the relief displacement can be computed by using the following equation:

$$d_r = \frac{rh}{H'}$$

where d_r is the relief displacement, r is the radial distance from the center of the

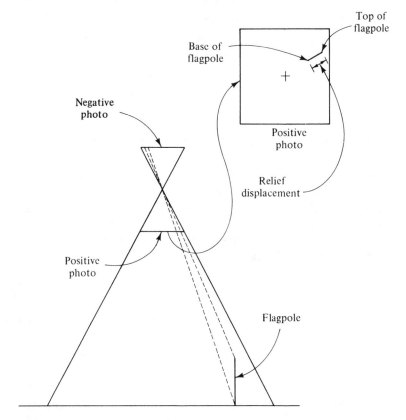

Figure 24-4   Schematic aerial photograph of a flagpole illustrating the concept of relief displacement.

photograph to the displaced image point, h is the height of the object, and H' is the height of the aircraft above the ground. In using this equation, the units of the relief displacement will be the same as the units of the radial distance measurement, while h and H' can be in any units as long as the same units are used.

The relief displacement equation can be used to determine the height of objects for certain situations in which the relief displacement can be measured on the aerial photograph. This condition exists when the image of both the top and bottom of an object can be observed on the photograph and the top of the object occurs vertically above the bottom. Features such as flagpoles, utility poles, some transmission towers, some smokestacks, and corners of buildings fulfill these criteria. The use of this approach to determine the height of objects is illustrated in Example 24-2.

Example 24-2

The relief displacement of the corner of a building was measured on a vertical aerial photograph and found to be 0.18 in. The top of the building was measured to be 2.95 in. from the center of the photograph. The photograph was taken from a height of 3000 ft above the ground. Compute the height of the building in feet.

Solution   The relief displacement equation can be solved for h to give the expression: $h = d_r H'/r$. Therefore, the height of the building is

$$h = \frac{d_r H'}{r} = \frac{(0.18 \text{ in.})(3000 \text{ ft})}{2.95 \text{ in.}} = \textbf{183 ft}$$

24-6 STEREOSCOPIC VIEWING OF AERIAL PHOTOGRAPHS

One of the most important concepts of photogrammetry is the concept of stereoscopic viewing of aerial photographs. When two aerial photographs taken under the proper conditions are viewed in a certain specific manner, the viewer perceives a three-dimensional image of the terrain in the photograph. The three-dimensional view permits the study of the shape of the terrain as it relates to engineering projects and provides the basis for topographic mapping by photogrammetric techniques.

There are three requirements for stereoscopic viewing:

1. The two aerial photographs must provide two views of the terrain taken from different positions.
2. The two aerial photographs must be oriented properly for viewing.
3. The viewer must have normal binocular vision.

The first requirement is usually fulfilled by the procedure that is used to take aerial photographs of project areas. As the aircraft flies along a particular line (the flight line) over the area being photographed, aerial photographs are taken at periodic intervals such that there is at least 50% overlap between adjacent photographs along the flight line, as shown in Fig. 24-5. To ensure that at least 50% overlap is achieved, the time interval between photographs is usually selected so that about 60 to 65% overlap is obtained under normal conditions.

The second requirement is fulfilled by orienting the two aerial photographs being viewed stereoscopically in the same relative position that the camera existed in the field at the time the aerial photographs were taken. This means that the two photographs must be overlapped along the flight line with the overlapping images superimposed. Then the two overlapping images are separated along the flight line by the proper amount to satisfy the requirements for the type of stereoscopic viewing device being used.

The third requirement is fulfilled by anyone who has vision in both eyes, even if the vision must be corrected by glasses. Vision in both eyes is necessary so that the two different images of the terrain contained in the separate overlapping aerial photographs can be transmitted to the brain by the two eyes of the viewer. The stereo model of the terrain is then perceived by the brain based on the slight differences in the two views of the terrain recorded on the aerial photographs.

A special viewing device is commonly used as an aid in the stereoscopic viewing of aerial photographs. The primary purpose of stereoscopic viewers is to ensure that the left eye sees only the aerial photograph on the left and the right eye sees only the aerial photograph on the right. Two common types of stereoscopic viewers are the small lens or pocket stereoscope, which uses refraction to produce the desired effect, and the larger mirror stereoscope, which uses reflection to produce the desired result. Aerial photograph users can also teach themselves to see the stereoscopic view without the aid of a stereoscope by learning to fix the image perceived by each eye on the two separate overlapping aerial

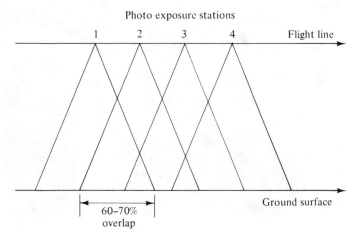

Figure 24-5 Overlapping of aerial photographs along the flight line to permit stereoscopic viewing.

Figure 24-6 Stereogram prepared from a pair of overlapping large-scale aerial photographs. (Courtesy of the Department of Civil Engineering, Clemson University.)

photographs, a process called *stereoscopic fusion*. However, considerable practice is necessary to achieve this capability.

An example stereogram that has been prepared from a pair of overlapping aerial photographs is shown in Fig. 24-6. The stereogram can be viewed with a pocket stereoscope or by stereoscopic fusion to observe the three-dimensional shape of the terrain covered by the photographs. The stereoscopic view that is obtained from this stereogram illustrates a common characteristic of such views, the vertical exaggeration that occurs when aerial photographs are used to produce a three-dimensional image.

24-7 PLANNING AERIAL PHOTOGRAPHY MISSIONS

Careful planning must be exercised in developing the specifications and flight plans for aerial photography missions. The parameters that are chosen for the aerial photographs are very important because these values influence the accuracy of the various photogrammetry operations that use the photographs. There are a number of interrelated variables that must be selected to define the characteristics of the aerial photographs that will be produced by a particular aerial photography mission. Also, a detailed flight plan must be prepared to define the area of coverage and the spacing between flight lines and between photographs.

The scale of the aerial photographs must be selected based on the use that is to be made of the photographs. For example, if topographic maps at a particular scale are to be prepared by photogrammetric techniques using a specific type of stereoscopic plotter (see Section 24-10), the scale of the aerial photographs to be used will be generally defined. Once the scale is selected, the combination of aerial camera focal length and flying altitude above the ground can be determined based on a number of considerations.

Two important parameters that must be selected are end-lap between adjacent aerial photographs along the flight line and side-lap between adjacent flight lines. These values are usually expressed in percents and can vary over a relatively narrow range depending on several considerations. End-lap values of between 60 and 70% are common and side-lap values of 20 to 40% are typical. The end-lap is an important variable in determining the photograph spacing along the flight line and the side-lap is important in determining the spacing between flight lines.

Figure 24-7 shows a typical flight plan for an aerial photography mission. Project areas are usually flown in a series of parallel north–south flight lines. One exception to this general rule is the case of rectangular project areas that are long in the east–west direction and short in the north–south direction, in which case east–west flight lines are used. For very irregular project areas, other flight line orientations may be adopted. The cost of flying the mission is lowest when the number of flight lines is minimized so that time and aircraft operational expense involved in the turning operation between flight lines is reduced.

The base map that is used for the flight map must be chosen carefully. The primary purpose of the flight map is to ensure that the flight crew can maintain

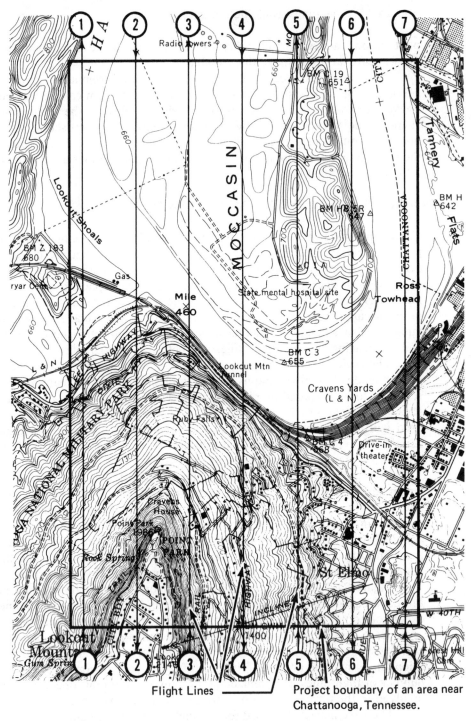

Flight Lines —————— Project boundary of an area near
Chattanooga, Tennessee.

Figure 24-7 Example of a flight line map of an area near Chattanooga, Tennesse.
(Base map courtesy of the Tennessee Valley Authority.)

the flight line alignment and start and stop the camera at the proper location. Therefore, the flight line base map must show a sufficient amount of natural and man-made detail that can be identified from the air to orient the crew adequately. In developed areas, the flight crew can usually orient the flight lines with reference to highways, railroads, building, and other man-made features. However, in undeveloped areas the flight crew may have trouble relating the base map features to the view observed from the air. In these instances, a flight line base map that consists of an aerial mosaic or other available small-scale aerial photographs may be more suitable.

Another important aspect of flight planning is preparation of a set of specifications for the aerial photographs. These specifications define the parameters that the aerial photographs should have and provide information on the limits in these values that will be permitted before the aerial photographs would be rejected. The specifications cover a number of aerial photograph characteristics including scale, camera focal length, percent of end-lap, percent of side-lap, tilt, and other parameters. The tilt values are particularly important since the assumption of vertical aerial photographs may not be applicable if the tilt values are too high.

There are a number of other considerations in flight planning that are beyond the scope of this chapter. Several of the references at the end of this chapter provide more detailed information on flight planning for aerial photography missions.

24-8 PHOTOGRAMMETRIC CONTROL SURVEYS

Probably the most direct relationship between surveying and photogrammetry is the situation in which field surveying techniques are used to provide the necessary ground control data for photogrammetric operations. Many photogrammetric operations require that the coordinates and elevations of a certain number of points be known. Ground control data requirements are usually given in terms of the number of horizontal control points for which map coordinates are needed and the number of vertical control points for which the elevations are needed. Although a primary advantage of photogrammetric techniques is the reduction in total project costs by reducing the amount of field surveying required, there will always be a need for some minimum amount of surveying to generate the needed ground control data. The cost of photogrammetric control surveys generally ranges from 20 to 50% of the total cost of photogrammetric mapping projects. Therefore, special attention must be given to the planning and supervision of photogrammetric control surveys to minimize the cost of these operations.

The surveying techniques used to develop horizontal ground control points require the measurement of distances and angles so that the coordinates of the points can be computed. The most common coordinate system that is used for horizontal ground control points is the state plane coordinate system (see Chapter

21), although latitude and longitude or other coordinate systems can be used. The types of surveying operations required to produce the horizontal ground control data can consist of triangulation, traverses, trilateration (see Chapter 20), or similar work. The surveying technique selected for the horizontal control surveys depends on a number of factors, including the topography, extent of the highway and railroad network in the project area, and density of vegetation. It is important to select the surveying technique that will provide the horizontal control data with the required accuracy at the lowest possible cost. Electronic distance measurement equipment has become widely accepted and used for horizontal control surveys.

Differential or spirit leveling is the usual surveying technique used for vertical control surveys. The leveling circuit should begin and end at bench marks with known elevations so that error adjustments can be made. In rugged terrain, trigonometric leveling is sometimes employed. For small-scale mapping projects in which the accuracy of the vertical control points is not critical, barometric leveling and other techniques have been used satisfactorily. New electronic devices that can provide three-dimensional coordinates of points based on signals from orbiting satellites have been developed that could revolutionize photogrammetric control surveys. (See Chapter 15.)

It is critical that surveyors who are performing photogrammetric control surveys understand how the ground control data are to be used. With a knowledge of how the ground control data are applied in various photogrammetric operations, the surveyor can locate the ground control points in the optimum location, ensure that the required accuracy is provided, and select points in such a manner that errors can be minimized. It is particularly important that the surveyor understand the critical influence that the ground control data has on the accuracy of photogrammetric operations. The equipment and surveying techniques employed should be selected to provide the necessary accuracy at the lowest possible cost. Providing an unrealistic level of accuracy in the ground control data is uneconomical.

The photogrammetrist should also have an understanding of surveying equipment and techniques used for control surveys. By having a knowledge of surveying procedures, the photogrammetrist can avoid requesting unnecessary data and making unrealistic accuracy demands that would significantly increase the cost of ground control surveys. By working closely and cooperatively, the surveyor and photogrammetrist can ensure that the best and most economical set of ground control data are obtained.

One very important consideration in ground control surveys is to ensure that the ground control points can be located and defined very accurately on the aerial photographs being used. This usually means that clearly defined objects with sharp boundaries should be used for control points. In many areas there are sufficient numbers of man-made objects, such as sidewalk and highway intersections, utility poles, fence corners, isolated trees, and similar objects so that the required number of ground control points can be located in the proper location on the photographs. In undeveloped areas artificial targets are often used to define the control points. This procedure, known as premarking or paneling, involves placing special shapes on the ground at the proper location before the aerial photographs are taken. Although this procedure is somewhat expensive, it can be very effective in improving the system of ground control points and minimizing errors in locating ground control points that would reduce the accuracy of the photogrammetric operations.

Surveyors who plan to become involved in photogrammetric control surveys should consult a good photogrammetry textbook to learn about the control data requirements for various photogrammetry operations. A photogrammetry textbook will describe the number of horizontal and vertical control points required and the preferred location of the points within the aerial photographs for different photogrammetry operations.

24-9 MOSAICS AND ORTHOPHOTOS

One photogrammetric product that is useful for a variety of purposes is a *mosaic*. A mosaic is defined as an assembly of overlapping aerial photographs into a continuous picture of an area. Mosaics are often prepared of certain geographic areas of interest such as cities or project areas so that the area of concern is shown conveniently in a single image rather than having to handle several photographs. The characteristics of aerial mosaics vary over a wide range, depending on the time and effort that is expended in preparing the aerial photographs and assembling the mosaic.

Aerial mosaics are commonly divided into three classes: uncontrolled, semicontrolled, and controlled. The accuracy of measurements made on mosaics are lowest for uncontrolled mosaics, intermediate for semicontrolled mosaics, and highest for controlled mosaics.

Uncontrolled mosaics are prepared by assembling contact print aerial photographs while matching the photographic detail of adjacent photographs as carefully as possible. Because of errors due to scale variation between photographs, tilt, aerial camera lens distortion, and other sources, the photographic detail does not match exactly and errors are introduced into the mosaic. No attempt is made to match specific features on the aerial photographs to known ground control points in an effort to improve the accuracy of uncontrolled mosaics. Uncontrolled mosaics are satisfactory for many applications in which high accuracy is not of critical importance.

Semicontrolled mosaics represent the next highest level in accuracy and cost of preparation. Semicontrolled mosaics can be prepared by using aerial photographs that have been rectified to correct for tilt and projection printed to a uniform scale to correct for scale variation between photographs. Alternatively, semicontrolled mosaics can be prepared by matching features on nonrectified and nonratioed aerial photographs to plotted ground control points. Semicontrolled mosaics represent a significant improvement in accuracy over uncontrolled mosaics, but the cost of preparation is considerably higher.

Controlled mosaics represent the highest quality of preparation and the highest level of accuracy is produced. First, the individual aerial photographs are rectified to remove errors due to tilt, and the photographs are ratioed or printed to a uniform scale to remove scale variation errors. Also, the coordinates of known ground control points are plotted on the base map, and the features corresponding to these points are matched to the plotted points. The result is a mosaic in which the photographic detail matches very closely between photographs. Therefore, accurate measurements are possible on controlled mosaics. However, the cost of preparing controlled mosaics is also high. It should be noted that the accuracy of mosaics is limited by the fact that each individual photograph contains relief distortions.

Only the central portion of each aerial photograph is used in mosaic preparation in order to improve the matching of photographic detail between adjacent photographs. By trimming and discarding the outer portions of each photograph, the portions of the photographs where tilt errors, lens distortion errors, and other errors are most pronounced are eliminated. The individual photographs are glued to a rigid material such as plywood or masonite board to provide support. The resulting continuous photograph of an area can be used for a variety of purposes.

A special type of photogrammetric product that has become more common in recent years is the *orthophoto*. Orthophotos show the photographic images of objects in the terrain in the form of an orthographic perspective, as contrasted with the central perspective view provided in aerial photographs. The most significant characteristic of orthophotos is that the relief distortions have been removed. Therefore, accurate measurements of distances, directions, and areas can be made directly on the orthophotos because image displacements caused by relief distortions do not occur. The preparation of orthophotos involves the differential rectification of aerial photographs using techniques that are beyond the scope of this discussion.

Orthophotos can be used individually or, more commonly, assembled into mosaics of several orthophotos. Such mosaics are known as *orthophotomosaics*. Since orthophotos and orthophotomosaics show the photographic detail in its true map position, contours (also produced by photogrammetric techniques) showing the shape of the terrain can be superimposed on the mosaic to offer another dimension of terrain information. Orthophotos and orthophotomosaics have the important advantages of showing the complete photographic detail in correct position and including the superimposed contours for terrain elevation data. These products are beginning to be widely accepted as a map base for a variety of applications. Many engineering projects, such as route selection, drainage design, and transmission-line layout, are now accomplished using orthophoto products for base maps. This product will continue to replace the more common topographic mapping techniques for a wide variety of mapping projects because of the inherent advantages of orthophotos.

24-10 STEREOSCOPIC PLOTTERS AND TOPOGRAPHIC MAPPING

The most important application of photogrammetry is the production of mapping products from overlapping aerial photographs. Although photogrammetric techniques are sometimes used to produce planimetric maps that show the true map positions of natural and man-made features, the most common photogrammetric product is the topographic map, which shows the shape of the terrain by the use of contours and also includes planimetric detail. The photogrammetric plotters that are used for map production are sophisticated instruments that employ the concept of stereoscopic viewing of aerial photographs to provide a three-dimensional view of the terrain to allow contour lines of equal elevation to be plotted. Figure 24-8 schematically illustrates the concept of a stereoscopic plotter.

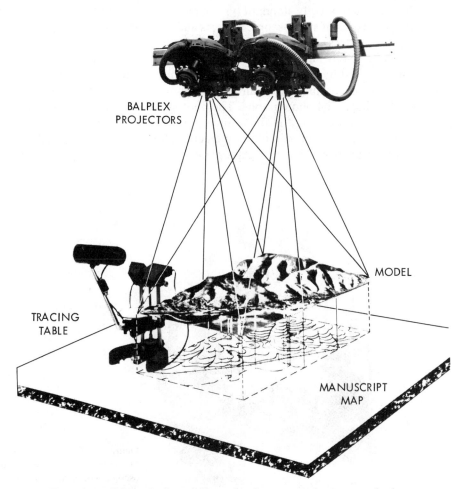

Figure 24-8   Schematic diagram illustrating the concept of a stereoscopic plotter. (Courtesy of Bausch & Lomb.)

The widespread substitution of photogrammetric techniques using stereoscopic plotters for field topographic mapping methods has occurred because of the lower cost of the photogrammetric approach on large mapping projects. Many aerial survey firms, engineering consulting firms, and government agencies have developed topographic mapping programs by purchasing stereoscopic plotters and training employees in photogrammetric mapping techniques. Several U.S. government agencies that have topographic mapping functions such as the U.S. Geological Survey and the Tennessee Valley Authority have been particularly aggressive in adopting and improving photogrammetric mapping procedures.

There are a number of different types of stereoscopic plotters that have been developed by various equipment manufacturers in the United States and Europe. These plotters vary considerably in cost, accuracy, and operating procedures. However, a relatively standard type of plotter can be used to describe the common characteristics of stereoscopic plotters and photogrammetric mapping techniques. Typical plotters used widely by aerial survey firms and government agencies in the United States are the Kelsh plotter and the ER-55 plotter and Balplex plotter manufactured by Bausch and Lomb. Figure 24-9 shows a modern Kelsh plotter.

A basic component of all photogrammetric mapping techniques is a pair of overlapping aerial photographs. The photographs are usually taken with about 60 to 70% overlap along the flight line so that they can be viewed stereoscopically in the plotter. For most plotters the aerial photographs are printed in the form of positive transparencies on a glass plate. This form of aerial photograph is known

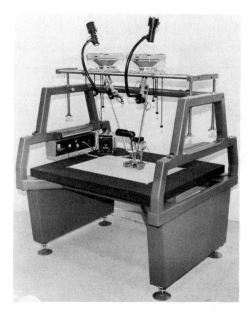

Figure 24-9 Modern Kelsh plotter, Model 5030-B, a common type of stereoscopic plotter for topographic map production. (Courtesy of the Kelsh Instrument Division of Danko Arlington, Inc.)

as a diapositive and is used to provide a stable and accurate photograph. The glass plates are very flat and are not affected significantly by changes in temperature and moisture conditions. Some plotters use diapositives that are the same size as the original aerial photograph, whereas other plotters use reduced-size diapositives that are produced by projection printing.

A typical stereoscopic plotter is composed of three primary systems: the projection system, the viewing system, and the measuring and plotting system. The projection system is composed of the lighting and optical components that are necessary to project images of the aerial photographs contained in the diapositives in such a manner that a stereoscopic view of the terrain is produced. The projection system usually consists of a light source placed above the diapositive and a lens system to ensure that the projected image is focused properly.

The viewing system is the mechanism that permits the operator to observe the stereoscopic view produced by the projection system. The primary function of the viewing system is to force the left eye of the operator to view the image projected from the left diapositive and the right eye to view the image projected from the right diapositive. This permits the operator to perceive the three-dimensional shape of the terrain, generally referred to as a stereomodel. Several different viewing systems are available, and considerable changes in viewing systems have occurred in recent years.

For many years the most common type of viewing system was the anaglyphic system. In this viewing system, a red filter is placed in the left projector and a blue-green filter is placed in the right projector so that a red image is projected from the left diapositive and a blue-green image is projected from the right diapositive. When the operator wears a pair of glasses composed of a red filter glass over the left eye and a blue-green filter glass over the right eye, the operator perceives the stereomodel. Because of the colors imparted to the projected images, the anaglyphic viewing system cannot be used with color aerial photographs. Also, the filters reduce the amount of light projected and serve to darken the model. However, this viewing system is relatively simple and inexpensive and has been used in a variety of stereoscopic plotters over the years.

In recent years two new viewing systems have been developed and widely adopted because of the inherent advantages of permitting the use of color diapositives and having much less light loss than the anaglyphic viewing system. One of the modern viewing systems is known as the stereo-image alternator or SIA system. This system uses rotating synchronized shutters which alternatively block and permit the passage of light from the projection system to the operator. One pair of shutters is placed in the projection system and the operator views the projected image through a second pair of shutters. By synchronizing the rotating shutters such that the left projection shutter is open when the left viewing shutter is open, the operator observes first the image from the left diapositive and then the image from the right diapositive. The shutters open and close so rapidly that the operator is not aware of the alternating images and the stereomodel is perceived. Stereoscopic plotters that used anaglyphic viewing systems previously can be retrofitted with SIA viewing systems, and many such conversions have been made.

Another modern viewing system is the polarized platen viewing system or PPV system. This system uses polarized light filters rather than the colored filters used with the anaglyphic viewing system. A pair of polarized filters of opposite polarity are used in the projectors and the operator views the projected images on the platen through a pair of eyeglasses with polarized filters over each eye. The PPV viewing system can also be retrofitted on plotters that originally used the anaglyphic viewing system and many such modifications have been made.

The measuring and plotting system provides a means of locating a specific point in the stereomodel and plotting this position on a map manuscript. The system permits planimetric features to be plotted in the correct map position and allows contours to be plotted to define the topography of the terrain. The most common type of measuring and plotting system consists of a tracing table composed of a platen, a floating dot reference mark in the center of the platen, a geared device for recording the vertical position of the platen (equivalent to elevations in the stereoscopic model), and a pencil to plot the position of the tracing table on the map manuscript. The platen is a white disk on which the diapositive images are projected so that a small area of the stereomodel can be viewed. The floating mark, which consists of a very small dot of light, can be made to rise or drop until it is in contact with the three-dimensional surface of the stereomodel by raising or lowering the platen. By setting the elevation dial at a specific even value and keeping the floating dot in contact with the surface of the stereomodel, a particular contour can be located and plotted by moving the tracing table with the plotting pencil extended to contact the manuscript. In a similar manner, the true map position of planimetric features can be plotted by following the outlines of the features while maintaining the floating dot in contact with the three-dimensional surface of the stereomodel and keeping the plotting pencil in contact with the manuscript. There are many other detailed aspects of stereoscopic plotters that cannot be covered herein because of space limitations. The student interested in learning more about photogrammetric techniques for topographic mapping should consult one of the photogrammetry textbooks in the list of references at the end of this chapter.

The use of stereoscopic plotters to produce topographic maps is a very versatile approach. A wide range of map scales can be accommodated by taking aerial photographs at various altitudes with different focal length cameras. For instance, topographic maps with scales as large as 1 in. equals 20 ft and a contour interval of 1 ft are often prepared in densely developed urban areas. On the other hand, topographic maps with scales as small as 1 in. equals 2000 ft and contour intervals of 10 or 20 ft are common for the 7½-minute quadrangle series of maps. More common scales for engineering projects range from 1 in. equals 100 ft with a 2-ft contour interval to 1 in. equals 500 ft with a 10-ft contour interval. It is important to recognize that the proper contour must be matched with the proper scale when producing topographic maps by photogrammetric techniques.

24-11 SOURCES OF AERIAL PHOTOGRAPHS

Although many photogrammetric projects require that a new set of aerial photographs be taken to fulfill the specific needs of the project, other projects can be accomplished with existing aerial photographs. In particular, aerial photographic interpretation projects can often use existing aerial photographs. If existing aerial photographs with the proper characteristics can be identified and used, a considerable savings can usually be realized. If a new set of aerial photographs must be taken, the total cost of the aerial photography mission must be borne by the purchaser. On the other hand, existing aerial photographs can usually be purchased for the cost of reproduction or only a slightly higher cost.

The user of aerial photographs should realize that a number of aerial survey firms and government agencies take aerial photographs for a variety of purposes. Therefore, most areas in the United States have been photographed several times by different organizations. In many instances one or more of the existing coverages will have a combination of dates, scales, film type, and overlap that are suitable for a particular application.

The organization with the largest inventory of aerial photographs is the Agricultural Stabilization and Conservation Service (ASCS) of the U.S. Department of Agriculture. This agency and its predecessor agencies produced aerial photographs of most counties in the United States that had a significant amount of agricultural land use on approximately a five-year repeat cycle over the past 40 years. Other U.S. Department of Agriculture agencies that have also produced significant amounts of aerial photography are the Soil Conservation Service and the U.S. Forest Service. Various U.S. government agencies that have significant aerial photography inventories are listed in Table 24-1.

Many state governments also have agencies with active aerial photography programs. The departments of transportation and highway agencies in many states have a photogrammetry department that has produced aerial photographs of large

TABLE 24-1   U.S. Government Agencies with Large Inventories of Aerial Photography

---

U.S. Department of Agriculture
    Agricultural Stabilization and Conservation Service
    Soil Conservation Service
    Forest Service
U.S. Geological Survey
National Ocean Survey (formerly U.S. Coast and Geodetic Survey)
National Aeronautics and Space Administration (NASA)
Tennessee Valley Authority
Bureau of Land Management
Bureau of Reclamation
National Park Service
U.S. Army Corps of Engineers

---

areas in the state. In some states, the state natural resources agency may have a large inventory of aerial photographs. Many county or municipal governments have a planning, engineering, or public works agency that maintains files of aerial photographs.

Many aerial survey firms have extensive inventories of aerial photographs in their files. These inventories vary widely in scale, film type, and other pertinent characteristics. In particular, aerial survey firms tend to accumulate large quantities of aerial photographs in the general vicinity of their home office and branch offices.

One problem that is frequently encountered in trying to determine the amount and characteristics of available aerial photographs is that many different agencies must be contacted. An organization that can be very helpful in this effort is the National Cartographic Information Center, an agency of the U.S. Geological Survey. This agency maintains current records of aerial photography coverage (as well as map coverage) that can be used as a source of information on which government agencies and aerial survey firms have aerial photographs of a specific geographic area. This program is known as the Aerial Photography Summary Record System, and use of this central data source can save considerable time and effort in gathering information on existing aerial photography. Inquiries to the National Cartographic Information Center can be made by writing to the agency at 507 National Center, 12001 Sunrise Valley Drive, Reston, VA 22092.

24-12 PHOTOGRAMMETRY ORGANIZATIONS

The primary professional organization in the photogrammetry field in the United States is the American Society for Photogrammetry and Remote Sensing. The national headquarters for this organization is located at 5410 Grosvenor Lane, Suite 210, Bethesda, Maryland 20814-2160. One of the important functions of the American Society for Photogrammetry and Remote Sensing is the publishing of the monthly technical journal entitled *Photogrammetric Engineering and Remote Sensing*. This magazine contains articles on all aspects of photogrammetry and aerial remote sensing, advertisements for photogrammetry equipment manufacturers and aerial survey firms, news articles of interest to the photogrammetry profession, and announcements of job opportunities in the photogrammetry field. Over the past several years, the American Society for Photogrammetry and Remote Sensing has prepared and published a number of important professional manuals including the *Manual of Photogrammetry, Manual of Remote Sensing, Manual of Photographic Interpretation*, and *Manual of Color Aerial Photography*. A student membership grade is available which allows students to receive the monthly journal and obtain discounts on other publications produced by the organization. Other professional organizations that provide information on photogrammetric techniques and the applications of photogrammetry are the Surveying and Mapping Division of the American Society of Civil Engineers; the American

Congress on Surveying and Mapping; the Photogrammetric Society, which has its headquarters in London; and the International Society of Photogrammetry, an association of 60 national societies of photogrammetry.

SELECTED REFERENCES

American Society of Photogrammetry. *Manual of Photogrammetry*, 4th ed. American Society of Photogrammetry, Falls Church, Va., 1980.

AVERY, THOMAS EUGENE. *Interpretation of Aerial Photographs*, 3rd ed. Burgess Publishing Co., Minneapolis, Minn., 1977.

CHURCH, EARL F. *Elements of Photogrammetry*. Syracuse University Press, Syracuse, N.Y., 1944.

CHURCH, EARL F., and ALFRED O. QUINN. *Elements of Photogrammetry*. Syracuse University Press, Syracuse, N.Y., 1948.

COLWELL, ROBERT N., editor. *Manual of Remote Sensing*, 2nd ed. American Society of Photogrammetry, Falls Church, Va., 1983.

CRONE, D. R. *Elementary Photogrammetry*. Frederick Ungar Publishing Co., New York, 1968.

HALLERT, B. *Photogrammetry*. McGraw-Hill Book Comparny, New York, 1960.

MOFFITT, FRANCIS H. and EDWARD M. MIKHAIL. *Photogrammetry*, 3rd ed. Harper & Row, Publishers, Inc., New York, 1980.

SCHWIDEFSKY, K. *An Outline of Photogrammetry*. Pitman Publishing Corp., New York, 1959.

SMITH, JOHN T., JR., editor. *Manual of Color Aerial Photography*. American Society of Photogrammetry, Falls Church, Va., 1968.

SPURR, STEPHEN H. *Photogrammetry and Photo-Interpretation*, 2nd ed. The Ronald Press Company, New York, 1960.

U.S. Dept. of Transportation, Federal Highway Administration. *Reference Guide Outline, Specifications for Aerial Surveys and Mapping by Photogrammetric Methods for Highways*. Washington, D.C., 1968.

WOLF, PAUL R. *Elements of Photogrammetry*, 2nd ed. McGraw-Hill Book Company, New York, 1983.

ZELLER, D. M. *Textbook of Photogrammetry*. H. K. Lewis and Co. Ltd., London, 1952.

PROBLEMS

24-1. Compute the scale (as a representative fraction) of a set of aerial photographs known to have an engineer's scale of 1 in. equals 2500 ft.

(*Ans.:* 1/30,000)

24-2. Carefully define the terms *metric photogrammetry, interpretative photogrammetry,* and *aerial remote sensing.*

24-3. Compute the scale (as a representative fraction) of a set of aerial photographs taken from an altitude of 3750 ft with a camera having a nominal focal length of 6 in. over a project area that has an average elevation of 750 ft.

(Ans.: 1/6000)

24-4. Describe the primary characteristics of vertical and oblique aerial photographs.

24-5. Compute the engineering scale of a vertical aerial photograph that shows two public land survey lines known to be 1 mile apart on the ground to be 5.28 in. apart on the photograph.

(Ans.: 1000 ft/in.)

24-6. Compute the scale (as a representative fraction) of a vertical aerial photograph that contains an image of a drag strip that is known to be $\frac{1}{4}$ mile in length if the image of the drag strip is 6.6 in. long.

24-7. Compute the engineer's scale of a set of aerial photographs taken with an aerial camera having a nominal focal length of $3\frac{1}{2}$ in. from an altitude of 1800 ft over a flat area that has an elevation of 400 ft.

(Ans.: 400 ft/in.)

24-8. A set of aerial photographs has been taken from an altitude of 15,000 ft with an aerial camera having a nominal 8.25 in. focal length lens. What is the maximum terrain elevation that can exist in the area if the aerial photographs scale at any point must meet the scale specification of $1:20,000 \pm 5\%$?

24-9. An aerial photograph of the central business district in a city shows the image of a large parking lot in which the painted parking stalls are clearly visible. One parking bay that contains 50 parking spaces which are known to be 9 ft wide on the ground measures 2.25 in. on the photograph. What is the engineer's scale of the aerial photograph?

(Ans.: 200 ft/in.)

24-10. List the three requirements that must be fulfilled to accomplish stereoscopic viewing.

24-11. The vertical component of an electric power transmission tower has a relief displacement that measures 0.21 in. on a vertical aerial photograph. The photograph was taken from a height of 4400 ft above the ground and the top of the tower is located at a distance of 4.86 in. from the center of the photograph. Compute the height of the transmission tower.

(Ans.: 190 ft)

24-12. Describe the three primary types of aerial mosaics.

24-13. Compute the relief displacement that would be expected to exist for a corner of the World Trade Center on an aerial photograph taken from an altitude of 10,000 ft above the ground if the top corner of the building is located at a distance of 3.5 in. from the center of the photograph. The World Trade Center is 1350 ft high.

(Ans.: 0.47 in.)

24-14. List and briefly describe the three primary components of a stereoscopic plotter.

24-15. An aerial photograph has the image of the top of a vertical flagpole located at a distance of 4.2 in. from the center of the photograph and the relief displacement is measured to be 0.38 in. The photograph was taken from an altitude of 1500 ft above the ground. What is the height of the flagpole?

(Ans.: 136 ft)

24-16. Describe the primary characteristics of orthophotos.

CHAPTER TWENTY-FIVE

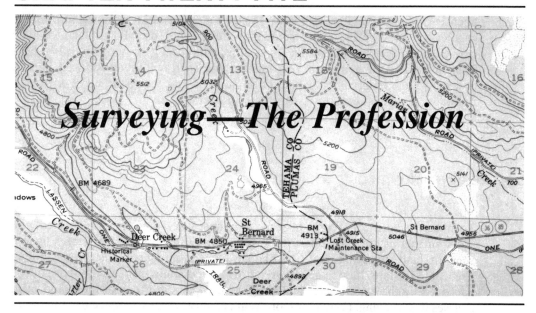

Surveying—The Profession

25-1 SURVEYING LICENSES

The purpose of building codes, medical codes, surveying registration requirements, and so on, is to protect the public from unqualified and/or unscrupulous people in these fields. Perhaps the first regulations of this type were contained in the code of laws of Hammurabi, who was king of Babylon in the eighteenth century B.C. His code covered many different subjects (including prohibition), but only his building code is mentioned here. It was famous for its "eye for an eye and tooth for a tooth" section. In general, the code said that if a builder constructed a house that collapsed and killed the owner, the builder would be killed. If the collapse caused the death of the owner's son, the builder's son would be killed, and so on. These laws were presumably very effective in making the builder put forth his best efforts.

Through the centuries since the time of Hammurabi many codes and laws have been established to guide the various professions, but it was not until 1883 that registration requirements reached the United States. At that time registration of dentists began and as the years went by doctors, lawyers, pharmacists, and others were required to obtain licenses before they could practice their professions.

Licensing of surveyors is not new. George Washington and Abraham Lincoln held surveyor's licenses, but licensing as it is known today for engineers and surveyors was begun in Wyoming in 1907. Today, the law in all 50 states, as well as Canada, Puerto Rico, and Guam, requires that a person must meet certain licensing requirements before practicing land surveying. For the purposes of such licensing, land surveying is generally said to include the determination of areas of tracts of land, the surveying needed for preparing descriptions of land for deed conveyance, the surveying necessary for establishing or reestablishing land boundaries, and the preparation of plats for land tracts and subdivisions. A license is usually not required for construction surveys and for route surveys (roads, railroads, pipelines, etc.) unless property corners are set.

25-2 REGISTRATION REQUIREMENTS

To obtain a land surveyor's license, it is necessary for a person to meet the requirements of the state board of engineering examiners. (A few states have a separate land surveying board.) Although these requirements vary considerably from state to state, they in general require (1) graduation from a school or college approved by the state board, including the completion of one or more approved courses in surveying, (2) two or more years of surveying experience of a character suitable to the board, and (3) successful completion of a written examination under the supervision of the state board. If the applicant is unable to meet the formal education requirements listed in (1), he or she will probably be required to obtain additional experience (perhaps as much as four or more years) before he or she will be permitted to take the written exam.

In 1973 the National Council of Examiners for Engineering and Surveying (NCEES) began offering a semiannual national surveying exam. This fundamental exam, which is of a basic and general type, consists of a half-day of multiple-choice questions and a half-day of more detailed problems. Since the laws pertaining to land description (metes and bounds, coordinates, public land surveys, etc.) are different from state to state, the individual states using the exam have different requirements. The so-called colonial or metes and bounds states have an approximately 7-hour exam for their types of surveys, followed by approximately a 1-hour exam concerning surveying in the particular state in which the exam is being given. Similarly, the public land survey states require approximately a 7-hour exam concerning their type of work plus the 1-hour individual state exam.

Almost every state uses the national exam, with the result that it is easier for a person registered in one state to become registered in another. Eventually, it is hoped that a surveyor who passes the national exam in one state will be permitted to obtain with little or no delay and few additional requirements a license in another state. Such reciprocity is a reality for professional engineering registration as a result of the national exams given by the NCEE, in which nearly all states participate.

The earliest known code of ethics is the Hippocratic oath of the medical profession, which is attributed to Hippocrates, a Greek physician (460–377 B.C.). The present form of the Hippocratic oath dates from approximately 300 A.D., but a detailed code of ethics for the medical profession did not appear in the United States until 1912. The code of ethics for the American Society of Civil Engineers was adopted in 1914.

A code of ethics is not intended to be a lengthy detailed statement of "thou shall nots," but rather a few general statements of noble motives expressing concern for the welfare of others and the standing of the profession as a whole. The classic comparison of law and ethics is made by relating them to medicine and hygiene. The object of medicine is to cure diseases, and the object of law is to cure or repress evil. The object of hygiene is to prevent illness, and the object of ethics is to prevent evil by raising the moral plane on which people deal with each other.

Another classic comparison of law and ethics is frequently made with relation to the Old and New Testaments. Essentially, the Old Testament is a list of "thou shall nots," such as "thou shall not kill, thou shall not steal," and so on; the New Testament is based on a concern for the welfare of our fellows and a desire to do good because it is a privilege and not a necessity. Many groups of land surveyors, primarily state societies, have published codes of ethics. In general, these codes are similar to each other and to the codes of the various engineering societies.

Perhaps the purpose and heart of a code of ethics may be summarized in a few sentences: The surveyor must faithfully and impartially perform work with fidelity to clients, employer, and the public. (For instance, when marking a property corner, its location will be the same regardless of which property owner is paying the surveyor.) The person will be seriously concerned with the standing of the profession in the public's eye and will not only strive to live and work according to a high standard of behavior, but, in addition, will avoid association with persons or enterprises of questionable character. In other words, the person will not only be concerned with evil but with the very appearance of evil. Further he or she will be actively interested in the welfare of the public and will always be ready to apply his or her knowledge for the benefit of humankind.

When a Roman soldier encountered a Jew in biblical times, he could require the Jew to carry his pack for a mile. Jesus said (Matthew 5:41) that he should carry it for 2 miles (thus the origin of the expression "go the extra mile"). This idea is the theme of a code of ethics.

Instead of reproducing one of the codes of ethics, the author is devoting the remainder of this section to the following general statements that summarize what a code of ethics means to him:

1. The surveyor must not place monetary values above other values. Although it may seem difficult to apply this to specific cases, it simply means that the

surveyor should never recommend to a client a course of action that is based on the amount of money that the surveyor will thereby receive.

2. In the course of work the surveyor may very well acquire knowledge that could be detrimental to the client if it were revealed to others. The surveyor's responsibility to the client goes beyond the immediate job and must not reveal private information concerning the client's business without the client's permission.

3. In concern for the reputation of the profession, the surveyor must refrain from speaking badly of other surveyors, or will be lowering the profession in the eyes of the public. This does not mean that there is not a time and place for an honest appraisal of other surveyors. A surveyor is far better able to judge the work of a fellow surveyor than is anyone else, whether lawyer, judge, or layperson.

4. In further concern for the standing of the profession, the surveyor must not become professionally associated with surveyors who do not conform to the standards of ethical practice discussed in this section. Furthermore, the surveyor will not become involved in any partnership, corporation, or other business group that is a cloak for unethical behavior. The surveyor must accept full responsibility for his or her work.

5. The surveyor will not be too proud to admit that he or she needs outside advice in order to solve a particular problem.

6. The surveyor will admit and accept his or her own mistakes.

7. When on a salaried job the surveyor will not do outside work to the detriment of his or her regular job. Further, the surveyor will not use such a job to compete unfairly with surveyors in private practice.

8. The surveyor will not agree to perform free surveying (except for community service) because he or she would be taking work away from the profession.

9. The surveyor will not advertise in a self-laudatory or blatant manner or in any other way that might be detrimental to the dignity of the profession.

10. The surveyor has a duty to increase the effectiveness of the profession by cooperating in the exchange of information and experience with other surveyors and students, by contributing to the work of surveying societies and by doing all he or she can to further the public's knowledge of surveying.

11. The surveyor will encourage employees to further their education, to attend and participate in professional meetings, and to become registered. The surveyor will do all he or she can to provide opportunities for the professional development and advancement of surveyors under his or her supervision.

12. The surveyor will not review the work of another surveyor without the knowledge or consent of that surveyor or unless the work has been terminated and the other surveyor has been paid for same.

13. The surveyor will always be greatly concerned with the safety and welfare of the public and his or her employees.

14. The surveyor will conform with the registration laws of the state.

Many people say that codes of ethics have not accomplished their purpose. The author likes to think that they have helped because they give surveyors a goal to strive for and because they make them look toward a higher moral plane. The author attended a speech by the late Winston Churchill in Boston in 1949 in which he said that the flame of religious ethics was still our highest guide and that to guard it and cherish it was our first interest in the world.

25-7 TO BE CLASSED AS A PROFESSIONAL

For surveying to be truly classed as a profession, the average standing of the group as a whole must be raised in the eyes of the public. This can only be done by having the title bestowed upon the group by the public because the public recognizes the group's high level of technical and ethical performance.

Although surveyors and engineers are highly respected by the community, they have not fully arrived as a profession in everyone's eyes. As a matter of fact, only doctors, lawyers, and clergymen have achieved this status. These three groups do not have to call themselves professional doctors or professional lawyers or professional clergymen because everyone recognizes them as professionals. But surveyors and engineers have persuaded government bodies to legislate titles for them as "professional surveyors" and "professional engineers."

No amount of self-proclamation or legislative action can achieve a true title as a profession. It can only be obtained by the actions of the group over a long period of time.

25-8 CONCLUSION

This book is an introduction to surveying and those planning to follow the subject as a career must continue their studies. Only a small percentage (if any at all) of those entering the surveying field have a completely adequate background for the work that they will face. The answer for most lies in many hours of *self-study*.

APPENDIX A

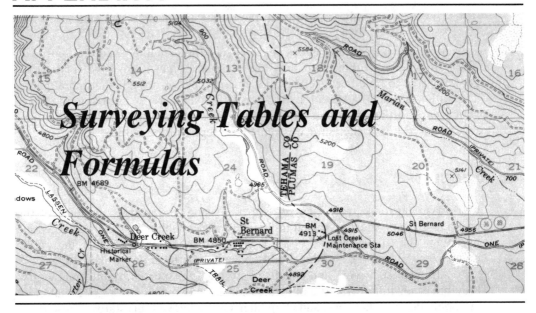

Surveying Tables and Formulas

TABLE A STADIA REDUCTIONS

Minutes	0° Hor. Dist.	0° Diff. Elev.	1° Hor. Dist.	1° Diff. Elev.	2° Hor. Dist.	2° Diff. Elev.	3° Hor. Dist.	3° Diff. Elev.
0	100.00	.00	99.97	1.74	99.88	3.49	99.73	5.23
2	100.00	.06	99.97	1.80	99.87	3.55	99.72	5.28
4	100.00	.12	99.97	1.86	99.87	3.60	99.71	5.34
6	100.00	.17	99.96	1.92	99.87	3.66	99.71	5.40
8	100.00	.23	99.96	1.98	99.86	3.72	99.70	5.46
10	100.00	.29	99.96	2.04	99.86	3.78	99.69	5.52
12	100.00	.35	99.96	2.09	99.85	3.84	99.69	5.57
14	100.00	.41	99.95	2.15	99.85	3.89	99.68	5.63
16	100.00	.47	99.95	2.21	99.84	3.95	99.68	5.69
18	100.00	.52	99.95	2.27	99.84	4.01	99.67	5.75
20	100.00	.58	99.95	2.33	99.83	4.07	99.66	5.80
22	100.00	.64	99.94	2.38	99.83	4.13	99.66	5.86
24	100.00	.70	99.94	2.44	99.82	4.18	99.65	5.92
26	99.99	.76	99.94	2.50	99.82	4.24	99.64	5.98
28	99.99	.81	99.93	2.56	99.81	4.30	99.63	6.04
30	99.99	.87	99.93	2.62	99.81	4.36	99.63	6.09
32	99.99	.93	99.93	2.67	99.80	4.42	99.62	6.15
34	99.99	.99	99.93	2.73	99.80	4.47	99.61	6.21
36	99.99	1.05	99.92	2.79	99.79	4.53	99.61	6.27
38	99.99	1.11	99.92	2.85	99.79	4.59	99.60	6.32
40	99.99	1.16	99.92	2.91	99.78	4.65	99.59	6.38
42	99.99	1.22	99.91	2.97	99.78	4.71	99.58	6.44
44	99.98	1.28	99.91	3.02	99.77	4.76	99.58	6.50
46	99.98	1.34	99.90	3.08	99.77	4.82	99.57	6.56
48	99.98	1.40	99.90	3.14	99.76	4.88	99.56	6.61
50	99.98	1.45	99.90	3.20	99.76	4.94	99.55	6.67
52	99.98	1.51	99.89	3.26	99.75	4.99	99.55	6.73
54	99.98	1.57	99.89	3.31	99.74	5.05	99.54	6.79
56	99.97	1.63	99.89	3.37	99.74	5.11	99.53	6.84
58	99.97	1.69	99.88	3.43	99.73	5.17	99.52	6.90
60	99.97	1.74	99.88	3.49	99.73	5.23	99.51	6.96
*C = .75	.75	.01	.75	.02	.75	.03	.75	.05
C = 1.00	1.00	.01	1.00	.03	1.00	.04	1.00	.06
C = 1.25	1.25	.02	1.25	.03	1.25	.05	1.25	.08

* These C values are discussed in Section 14.9.

TABLE A STADIA REDUCTIONS (*continued*)

Minutes	4° Hor. Dist.	4° Diff. Elev.	5° Hor. Dist.	5° Diff. Elev.	6° Hor. Dist.	6° Diff. Elev.	7° Hor. Dist.	7° Diff. Elev.
0	99.51	6.96	99.24	8.68	98.91	10.40	98.51	12.10
2	99.51	7.02	99.23	8.74	98.90	10.45	98.50	12.15
4	99.50	7.07	99.22	8.80	98.88	10.51	98.49	12.21
6	99.49	7.13	99.21	8.85	98.87	10.57	98.47	12.27
8	99.48	7.19	99.20	8.91	98.86	10.62	98.46	12.32
10	99.47	7.25	99.19	8.97	98.85	10.68	98.44	12.38
12	99.46	7.30	99.18	9.03	98.83	10.74	98.43	12.43
14	99.46	7.36	99.17	9.08	98.82	10.79	98.41	12.49
16	99.45	7.42	99.16	9.14	98.81	10.85	98.40	12.55
18	99.44	7.48	99.15	9.20	98.80	10.91	98.39	12.60
20	99.43	7.53	99.14	9.25	98.78	10.96	98.37	12.66
22	99.42	7.59	99.13	9.31	98.77	11.02	98.36	12.72
24	99.41	7.65	99.11	9.37	98.76	11.08	98.34	12.77
26	99.40	7.71	99.10	9.43	98.74	11.13	98.33	12.83
28	99.39	7.76	99.09	9.48	98.73	11.19	98.31	12.88
30	99.38	7.82	99.08	9.54	98.72	11.25	98.30	12.94
32	99.38	7.88	99.07	9.60	98.71	11.30	98.28	13.00
34	99.37	7.94	99.06	9.65	98.69	11.36	98.27	13.05
36	99.36	7.99	99.05	9.71	98.68	11.42	98.25	13.11
38	99.35	8.05	99.04	9.77	98.67	11.47	98.24	13.17
40	99.34	8.11	99.03	9.83	98.65	11.53	98.22	13.22
42	99.33	8.17	99.01	9.88	98.64	11.59	98.20	13.28
44	99.32	8.22	99.00	9.94	98.63	11.64	98.19	13.33
46	99.31	8.28	98.99	10.00	98.61	11.70	98.17	13.39
48	99.30	8.34	98.98	10.05	98.60	11.76	98.16	13.45
50	99.29	8.40	98.97	10.11	98.58	11.81	98.14	13.50
52	99.28	8.45	98.96	10.17	98.57	11.87	98.13	13.56
54	99.27	8.51	98.94	10.22	98.56	11.93	98.11	13.61
56	99.26	8.57	98.93	10.28	98.54	11.98	98.10	13.67
58	99.25	8.63	98.92	10.34	98.53	12.04	98.08	13.73
60	99.24	8.68	98.91	10.40	98.51	12.10	98.06	13.78
C = .75	.75	.06	.75	.07	.75	.08	.74	.10
C = 1.00	1.00	.08	1.00	.10	.99	.11	.99	.13
C = 1.25	1.25	.10	1.24	.12	1.24	.14	1.24	.16

TABLE A   STADIA REDUCTIONS (*continued*)

Minutes	8° Hor. Dist.	8° Diff. Elev.	9° Hor. Dist.	9° Diff. Elev.	10° Hor. Dist.	10° Diff. Elev.	11° Hor. Dist.	11° Diff. Elev.
0	98.06	13.78	97.55	15.45	96.98	17.10	96.36	18.73
2	98.05	13.84	97.53	15.51	96.96	17.16	96.34	18.78
4	98.03	13.89	97.52	15.56	96.94	17.21	96.32	18.84
6	98.01	13.95	97.50	15.62	96.92	17.26	96.29	18.89
8	98.00	14.01	97.48	15.67	96.90	17.32	96.27	18.95
10	97.98	14.06	97.46	15.73	96.88	17.37	96.25	19.00
12	97.97	14.12	97.44	15.78	96.86	17.43	96.23	19.05
14	97.95	14.17	97.43	15.84	96.84	17.48	96.21	19.11
16	97.93	14.23	97.41	15.89	96.82	17.54	96.18	19.16
18	97.92	14.28	97.39	15.95	96.80	17.59	96.16	19.21
20	97.90	14.34	97.37	16.00	96.78	17.65	96.14	19.27
22	97.88	14.40	97.35	16.06	96.76	17.70	96.12	19.32
24	97.87	14.45	97.33	16.11	96.74	17.76	96.09	19.38
26	97.85	14.51	97.31	16.17	96.72	17.81	96.07	19.43
28	97.83	14.56	97.29	16.22	96.70	17.86	96.05	19.48
30	97.82	14.62	97.28	16.28	96.68	17.92	96.03	19.54
32	97.80	14.67	97.26	16.33	96.66	17.97	96.00	19.59
34	97.78	14.73	97.24	16.39	96.64	18.03	95.98	19.64
36	97.76	14.79	97.22	16.44	96.62	18.08	95.96	19.70
38	97.75	14.84	97.20	16.50	96.60	18.14	95.93	19.75
40	97.73	14.90	97.18	16.55	96.57	18.19	95.91	19.80
42	97.71	14.95	97.16	16.61	96.55	18.24	95.89	19.86
44	97.69	15.01	97.14	16.66	96.53	18.30	95.86	19.91
46	97.68	15.06	97.12	16.72	96.51	18.35	95.84	19.96
48	97.66	15.12	97.10	16.77	96.49	18.41	95.82	20.02
50	97.64	15.17	97.08	16.83	96.47	18.46	95.79	20.07
52	97.62	15.23	97.06	16.88	96.45	18.51	95.77	20.12
54	97.61	15.28	97.04	16.94	96.42	18.57	95.75	20.18
56	97.59	15.34	97.02	16.99	96.40	18.62	95.72	20.23
58	97.57	15.40	97.00	17.05	96.38	18.68	95.70	20.28
60	97.55	15.45	96.98	17.10	96.36	18.73	95.68	20.34
$C = .75$	.74	.11	.74	.12	.74	.14	.73	.15
$C = 1.00$	.99	.15	.99	.17	.98	.18	.98	.20
$C = 1.25$	1.24	.18	1.23	.21	1.23	.23	1.22	.25

TABLE A STADIA REDUCTIONS (*continued*)

Minutes	12° Hor. Dist.	12° Diff. Elev.	13° Hor. Dist.	13° Diff. Elev.	14° Hor. Dist.	14° Diff. Elev.	15° Hor. Dist.	15° Diff. Elev.
0	95.68	20.34	94.94	21.92	94.15	23.47	93.30	25.00
2	95.65	20.39	94.91	21.97	94.12	23.52	93.27	25.05
4	95.63	20.44	94.89	22.02	94.09	23.58	93.24	25.10
6	95.61	20.50	94.86	22.08	94.07	23.63	93.21	25.15
8	95.58	20.55	94.84	22.13	94.04	23.68	93.18	25.20
10	95.56	20.60	94.81	22.18	94.01	23.73	93.16	25.25
12	95.53	20.66	94.79	22.23	93.98	23.78	93.13	25.30
14	95.51	20.71	94.76	22.28	93.95	23.83	93.10	25.35
16	95.49	20.76	94.73	22.34	93.93	23.88	93.07	25.40
18	95.46	20.81	94.71	22.39	93.90	23.93	93.04	25.45
20	95.44	20.87	94.68	22.44	93.87	23.99	93.01	25.50
22	95.41	20.92	94.66	22.49	93.84	24.04	92.98	25.55
24	95.39	20.97	94.63	22.54	93.82	24.09	92.95	25.60
26	95.36	21.03	94.60	22.60	93.79	24.14	92.92	25.65
28	95.34	21.08	94.58	22.65	93.76	24.19	92.89	25.70
30	95.32	21.13	94.55	22.70	93.73	24.24	92.86	25.75
32	95.29	21.18	94.52	22.75	93.70	24.29	92.83	25.80
34	95.27	21.24	94.50	22.80	93.67	24.34	92.80	25.85
36	95.24	21.29	94.47	22.85	93.65	24.39	92.77	25.90
38	95.22	21.34	94.44	22.91	93.62	24.44	92.74	25.95
40	95.19	21.39	94.42	22.96	93.59	24.49	92.71	26.00
42	95.17	21.45	94.39	23.01	93.56	24.55	92.68	26.05
44	95.14	21.50	94.36	23.06	93.53	24.60	92.65	26.10
46	95.12	21.55	94.34	23.11	93.50	24.65	92.62	26.15
48	95.09	21.60	94.31	23.16	93.47	24.70	92.59	26.20
50	95.07	21.66	94.28	23.22	93.45	24.75	92.56	26.25
52	95.04	21.71	94.26	23.27	93.42	24.80	92.53	26.30
54	95.02	21.76	94.23	23.32	93.39	24.85	92.49	26.35
56	94.99	21.81	94.20	23.37	93.36	24.90	92.46	26.40
58	94.97	21.87	94.17	23.42	93.33	24.95	92.43	26.45
60	94.94	21.92	94.15	23.47	93.30	25.00	92.40	26.50
$C = .75$	.73	.16	.73	.18	.73	.19	.72	.20
$C = 1.00$	.98	.22	.97	.23	.97	.25	.96	.27
$C = 1.25$	1.22	.27	1.22	.29	1.21	.31	1.20	.33

TABLE A STADIA REDUCTIONS (*continued*)

Minutes	16° Hor. Dist.	16° Diff. Elev.	17° Hor. Dist.	17° Diff. Elev.	18° Hor. Dist.	18° Diff. Elev.	19° Hor. Dist.	19° Diff. Elev.
0	92.40	26.50	91.45	27.96	90.45	29.39	89.40	30.78
2	92.37	26.55	91.42	28.01	90.42	29.44	89.36	30.83
4	92.34	26.59	91.39	28.06	90.38	29.48	89.33	30.87
6	92.31	26.64	91.35	28.10	90.35	29.53	89.29	30.92
8	92.28	26.69	91.32	28.15	90.31	29.58	89.26	30.97
10	92.25	26.74	91.29	28.20	90.28	29.62	89.22	31.01
12	92.22	26.79	91.26	28.25	90.24	29.67	89.18	31.06
14	92.19	26.84	91.22	28.30	90.21	29.72	89.15	31.10
16	92.15	26.89	91.19	28.34	90.18	29.76	89.11	31.15
18	92.12	26.94	91.16	28.39	90.14	29.81	89.08	31.19
20	92.09	26.99	91.12	28.44	90.11	29.86	89.04	31.24
22	92.06	27.04	91.09	28.49	90.07	29.90	89.00	31.28
24	92.03	27.09	91.06	28.54	90.04	29.95	88.97	31.33
26	92.00	27.13	91.02	28.58	90.00	30.00	88.93	31.38
28	91.97	27.18	90.99	28.63	89.97	30.04	88.89	31.42
30	91.93	27.23	90.96	28.68	89.93	30.09	88.86	31.47
32	91.90	27.28	90.92	28.73	89.90	30.14	88.82	31.51
34	91.87	27.33	90.89	28.77	89.86	30.18	88.78	31.56
36	91.84	27.38	90.86	28.82	89.83	30.23	88.75	31.60
38	91.81	27.43	90.82	28.87	89.79	30.28	88.71	31.65
40	91.77	27.48	90.79	28.92	89.76	30.32	88.67	31.69
42	91.74	27.52	90.76	28.96	89.72	30.37	88.64	31.74
44	91.71	27.57	90.72	29.01	89.69	30.41	88.60	31.78
46	91.68	27.62	90.69	29.06	89.65	30.46	88.56	31.83
48	91.65	27.67	90.66	29.11	89.61	30.51	88.53	31.87
50	91.61	27.72	90.62	29.15	89.58	30.55	88.49	31.92
52	91.58	27.77	90.59	29.20	89.54	30.60	88.45	31.96
54	91.55	27.81	90.55	29.25	89.51	30.65	88.41	32.01
56	91.52	27.86	90.52	29.30	89.47	30.69	88.38	32.05
58	91.48	27.91	90.49	29.34	89.44	30.74	88.34	32.09
60	91.45	27.96	90.45	29.39	89.40	30.78	88.30	32.14
$C = .75$	.72	.21	.72	.23	.71	.24	.71	.25
$C = 1.00$	.96	.28	.95	.30	.95	.32	.94	.33
$C = 1.25$	1.20	.36	1.19	.38	1.19	.40	1.18	.42

TABLE A   STADIA REDUCTIONS (*continued*)

Minutes	20° Hor. Dist.	20° Diff. Elev.	21° Hor. Dist.	21° Diff. Elev.	22° Hor. Dist.	22° Diff. Elev.	23° Hor. Dist.	23° Diff. Elev.
0	88.30	32.14	87.16	33.46	85.97	34.73	84.73	35.97
2	88.26	32.18	87.12	33.50	85.93	34.77	84.69	36.01
4	88.23	32.23	87.08	33.54	85.89	34.82	84.65	36.05
6	88.19	32.27	87.04	33.59	85.85	34.86	84.61	36.09
8	88.15	32.32	87.00	33.63	85.80	34.90	84.57	36.13
10	88.11	32.36	86.96	33.67	85.76	34.94	84.52	36.17
12	88.08	32.41	86.92	33.72	85.72	34.98	84.48	36.21
14	88.04	32.45	86.88	33.76	85.68	35.02	84.44	36.25
16	88.00	32.49	86.84	33.80	85.64	35.07	84.40	36.29
18	87.96	32.54	86.80	33.84	85.60	35.11	84.35	36.33
20	87.93	32.58	86.77	33.89	85.56	35.15	84.31	36.37
22	87.89	32.63	86.73	33.93	85.52	35.19	84.27	36.41
24	87.85	32.67	86.69	33.97	85.48	35.23	84.23	36.45
26	87.81	32.72	86.65	34.01	85.44	35.27	84.18	36.49
28	87.77	32.76	86.61	34.06	85.40	35.31	84.14	36.53
30	87.74	32.80	86.57	34.10	85.36	35.36	84.10	36.57
32	87.70	32.85	86.53	34.14	85.31	35.40	84.06	36.61
34	87.66	32.89	86.49	34.18	85.27	35.44	84.01	36.65
36	87.62	32.93	86.45	34.23	85.23	35.48	83.97	36.69
38	87.58	32.98	86.41	34.27	85.19	35.52	83.93	36.73
40	87.54	33.02	86.37	34.31	85.15	35.56	83.89	36.77
42	87.51	33.07	86.33	34.35	85.11	35.60	83.84	36.80
44	87.47	33.11	86.29	34.40	85.07	35.64	83.80	36.84
46	87.43	33.15	86.25	34.44	85.02	35.68	83.76	36.88
48	87.39	33.20	86.21	34.48	84.98	35.72	83.72	36.92
50	87.35	33.24	86.17	34.52	84.94	35.76	83.67	36.96
52	87.31	33.28	86.13	34.57	84.90	35.80	83.63	37.00
54	87.27	33.33	86.09	34.61	84.86	35.85	83.59	37.04
56	87.24	33.37	86.05	34.65	84.82	35.89	83.54	37.08
58	87.20	33.41	86.01	34.69	84.77	35.93	83.50	37.12
60	87.16	33.46	85.97	34.73	84.73	35.97	83.46	37.16
$C = .75$	.70	.26	.70	.27	.69	.29	.69	.30
$C = 1.00$	.94	.35	.93	.37	.92	.38	.92	.40
$C = 1.25$	1.17	.44	1.16	.46	1.15	.48	1.15	.50

TABLE A   STADIA REDUCTIONS (*continued*)

Minutes	24° Hor. Dist.	24° Diff. Elev.	25° Hor. Dist.	25° Diff. Elev.	26° Hor. Dist.	26° Diff. Elev.	27° Hor. Dist.	27° Diff. Elev.
0	83.46	37.16	82.14	38.30	80.78	39.40	79.39	40.45
2	83.41	37.20	82.09	38.34	80.74	39.44	79.34	40.49
4	83.37	37.23	82.05	38.38	80.69	39.47	79.30	40.52
6	83.33	37.27	82.01	38.41	80.65	39.51	79.25	40.55
8	83.28	37.31	81.96	38.45	80.60	39.54	79.20	40.59
10	83.24	37.35	81.92	38.49	80.55	39.58	79.15	40.62
12	83.20	37.39	81.87	38.53	80.51	39.61	79.11	40.66
14	83.15	37.43	81.83	38.56	80.46	39.65	79.06	40.69
16	83.11	37.47	81.78	38.60	80.41	39.69	79.01	40.72
18	83.07	37.51	81.74	38.64	80.37	39.72	78.96	40.76
20	83.02	37.54	81.69	38.67	80.32	39.76	78.92	40.79
22	82.98	37.58	81.65	38.71	80.28	39.79	78.87	40.82
24	82.93	37.62	81.60	38.75	80.23	39.83	78.82	40.86
26	82.89	37.66	81.56	38.78	80.18	39.86	78.77	40.89
28	82.85	37.70	81.51	38.82	80.14	39.90	78.73	40.92
30	82.80	37.74	81.47	38.86	80.09	39.93	78.68	40.96
32	82.76	37.77	81.42	38.89	80.04	39.97	78.63	40.99
34	82.72	37.81	81.38	38.93	80.00	40.00	78.58	41.02
36	82.67	37.85	81.33	38.97	79.95	40.04	78.54	41.06
38	82.63	37.89	81.28	39.00	79.90	40.07	78.49	41.09
40	82.58	37.93	81.24	39.04	79.86	40.11	78.44	41.12
42	82.54	37.96	81.19	39.08	79.81	40.14	78.39	41.16
44	82.49	38.00	81.15	39.11	79.76	40.18	78.34	41.19
46	82.45	38.04	81.10	39.15	79.72	40.21	78.30	41.22
48	82.41	38.08	81.06	39.18	79.67	40.24	78.25	41.26
50	82.36	38.11	81.01	39.22	79.62	40.28	78.20	41.29
52	82.32	38.15	80.97	39.26	79.58	40.31	78.15	41.32
54	82.27	38.19	80.92	39.29	79.53	40.35	78.10	41.35
56	82.23	38.23	80.87	39.33	79.48	40.38	78.06	41.39
58	82.18	38.26	80.83	39.36	79.44	40.42	78.01	41.42
60	82.14	38.30	80.78	39.40	79.39	40.45	77.96	41.45
$C = .75$	.68	.31	.68	.32	.67	.33	.67	.35
$C = 1.00$	.91	.41	.90	.43	.89	.45	.89	.46
$C = 1.25$	1.14	.52	1.13	.54	1.12	.56	1.11	.58

TABLE A   STADIA REDUCTIONS (*continued*)

Minutes	28° Hor. Dist.	28° Diff. Elev.	29° Hor. Dist.	29° Diff. Elev.	30° Hor. Dist.	30° Diff. Elev.
0	77.96	41.45	76.50	42.40	75.00	43.30
2	77.91	41.48	76.45	42.43	74.95	43.33
4	77.86	41.52	76.40	42.46	74.90	43.36
6	77.81	41.55	76.35	42.49	74.85	43.39
8	77.77	41.58	76.30	42.53	74.80	43.42
10	77.72	41.61	76.25	42.56	74.75	43.45
12	77.67	41.65	76.20	42.59	74.70	43.47
14	77.62	41.68	76.15	42.62	74.65	43.50
16	77.57	41.71	76.10	42.65	74.60	43.53
18	77.52	41.74	76.05	42.68	74.55	43.56
20	77.48	41.77	76.00	42.71	74.49	43.59
22	77.42	41.81	75.95	42.74	74.44	43.62
24	77.38	41.84	75.90	42.77	74.39	43.65
26	77.33	41.87	75.85	42.80	73.34	43.67
28	77.28	41.90	75.80	42.83	74.29	43.70
30	77.23	41.93	75.75	42.86	74.24	43.73
32	77.18	41.97	75.70	42.89	74.19	43.76
34	77.13	42.00	75.65	42.92	74.14	43.79
36	77.09	42.03	75.60	42.95	74.09	43.82
38	77.04	42.06	75.55	42.98	74.04	43.84
40	76.99	42.09	75.50	43.01	73.99	43.87
42	76.94	42.12	75.45	43.04	73.93	43.90
44	76.89	42.15	75.40	43.07	73.88	43.93
46	76.84	42.19	75.35	43.10	73.83	43.95
48	76.79	42.22	75.30	43.13	73.78	43.98
50	76.74	42.25	75.25	43.16	73.73	44.01
52	76.69	42.28	75.20	43.18	73.68	44.04
54	76.64	42.31	75.15	43.21	73.63	44.07
56	76.59	42.34	75.10	43.24	73.58	44.09
58	76.55	42.37	75.05	43.27	73.52	44.12
60	76.50	42.40	75.00	43.30	73.47	44.15
C = .75	.66	.36	.65	.37	.65	.38
C = 1.00	.88	.48	.87	.49	.86	.51
C = 1.25	1.10	.60	1.09	.62	1.08	.63

TABLE B   TRIGONOMETRIC FORMULAS FOR THE SOLUTION OF RIGHT TRIANGLES

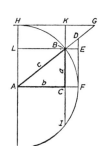

Let A = angle BAC = arc BF, and let radius $AF = AB = AH = 1$. Then,

$\sin A = BC$	$\csc A = AG$
$\cos A = AC$	$\sec A = AD$
$\tan A = DF$	$\cot A = HG$
vers $A = CF = BE$	covers $A = BK = LH$
exsec $A = BD$	coexsec $A = BG$
chord $A = BF$	chord $2A = BI = 2BC$

In the right-angled triangle ABC, let $AB = c$, $BC = a$, $CA = b$. Then,

1. $\sin A = \dfrac{a}{c}$

2. $\cos A = \dfrac{b}{c}$

3. $\tan A = \dfrac{a}{b}$

4. $\cot A = \dfrac{b}{a}$

5. $\sec A = \dfrac{c}{b}$

6. $\csc A = \dfrac{c}{a}$

7. vers $A = 1 - \cos A = \dfrac{c - b}{c}$ = covers B

8. exsec $A = \sec A - 1 = \dfrac{c - b}{b}$ = coexsec B

9. covers $A = \dfrac{c - a}{c}$ = vers B

10. coexsec $A = \dfrac{c - a}{a}$ = exsec B

11. $a = c \sin A = b \tan A$

12. $b = c \cos A = a \cot A$

13. $c = \dfrac{a}{\sin A} = \dfrac{b}{\cos A}$

14. $a = c \cos B = b \cot B$

15. $b = c \sin B = a \tan B$

16. $c = \dfrac{a}{\cos B} = \dfrac{b}{\sin B}$

17. $a = \sqrt{c^2 - b^2} = \sqrt{(c - b)(c + b)}$

18. $b = \sqrt{c^2 - a^2} = \sqrt{(c - a)(c + a)}$

19. $c = \sqrt{a^2 + b^2}$

20. $C = 90° = A + B$

21. Area = $\tfrac{1}{2}ab$

TABLE C TRIGONOMETRIC FORMULAS FOR THE SOLUTION OF OBLIQUE TRIANGLES

Given	Sought	Formula
A, B, a	C, b, c	$C = 180° - (A + B)$
		$b = \dfrac{a}{\sin A} \times \sin B$
		$c = \dfrac{a}{\sin A} \times \sin (A + B) = \dfrac{a}{\sin A} \times \sin C$
	Area	$\text{Area} = \tfrac{1}{2}ab \sin C = \dfrac{a^2 \sin B \sin C}{2 \sin A}$
A, a, b	B, C, c	$\sin B = \dfrac{\sin A}{a} \times b$
		$C = 180° - (A + B)$
		$c = \dfrac{a}{\sin A} \times \sin C$
	Area	$\text{Area} = \tfrac{1}{2}ab \sin C$
C, a, b	c	$c = \sqrt{a^2 + b^2 - 2ab \cos C}$
	$\tfrac{1}{2}(A + B)$	$\tfrac{1}{2}(A + B) = 90° - \tfrac{1}{2}C$
	$\tfrac{1}{2}(A - B)$	$\tan \tfrac{1}{2}(A - B) = \dfrac{a - b}{a + b} \times \tan \tfrac{1}{2}(A + B)$
	A, B	$A = \tfrac{1}{2}(A + B) + \tfrac{1}{2}(A - B)$
		$B = \tfrac{1}{2}(A + B) - \tfrac{1}{2}(A - B)$
	c	$c = (a + b) \times \dfrac{\cos \tfrac{1}{2}(A + B)}{\cos \tfrac{1}{2}(A - B)} = (a - b) \times \dfrac{\sin \tfrac{1}{2}(A + B)}{\sin \tfrac{1}{2}(A - B)}$
	Area	$\text{Area} = \tfrac{1}{2}ab \sin C$

TABLE C   TRIGONOMETRIC FORMULAS FOR THE SOLUTION OF OBLIQUE TRIANGLES
(*cont.*)

Given	Sought	Formula
a, b, c	A	Let $s = \dfrac{a + b + c}{2}$ $\sin \tfrac{1}{2} A = \sqrt{\dfrac{(s - b)(s - c)}{bc}}$ $\cos \tfrac{1}{2} A = \sqrt{\dfrac{s(s - a)}{bc}}$ $\tan \tfrac{1}{2} A = \sqrt{\dfrac{(s - b)(s - c)}{s(s - a)}}$ $\sin A = \dfrac{2\sqrt{s(s - a)(s - b)(s - c)}}{bc}$ $\cos A = \dfrac{b^2 + c^2 - a^2}{2bc}$
	Area	$\text{Area} = \sqrt{s(s - a)(s - b)(s - c)}$

TABLE D GREENWICH HOUR ANGLE FOR THE SUN AND POLARIS FOR 0 HOUR UNIVERSAL TIME

April 1990

Day	GHA (Sun)	Declination	Eq. of time appt-mean	Semi-diam.	GHA (Polaris)	Declination	Greenwich transit
	° ′ ″	° ′ ″	M S	′ ″	° ′ ″	° ′ ″	H M S
1 SU	178 58 39.0	4 21 21.8	− 04 05.40	16 01.9	154 08 01.8	89 13 29.39	13 41 13.
2 M	179 03 07.1	4 44 31.2	− 03 47.53	16 01.7	155 07 16.0	89 13 29.15	13 37 17.
TU	179 07 33.6	5 07 35.4	− 03 29.76	16 01.4	156 06 31.7	89 13 28.91	13 33 20.
4 W	179 11 58.3	5 30 33.9	− 03 12.11	16 01.1	156 05 48.7	89 13 28.66	13 29 24.
5 TH	179 16 21.0	5 53 26.5	− 02 54.60	16 00.9	158 05 06.5	89 13 28.39	13 25 27.
6 F	179 20 41.3	6 16 12.9	− 02 37.25	16 00.6	159 04 24.4	89 13 28.10	13 21 31.
7 SA	179 24 58.9	6 38 52.7	− 02 20.08	16 00.3	160 03 41.8	89 13 27.79	13 17 34.
8 SU	179 29 13.4	7 01 25.7	− 02 03.11	16 00.0	161 02 58.1	89 13 27.46	13 13 38.
9 M	179 33 24.7	7 23 51.4	− 01 46.36	15 59.8	162 02 12.8	89 13 27.12	13 09 41.
10 TU	179 37 32.3	7 46 09.6	− 01 29.85	15 59.5	163 01 25.8	89 13 26.77	13 05 45.
11 W	179 41 35.9	8 08 20.1	− 01 13.60	15 59.2	164 00 37.2	89 13 26.42	13 01 49.
12 TH	179 45 35.4	8 30 22.4	− 00 57.64	15 58.9	164 59 47.1	89 13 26.08	12 57 53.
13 F	179 49 30.2	8 52 16.2	− 00 41.98	15 58.7	165 58 55.9	89 13 25.75	12 53 57.
14 SA	179 53 20.3	9 14 01.3	− 00 26.65	15 58.4	166 58 04.1	89 13 25.43	12 50 01.
15 SU	179 57 05.3	9 35 37.2	− 00 11.65	15 58.1	167 57 12.2	89 13 25.13	12 46 05.
16 M	180 00 44.9	9 57 03.8	00 02.99	15 57.9	168 56 20.7	89 13 24.85	12 42 09.
17 TU	180 04 18.9	10 18 20.6	00 17.26	15 57.6	169 55 29.8	89 13 24.58	12 38 13.
18 W	180 07 47.0	10 39 27.3	00 31.13	15 57.3	170 54 39.8	89 13 24.31	12 34 17.
19 TH	180 11 09.0	11 00 23.5	00 44.60	15 57.0	171 53 50.7	89 13 24.04	12 30 21.
20 F	180 14 24.7	11 21 09.0	00 57.65	15 56.8	172 53 02.1	89 13 23.75	12 26 25.
21 SA	180 17 33.8	11 41 43.3	01 10.26	15 56.5	173 52 13.7	89 13 23.45	12 22.29.
22 SU	180 20 36.3	12 02 06.1	01 22.42	15 56.2	174 51 24.5	89 13 23.12	12 18 33.
23 M	180 23 31.8	12 22 17.1	01 34.12	15 56.0	175 50 33.6	89 13 22.77	12 14 37.
24 TU	180 26 20.3	12 42 15.9	01 45.36	15 55.7	176 49 40.1	89 13 22.41	12 10 41.
25 W	180 29 01.7	13 02 02.2	01 56.12	15 55.5	177 48 43.7	89 13 22.05	12 06 46.
26 TH	180 31 35.9	13 21 35.6	02 06.39	15 55.2	178 47 45.0	89 13 21.71	12 02 50.
27 F	180 34 02.8	13 40 55.7	02 16.19	15 55.0	179 46 45.0	89 13 21.39	11 58 55.
28 SA	180 36 22.3	14 00 02.3	02 25.49	15 54.7	180 45 45.1	89 13 21.11	11 55 00.
29 SU	180 38 34.4	14 18 55.0	02 34.30	15 54.5	181 44 46.2	89 13 20.85	11 51 04.
30 M	180 40 39.0	14 37 33.4	02 42.60	15 54.2	182 43 49.0	89 13 20.60	11 47 09.

TABLE D (*continued*)

May 1990

Day	GHA (Sun) ° ′ ″	Declination ° ′ ″	Eq. of time appt-mean M S	Semi-diam. ′ ″	GHA (Polaris) ° ′ ″	Declination ° ′ ″	Greenwich transit H M S
1 TU	180 42 36.0	14 55 57.3	02 50.40	15 54.0	183 42 53.4	89 13 20.35	11 43 13.
2 W	180 44 25.3	15 14 06.2	02 57.69	15 53.8	184 41 58.8	89 13 20.09	11 39 17.
3 TH	180 46 06.7	15 31 59.9	03 04.45	15 53.5	185 41 04.5	89 13 19.81	11 35 21.
4 F	180 47 40.3	15 49 38.0	03 10.69	15 53.3	186 40 09.8	89 13 19.50	11 31 26.
5 SA	180 49 05.8	16 07 00.4	03 16.38	15 53.1	187 39 14.1	89 13 19.18	11 27 30.
6 SU	180 50 23.1	16 24 06.7	03 21.54	15 52.9	188 38 17.0	89 13 18.85	11 23 35.
7 M	180 51 32.2	16 40 56.5	03 26.15	15 52.6	189 37 18.3	89 13 18.51	11 19 39.
8 TU	180 52 32.9	16 57 29.7	03 30.20	15 52.4	190 36 17.9	89 13 18.18	11 15 44.
9 W	180 53 25.3	17 13 46.0	03 33.68	15 52.2	191 35 16.0	89 13 17.85	11 11 49.
10 TH	180 54 09.0	17 29 45.0	03 36.60	15 52.0	192 34 13.0	89 13 17.54	11 07 53.
11 F	180 54 44.2	17 45 26.5	03 38.95	15 51.8	193 33 09.4	89 13 17.25	11 03 58.
12 SA	180 55 10.8	18 00 50.2	03 40.72	15 51.5	194 32 05.5	89 13 16.97	11 00 03.
13 SU	180 55 28.7	18 15 55.8	03 41.91	15 51.3	195 31 02.0	89 13 16.72	10 56 08.
14 M	180 55 37.8	18 30 43.1	03 42.52	15 51.1	196 29 59.1	89 13 16.47	10 52 13.
15 TU	180 55 38.2	18 45 11.8	03 42.54	15 50.9	197 28 57.1	89 13 16.24	10 48 18.
16 W	180 55 29.8	18 59 21.5	03 41.99	15 50.7	198 27 56.0	89 13 16.01	10 44 22.
17 TH	180 55 12.7	19 13 12.1	03 40.85	15 50.5	199 26 55.6	89 13 15.76	10 40 27.
18 F	180 54 46.9	19 26 43.1	03 39.12	15 50.3	200 25 55.5	89 13 15.51	10 36 32.
19 SA	180 54 12.4	19 39 54.3	03 36.83	15 50.1	201 24 54.9	89 13 15.24	10 32 36.
20 SU	180 53 29.3	19 52 45.5	03 33.96	15 49.9	202 23 53.2	89 13 14.95	10 28 41.
21 M	180 52 37.8	20 05 16.4	03 30.52	15 49.7	203 22 49.4	89 13 14.64	10 24 46.
22 TU	180 51 37.9	20 17 26.7	03 26.53	15 49.5	204 21 43.0	89 13 14.33	10 20 51.
23 W	180 50 29.9	20 29 16.1	03 21.99	15 49.3	205 20 34.0	89 13 14.04	10 16 56.
24 TH	180 49 13.9	20 40 44.5	03 16.92	15 49.2	206 19 23.2	89 13 13.77	10 13 02.
25 F	180 47 50.1	20 51 51.5	03 11.34	15 49.0	207 18 11.9	89 13 13.54	10 09 07.
26 SA	180 46 18.9	21 02 36.9	03 05.26	15 48.8	208 17 01.3	89 13 13.34	10 05 12.
27 SU	180 44 40.5	21 13 00.6	02 58.70	15 48.7	209 15 52.3	89 13 13.16	10 01 18.
28 M	180 42 55.1	21 23 02.2	02 51.67	15 48.5	210 14 45.3	89 13 12.99	9 57 23.
29 TU	180 41 03.0	21 32 41.5	02 44.20	15 48.4	211 13 39.8	89 13 12.81	9 53 28.
30 W	180 39 04.5	21 41 58.4	02 36.30	15 48.2	212 12 34.9	89 13 12.61	9 49 33.
31 TH	180 36 59.7	21 50 52.4	02 27.98	15 48.1	213 11 30.0	89 13 12.39	9 45 38.

TABLE D (*continued*)

November 1990

Day	GHA (Sun)	Declination	Eq. of time appt-mean	Semi-diam.	GHA (Polaris)	Declination	Greenwich transit
	° ′ ″	° ′ ″	M S	′ ″	° ′ ″	° ′ ″	H M S
1 TH	184 05 29.7	− 14 16 33.5	16 21.98	16 08.4	4 07 49.8	89 13 32.80	23 39 35.
2 F	184 05 55.9	− 14 35 45.9	16 23.73	16 08.6	5 06 53.0	89 13 33.13	23 35 40.
3 SA	184 06 10.2	− 14 54 44.3	16 24.68	16 08.9	6 05 54.5	89 13 33.46	23 31 44.
4 SU	184 06 12.3	− 15 13 28.1	16 24.82	16 09.1	7 04 55.0	89 13 33.81	23 27 49.
5 M	184 06 02.0	− 15 31 57.1	16 24.13	16 09.3	8 03 55.8	89 13 34.20	23 23 54.
6 TU	184 05 39.2	− 15 50 10.9	16 22.61	16 09.6	9 02 58.2	89 13 34.61	23 19 58.
7 W	184 05 03.7	− 16 08 09.0	16 20.25	16 09.8	10 02 02.9	89 13 35.04	23 16 02.
8 TH	184 04 15.4	− 16 25 51.1	16 17.02	16 10.1	11 01 10.2	89 13 35.47	23 12 07.
9 F	184 03 14.1	− 16 43 16.6	16 12.94	16 10.3	12 00 19.8	89 13 35.89	23 08 11.
10 SA	184 01 59.9	− 17 00 25.3	16 08.00	16 10.5	12 59 30.7	89 13 36.29	23 04 15.
11 SU	184 00 32.7	− 17 17 16.7	16 02.18	16 10.7	13 58 42.3	89 13 36.66	23 00 18.
12 M	183 58 52.5	− 17 33 50.3	15 55.50	16 11.0	14 57 53.7	89 13 37.01	22 56 22.
13 TU	183 56 59.2	− 17 50 05.8	15 47.95	16 11.2	15 57 04.6	89 13 37.34	22 52 26.
14 W	183 54 53.0	− 18 06 02.7	15 39.53	16 11.4	16 56 14.7	89 13 37.67	22 48 30.
15 TH	183 52 33.9	− 18 21 40.7	15 30.26	16 11.6	17 55 23.9	89 13 37.99	22 44 34.
16 F	183 50 01.9	− 18 36 59.3	15 20.13	16 11.8	18 54 32.4	89 13 38.32	22 40 38.
17 SA	183 57 17.3	− 18 51 58.2	15 09.15	16 12.0	19 53 40.7	89 13 38.66	22 36 42.
18 SU	183 44 20.0	− 19 06 36.9	14 57.33	16 12.2	20 52 49.1	89 13 39.01	22 32 47.
19 M	183 41 10.3	− 19 20 55.1	14 44.68	16 12.5	21 51 58.2	89 13 39.38	22 28 51.
20 TU	183 37 48.2	− 19 34 52.4	14 31.21	16 12.7	22 51 08.4	89 13 39.76	22 24 55.
21 W	183 34 14.0	− 19 48 28.4	14 16.94	16 12.9	23 50 20.0	89 13 40.15	22 20 58.
22 TH	183 30 27.9	− 20 01 42.8	14 01.86	16 13.1	24 49 33.2	89 13 40.54	22 17 02.
23 F	183 26 30.0	− 20 14 35.1	13 46.00	16 13.3	25 48 48.1	89 13 40.93	22 13 06.
24 SA	183 22 20.5	− 20 27 05.0	13 29.37	16 13.5	26 48 04.4	89 13 41.31	22 09 09.
25 SU	183 17 59.7	− 20 39 12.2	13 11.98	16 13.6	27 47 21.8	89 13 41.67	22 05 13.
26 M	183 13 27.8	− 20 50 56.3	12 53.85	16 13.8	28 46 39.8	89 13 42.02	22 01 16.
27 TU	183 08 45.0	− 21 02 17.1	12 35.00	16 14.0	29 45 57.6	89 13 42.34	21 57 20.
28 W	183 03 51.6	− 21 13 14.1	12 15.44	16 14.2	30 45 74.4	89 13 42.64	21 53 23.
29 TH	182 58 47.7	− 21 23 47.1	11 55.18	16 14.4	11 44 29.8	89 13 42.92	21 49 27.
30 F	182 53 33.7	− 21 33 55.8	11 34.24	16 14.5	32 43 43.4	89 13 43.21	21 45 31.

TABLE D (*continued*)

December 1990

Day	GHA (Sun) ° ′ ″	Declination ° ′ ″	Eq. of time appt-mean M S	Semi-diam. ′ ″	GHA (Polaris) ° ′ ″	Declination ° ′ ″	Greenwich transit H M S
1 SA	182 48 09.7	−21 43 39.9	11 12.64	16 14.7	33 42 55.6	89 13 43.52	21 41 34.
2 SU	182 42 35.9	−21 52 59.1	10 50.39	16 14.9	34 42 07.4	89 13 43.85	21 37 38.
3 M	182 36 52.7	−22 01 53.3	10 27.51	16 15.0	35 41 20.1	89 13 44.21	21 33 42.
4 TU	182 31 00.2	−22 10 22.1	10 04.02	16 15.2	36 40 34.8	89 13 44.59	21 29 46.
5 W	182 24 58.8	−22 18 25.3	09 39.92	16 15.3	37 39 52.3	89 13 44.97	21 25 49.
6 TH	182 18 48.7	−22 26 02.5	09 15.24	16 15.5	38 39 12.3	89 13 45.35	21 21 53.
7 F	182 12 30.2	−22 33 13.6	08 50.01	16 15.6	39 38 34.2	89 13 45.70	21 17 56.
8 SA	182 06 03.7	−22 39 58.3	08 24.25	16 15.7	40 37 57.0	89 13 46.03	21 13 59.
9 SU	181 59 29.6	−22 46 16.4	07 57.97	16 15.8	41 37 19.7	89 13 46.32	21 10 02.
10 M	181 52 48.3	−22 52 07.5	07 31.22	16 16.0	42 16 41.9	89 13 46.60	21 06 05.
11 TU	181 46 00.3	−22 57 31.6	07 04.02	16 16.1	43 36 03.2	89 13 46.86	21 02 08.
12 W	181 39 05.9	−23 02 28.5	06 36.40	16 16.2	44 35 23.6	89 13 47.11	20 58 12.
13 TH	181 32 05.8	−23 06 58.0	06 08.39	16 16.3	45 34 43.3	89 13 47.36	20 54 15.
14 F	181 25 00.4	−23 10 59.9	05 40.03	16 16.4	46 34 02.5	89 13 47.63	20 50 18.
15 SA	181 17 50.2	−23 14 34.1	05 11.35	16 16.5	47 33 21.8	89 13 47.90	20 46 22.
16 SU	181 10 35.8	−23 17 40.4	04 42.39	16 16.6	48 32 41.6	89 13 48.19	20 42 25.
17 M	181 03 17.7	−23 20 18.9	04 13.18	16 16.7	49 32 02.3	89 13 48.48	20 38 28.
18 TU	180 55 56.4	−23 22 29.3	03 43.76	16 16.7	50 31 24.1	89 13 48.79	20 34 32.
19 W	180 48 32.5	−23 24 11.6	03 14.17	16 16.8	51 30 47.9	89 13 49.09	20 30 35.
20 TH	180 41 06.6	−23 25 25.8	02 44.44	16 16.9	52 30 13.0	89 13 49.39	20 26 38.
21 F	180 33 39.3	−23 26 11.8	02 14.62	16 17.0	53 29 39.6	89 13 49.68	20 22 41.
22 SA	180 26 11.0	−23 26 29.5	01 44.74	16 17.1	54 29 07.3	89 13 49.95	20 16 43.
23 SU	180 18 42.5	−23 26 18.9	01 14.83	16 17.1	55 28 35.5	89 13 50.20	20 14 46.
24 M	180 11 14.1	−23 25 40.1	00 44.94	16 17.2	56 28 03.7	89 13 50.43	20 10 49.
25 TU	180 03 46.5	−23 24 33.1	00 15.10	16 17.2	57 27 31.2	89 13 50.63	20 06 52.
26 W	179 56 20.1	−23 22 57.8	−00 14.66	16 17.3	58 26 57.4	89 13 50.82	20 02 55.
27 TH	179 48 55.6	−23 20 54.4	−00 44.30	16 17.3	59 26 21.9	89 13 51.00	19 58 58.
28 F	179 41 33.3	−23 18 22.9	−01 13.78	16 17.4	60 25 44.8	89 13 51.18	19 55 01.
29 SA	179 34 13.8	−23 15 23.4	−01 43.08	16 17.4	61 25 06.7	89 13 51.39	19 51 04.
30 SU	179 26 57.5	−23 11 55.9	−02 12.17	16 17.5	62 24 28.8	89 13 51.62	19 47 07.
31 M	179 19 44.8	−23 08 00.7	−02 41.02	16 17.5	63 23 52.2	89 13 51.87	19 43 10.

TABLE E   LAMBERT PROJECTION FOR SOUTH CAROLINA—NORTH

Part I

Lat.	R (ft)	Y' Y Value on Central Meridian (ft)	Tabular Difference for 1 sec of Lat. (ft)	Scale in Units of 7th Place of Logs	Scale Expressed as a Ratio
33°00'	31,127,724.75	0	101.09167	+987.1	1.0002273
01	31,121,659.25	6,065.50	101.09117	+957.5	1.0002205
02	31,115,593.78	12,130.97	101.09083	+928.3	1.0002137
03	31,109,528.33	18,196.42	101.09033	+899.5	1.0002071
04	31,103,462.91	24,261.84	101.09000	+871.0	1.0002006
05	31,097,397.51	30,327.24	101.08967	+842.9	1.0001941
33°06'	31,091,332.13	36,392.62	101.08933	+815.1	1.0001877
07	31,085,266.77	42,457.98	101.08883	+787.7	1.0001814
08	31,079,201.44	48,523.31	101.08850	+760.6	1.0001751
09	31,073,136.13	54,588.62	101.08833	+733.9	1.0001690
10	31,067,070.83	60,653.92	101.08783	+707.6	1.0001629
33°11'	31,061,005.56	66,719.19	101.08750	+681.6	1.0001569
12	31,054,940.31	72,784.44	101.08733	+656.0	1.0001510
13	31,048,875.07	78,849.68	101.08683	+630.7	1.0001452
14	31,042,809.86	84,914.89	101.08667	+605.8	1.0001395
15	31,036,744.66	90,980.09	101.08633	+581.3	1.0001338
33°16'	31,030,679.48	97,045.27	101.08600	+557.1	1.0001283
17	31,024,614.32	103,110.43	101.08583	+533.3	1.0001228
18	31,018,549.17	109,175.58	101.08550	+509.8	1.0001174
19	31,012,484.04	115,240.71	101.08517	+486.7	1.0001121
20	31,006,418.93	121,305.82	101.08500	+464.0	1.0001068
33°21'	31,000,353.83	127,370.92	101.08483	+441.6	1.0001017
22	30,994,288.74	133,436.01	101.08450	+419.6	1.0000966
23	30,988,223.67	139,501.08	101.08433	+398.0	1.0000916
24	30,982,158.61	145,566.14	101.08400	+376.7	1.0000867
25	30,976,093.57	151,631.18	101.08400	+355.8	1.0000819
33°26'	30,970,028.53	157,692.22	101.08367	+335.2	1.0000772
27	30,963,963.51	163,761.24	101.08350	+315.0	1.0000725
28	30,957,898.50	169,826.25	101.08317	+295.2	1.0000680
29	30,951,833.51	175,891.24	101.08317	+275.7	1.0000635
30	30,945,768.52	181,956.23	101.08300	+256.6	1.0000591
33°31'	30,939,703.54	188,021.21	101.08283	+237.9	1.0000548
32	30,933,638.57	194,086.18	101.08267	+219.5	1.0000505
33	30,927,573.61	200,151.14	101.08250	+201.4	1.0000464
34	30,921,508.66	206,216.09	101.08233	+183.8	1.0000423
35	30,915,443.72	212,281.03	101.08233	+166.5	1.0000383
33°36'	30,909,378.78	218,345.97	101.08217	+149.6	1.0000344
37	30,903,313.85	224,410.90	101.08200	+133.0	1.0000306
38	30,897,248.93	230,475.82	101.08200	+116.8	1.0000269
39	30,891,184.01	236,540.74	101.08183	+100.9	1.0000232
40	30,885,119.10	242,605.65	101.08183	+ 85.4	1.0000197

TABLE E   LAMBERT PROJECTION FOR SOUTH CAROLINA—NORTH

Part I (*continued*)

Lat.	R (ft)	Y' Y Value on Central Meridian (ft)	Tabular Difference for 1 sec of Lat. (ft)	Scale in Units of 7th Place of Logs	Scale Expressed as a Ratio
33°41'	30,879,054.19	248,670.56	101.08167	+ 70.3	1.0000162
42	30,872,989.29	254,735.46	101.08167	+ 55.5	1.0000128
43	30,866,924.39	260,800.36	101.08150	+ 41.1	1.0000095
44	30,860,859.50	266,865.25	101.08150	+ 27.1	1.0000062
45	30,854,794.61	272,930.14	101.08150	+ 13.4	1.0000031
33°46'	30,848,729.72	278,995.03	101.08150	0.0	1.00000000
47	30,842,664.83	285,059.92	101.08150	− 12.9	0.9999970
48	30,836,599.94	291,124.81	101.08150	− 25.5	0.9999941
49	30,830,535.05	297,189.70	101.08133	− 37.7	0.9999913
50	30,824,470.17	303,254.58	101.08150	− 49.6	0.9999886
33°51'	30,818,405.28	309,319.47	101.08150	− 61.1	0.9999859
52	30,812,340.39	315,384.36	101.08133	− 72.3	0.9999834
53	30,806,275.51	321,449.24	101.08150	− 83.1	0.9999809
54	30,800,210.62	327,514.13	101.08167	− 93.5	0.9999785
55	30,794,145.72	333,570.03	101.08150	− 103.5	0.9999762
33°56'	30,788,080.83	339,643.92	101.08167	− 113.2	0.9999739
57	30,782,015.93	345,708.82	101.08167	− 122.5	0.9999718
58	30,775,951.03	351,773.72	101.08183	− 131.5	0.9999697
59	30,769,886.12	357,838.63	101.08183	− 140.1	0.9999677
34°00'	30,763,821.21	363,903.54	101.08200	− 148.3	0.9999659
34°01'	30,757,756.29	369,968.46	101.08217	− 156.2	0.9999640
02	30,751,691.36	376,033.39	101.08217	− 163.7	0.9999623
03	30,745,626.43	382,098.32	101.08217	− 170.8	0.9999607
04	30,739,561.50	388,163.25	101.08250	− 177.6	0.9999591
05	30,733,496.55	394,228.20	101.08250	− 184.0	0.9999576
34°06'	30,727,431.60	400,293.15	101.08267	− 190.0	0.9999563
07	30,721,366.64	406,358.11	101.08283	− 195.7	0.9999549
08	30,715,301.67	412,423.08	101.08300	− 201.0	0.9999537
09	30,709,236.69	418,488.06	101.08317	− 205.9	0.9999536
10	30,703,171.70	424,553.05	101.08333	− 210.5	0.9999515
34°11'	30,697,106.70	430,618.05	101.08350	− 214.7	0.9999506
12	30,691,041.69	436,683.06	101.08367	− 218.6	0.9999497
13	30,684,976.67	442,748.08	101.08400	− 222.1	0.9999489
14	30,678,911.63	448,813.12	101.08417	− 225.2	0.9999481
15	30,672,846.58	454,878.17	101.08433	− 227.9	0.9999475
34°16'	30,666,781.52	460,943.23	101.08450	− 230.3	
17	30,660,716.45	467,008.30	101.08483	− 232.3	0.9999470
18	30,654,651.36	473,073.39	101.08500	− 234.0	0.9999465
19	30,648,586.26	479,138.49	101.08533	− 235.3	0.9999458
20	30,642,521.14	485,203.61	101.08550	− 236.2	0.9999456

TABLE E   LAMBERT PROJECTION FOR SOUTH CAROLINA—NORTH

Part I (*continued*)

Lat.	R (ft)	Y' Y Value on Central Meridian (ft)	Tabular Difference for 1 sec of Lat. (ft)	Scale in Units of 7th Place of Logs	Scale Expressed as a Ratio
34°21'	30,636,456.01	491,268.74	101.08583	− 236.8	0.9999455
22	30,630,390.86	497,333.89	101.08617	− 237.0	0.9999454
23	30,624,325.69	503,399.06	101.08633	− 236.8	0.9999455
24	30,618,260.51	509,464.24	101.08667	− 236.3	0.9999456
25	30,612,195.31	515,529.44	101.08700	− 235.4	0.9999458
34°26'	30,606,130.09	521,594.66	101.08733	− 234.1	0.9999461
27	30,600.064.85	527,659.90	101.08767	− 232.5	0.9999465
28	30,593,999.59	533,725.16	101.08800	− 230.5	0.9999469
29	30,587,934.31	539,790.44	101.08817	− 228.1	0.9999475
30	30,581,869.02	545,855.73	101.08867	− 225.4	0.9999481
34°31'	30,575,803.70	551,921.05	101.08900	− 222.3	0.9999488
32	30,569,738.36	557,986.39	101.08933	− 218.9	0.9999496
33	30,563,673.00	564,051.75	101.08983	− 215.1	0.9999505
34	30,557,607.61	570,117.14	101.09000	− 210.9	0.9999514
35	30,551,542.21	576,182.54	101.09050	− 206.3	0.9999525
34°36'	30,545,476.78	582,247.97	101.09100	− 201.4	0.9999536
37	30,539,411.32	588,313.43	101.09133	− 196.1	0.9999548
38	30,533,345.84	594,378.91	101.09167	− 190.4	0.9999562
39	30,527,280.34	600,444.41	101.09217	− 184.4	0.9999575
40	30,521,214.81	606,509.94	101.09267	− 178.0	0.9999590
34°41'	30,515,149.25	612,575.50	101,09300	− 171.2	0.9999606
42	30,509,083.67	618,641.08	101.09350	− 164.1	0.9999622
43	30,503,018.06	624,706.69	101.09383	− 156.6	0.9999639
44	30,496,952.43	630,772.32	101.09450	− 148.7	0.9999658
45	30,490,886.76	636,837.99	101.09483	− 140.5	0.9999676
34°46'	30,484,821.70	642,903.68	101.09533	− 131.9	0.9999696
47	30,478,755.35	648,969.40	101.09600	− 122.9	0.9999717
48	30,472,689.59	655,035.16	101.09633	− 113.6	0.9999738
49	30,466,623.81	661,100.94	101.09683	− 103.9	0.9999761
50	30,460,558.00	667,166.75	101.09733	− 93.8	0.9999784
34°51'	30,454,492.16	673,232.59	101.09800	− 83.3	0.9999808
52	30,448,426.28	679,298.47	101.09850	− 72.5	0.9999833
53	30,442,360.37	685,364.38	101.09900	− 61.3	0.9999859
54	30,436,294.43	691,430.32	101.09950	− 49.8	0.9999885
55	30,430,228.46	697,496.29	101.10017	− 37.9	0.9999913
34°56'	30,424,162.45	703,562.30	101.10067	− 25.6	0.9999941
57	30,418,096.41	709,628.34	101.10117	− 13.0	0.9999970
58	30,412,030.34	715,694.41	101.10183	0.0	1.0000000
59	30,405,964.23	721,760.52	101.10250	+ 13.4	1.0000031
35°00	30,399,898.08	727,826.67	101.10317	+ 27.2	1.0000063

TABLE E   LAMBERT PROJECTION FOR SOUTH CAROLINA—NORTH

Part I (*continued*)

Lat.	R (ft)	Y' Y Value on Central Meridian (ft)	Tabular Difference for 1 sec of Lat. (ft)	Scale in Units of 7th Place of Logs	Scale Expressed as a Ratio
35°01′	30,393,831.89	733,892.86	101.10367	+ 41.3	1.0000095
02	30,387,765.67	739,959.08	101.10417	+ 55.8	1.0000128
03	30,381,699.42	746,025.33	101.10500	+ 70.7	1.0000163
04	30,375,633.12	752,091.63	101.10567	+ 85.9	1.0000198
05	30,369,566.78	758,157.97	101.10617	+101.5	1.0000234
35°06′	30,363,500.41	764,224.34	101.10683	+117.4	1.0000270
07	30,357,434.00	770,290.75	101.10750	+133.8	1.0000308
08	30,351,367.55	776,357.20	101.10833	+150.5	1.0000347
09	30,345,301.05	782,423.70	101.10883	+167.6	1.0000386
10	30,339,234.52	788,490.23	101.10950	+185.0	1.0000426
35°11′	30,333,167.95	794,556.80	101.11033	+202.8	1.0000467
12	30,327,101.33	800,623.42	101.11100	+221.0	1.0000509
13	30,321,034.67	806,690.08	101.11183	+239.6	1.0000552
14	30,314,967.96	812,756.79	101.11233	+258.5	1.0000595
15	30,308,901.22	818,823.53	101.11317	+277.8	1.0000640
35°16′	30,302,834.43	824,890.32	101.11400	+297.4	1.0000685
17	30,296,767.59	830,957.16	101.11467	+317.5	1.0000731
18	30,290,700.71	837,024.04	101.11550	+337.9	1.0000778
19	30,284,633.78	843,090.97	101.11617	+358.7	1.0000826
20	30,278,566.81	849,157.94	101.11700	+379.8	1.0000875
35°21′	30,272,499.79	855,224.96	101.11783	+401.3	1.0000924
22	30,266,432.72	861,292.03	101.11850	+423.2	1.0000974
23	30,260,365.61	867,359.14	101.11950	+445.5	1.0001026
24	30,254,298.44	873,426.31	101.12017	+468.1	1.0001078
25	30,248,231.23	879,493.52	101.12100	+491.1	1.0001131
35°26′	30,242,163.97	885,560.78	101.12183	+514.4	1.0001184
27	30,236,096.66	891,628.09	101.12267	+538.2	1.0001239
28	30,230,029.30	897,695.45	101.12350	+562.3	1.0001295
29	30,223,961.89	903,762.86	101.12450	+586.8	1.0001351
30	30,217,894.42	909,830.33		+611.6	1.0001408

TABLE F LAMBERT PROJECTION FOR SOUTH CAROLINA—NORTH

Part II

1″ of long. = 0″.56449738 of θ

Long.	θ	Long.	θ	Long.	θ
78°20′	+1°30′19″.1748				
21	+1 29 45.3050	79°01′	+1°07′10″.5113	79°41′	+0°44′35″.7176
22	+1 29 11.4352	02	+1 06 36.6415	42	+0 44 01.8477
23	+1 28 37.5653	03	+1 06 02.7716	43	+0 43 27.9779
24	+1 28 03.6955	04	+1 05 28.9018	44	+0 42 54.1081
25	+1 27 29.8256	05	+1 04 55.0319	45	+0 42 20.2382
78°26′	+1 26 55.9558	79°06′	+1 04 21.1621	79°46′	+0 41 46.3684
27	+1 26 22.0859	07	+1 03 47.2922	47	+0 41 12.4985
28	+1 25 48.2161	08	+1 03 13.4224	48	+0 40 38.6287
29	+1 25 14.3463	09	+1 02 39.5526	49	+0 40 04.7588
30	+1 24 40.4764	10	+1 02 05.6827	50	+0 39 30.8890
78°31′	+1 24 06.6066	79°11′	+1 01 31.8129	79°51′	+0 38 57.0192
32	+1 23 32.7367	12	+1 00 57.9430	52	+0 38 23.1493
33	+1 22 58.8669	13	+1 00 24.0732	53	+0 37 49.2795
34	+1 22 24.9970	14	+0 59 50.2033	54	+0 37 15.4096
35	+1 21 51.1272	15	+0 59 16.3335	55	+0 36 41.5398
78°36′	+1 21 17.2574	79°16′	+0 58 42.4637	79°56′	+0 36 07.6699
37	+1 20 43.3875	17	+0 58 08.5938	57	+0 35 33.8001
38	+1 20 09.5177	18	+0 57 34.7240	58	+0 34 59.9303
39	+1 19 35.6478	19	+0 57 00.8541	59	+0 34 26.0604
40	+1 19 01.7780	20	+0 56 26.9843	80°00′	+0 33 52.1906
78°41′	+1 18 27.9081	79°21′	+0°55′53″.1144	80°01′	+0 33 18.3207
42	+1 17 54.0383	22	+0 55 19.2446	02	+0 32 44.4509
43	+1 17 20.1685	23	+0 54 45.3748	03	+0 32 10.5810
44	+1 16 46.2986	24	+0 54 11.5049	04	+0 31 36.7112
45	+1 16 12.4288	25	+0 53 37.6351	05	+0 31 02.8414
78°46′	+1 15 38.5589	79°26′	+0 53 03.7652	80°06′	+0 30 28.9715
47	+1 15 04.6891	27	+0 52 29.8954	07	+0 29 55.1017
48	+1 14 30.8192	28	+0 51 56.0255	08	+0 29 21.2318
49	+1 13 56.9494	29	+0 51 22.1557	09	+0 28 47.3620
50	+1 13 23.0796	30	+0 50 48.2859	10	+0 28 13.4921
78°51′	+1 12 49.2097	79°31′	+0 50 14.4160	80°11′	+0 27 39.6223
52	+1 12 15.3399	32	+0 49 40.5462	12	+0 27 05.7525
53	+1 11 41.4700	33	+0 49 06.6763	13	+0 26 31.8826
54	+1 11 07.6002	34	+0 48 32.8065	14	+0 25 58.0128
55	+1 10 33.7304	35	+0 47 58.9366	15	+0 25 24.1429
78°56′	+1 09 59.8605	79°36′	+0 47 25.0668	80°16′	+0 24 50.2731
57	+1 09 25.9907	37	+0 46 51.1970	17	+0 24 16.4032
58	+1 08 52.1208	38	+0 46 17.3271	18	+0 23 42.5334
59	+1 08 18.2510	39	+0 45 43.4573	19	+0 23 08.6636
79°00′	+1 07 44.3811	40	+0 45 09.5874	20	+0 22 34.7937

TABLE F LAMBERT PROJECTION FOR SOUTH CAROLINA—NORTH

Part II (*continued*)

1″ of long. = 0″56449738 of θ

Long.	θ	Long.	θ	Long.	θ
80°21′	+0°22′00″9239	81°01′	−0°00′33″8698	81°41′	−0°23′08″6636
22	+0 22 27.0540	02	−0 01 07.7397	42	−0 23 42.5334
23	+0 20 53.1842	03	−0 01 41.6095	43	−0 24 16.4032
24	+0 20 19.3143	04	−0 02 15.4794	44	−0 24 50.2731
25	+0 19 45.4445	05	−0 02 49.3492	45	−0 25 24.1429
80°26′	+0 19 11.5747	81°06′	−0 03 23.2191	81°46′	−0 25 58.0128
27	+0 18 37.7048	07	−0 03 57.0889	47	−0 26 31.8826
28	+0 18 03.8350	08	−0 04 30.9587	48	−0 27 05.7525
29	+0 17 29.9651	09	−0 05 04.8286	49	−0 27 39.6223
30	+0 16 56.0953	10	−0 05 38.6984	50	−0 28 13.4921
80°31′	+0°16′22″2254	81°11′	−0 06 12.5683	81°51′	−0 28 47.3620
32	+0 15 48.3556	12	−0 06 46.4381	52	−0 29 21.2318
33	+0 15 14.4858	13	−0 07 20.3080	53	−0 29 55.1017
34	+0 14 40.6159	14	−0 07 54.1778	54	−0 30 28.9715
35	+0 14 06.7461	15	−0 08 28.0476	55	−0 31 02.8414
80°36′	+0 13 32.8762	81°16′	−0 09 01.9175	81°56′	−0 31 36.7112
37	+0 12 59.0064	17	−0 09 35.7873	57	−0 32 10.5810
38	+0 12 25.1365	18	−0 10 09.6572	58	−0 32 44.4509
39	+0 11 51.2667	19	−0 10 43.5270	59	−0 33 18.3207
40	+0 11 17.3969	20	−0 11 17.3969	82°00	−0 33 52.1906
80°41′	+0 10 43.5270	81°21′	−0 11 51.2667	82°01′	−0 34 26.0604
42	+0 10 09.6572	22	−0 12 25.1365	02	−0 34 59.9303
43	+0 09 35.7873	23	−0 12 59.0064	03	−0 35 33.8001
44	+0 09 01.9175	24	−0 13 32.8762	04	−0 36 07.6699
45	+0 08 28.0476	25	−0 14 06.7461	05	−0 36 41.5398
80°46′	+0 07 54.1778	81°26′	−0 14 40.6159	82°06′	−0 37 15.4096
47	+0 07 20.3080	27	−0 15 14.4858	07	−0 37 49.2795
48	+0 06 46.4381	28	−0 15 48.3556	08	−0 38 23.1493
49	+0 06 12.5683	29	−0 16 22.2254	09	−0 38 57.0192
50	+0 05 38.6984	30	−0 16 56.0953	10	−0 39 30.8890
80°51′	+0 05 04.8286	81°31′	−0 17 29.9651	82°11′	−0 40 04.7588
52	+0 04 30.9587	32	−0 18 03.8350	12	−0 40 38.6287
53	+0 03 57.0889	33	−0 18 37.7048	13	−0 41 12.4985
54	+0 03 23.2191	34	−0 19 11.5747	14	−0 41 46.3684
55	+0 02 49.3492	35	−0 19 45.445	15	−0 42 20.2382
80°56′	+0 02 15.4794	81°36′	−0 20 19.3143	82°16′	−0 42 54.1081
57	+0 01 41.6095	37	−0 20 53.1842	17	−0 43 27.9779
58	+0 01 07.7397	38	−0 21 27.0540	18	−0 44 01.8477
59	+0 00 33.8698	39	−0 22 00.9239	19	−0 44 35.7176
80°00	0 00 00.0000	40	−0 22 34.7937	20	−0 45 09.5874

TABLE F LAMBERT PROJECTION FOR SOUTH CAROLINA—NORTH

Part II (*continued*)

1″ of long. = 0″56449738 or θ

Long.	θ	Long.	θ	Long.	θ
82°21′	− 0°45′43″4573	82°46′	− 0°59′50″2033	83°11′	− 1°13′56″9494
22	− 0 46 17.3271	47	− 1 00 24.0732	12	− 1 14 30.8192
23	− 0 46 51.1970	48	− 1 00 57.9430	13	− 1 15 04.6891
24	− 0 47 25.0668	49	− 1 01 31.8129	14	− 1 15 38.5589
25	− 0 47 58.9366	50	− 1 02 05.6827	15	− 1 16 12.4288
82°26′	− 0 48 32.8065	82°51′	− 1°02′39″5526	83°16′	− 1 16 46.2986
27	− 0 49 06.6763	52	− 1 03 13.4224	17	− 1 17 20.1685
28	− 0 49 40.5462	53	− 1 03 47.2922	18	− 1 17 54.0383
29	− 0 50 14.4160	54	− 1 04 21.1621	19	− 1 18 27.9081
30	− 0 50 48.2859	55	− 1 04 55.0319	20	− 1 19 01.7780
82°31′	− 0 51 22.1557	82°56′	− 1 05 28.9018	82°21′	− 1 19 35.6478
32	− 0 51 56.0255	57	− 1 06 02.7716	22	− 1 20 09.5177
33	− 0 52 29.8954	58	− 1 06 36.6415	23	− 1 20 43.3875
34	− 0 53 03.7652	59	− 1 07 10.5113	24	− 1 21 17.2574
35	− 0 53 37.6351	83°00′	− 1 07 44.3811	25	− 1 21 51.1272
82°36′	− 0 54 11.5049	83°01′	− 1 08 18.2510	83°26′	− 1 22 24.9970
37	− 0 54 45.3748	02	− 1 08 52.1208	27	− 1 22 58.8669
38	− 0 55 19.2446	03	− 1 09 25.9907	28	− 1 23 32.7367
39	− 0 55 53.1144	04	− 1 09 59.8605	29	− 1 24 06.6066
40	− 0 56 26.9843	05	− 1 10 33.7304	30	− 1 24 40.4764
82°41′	− 0 57 00.8541	83°06′	− 1 11 07.6002	83°31′	− 1 25 14.3463
42	− 0 57 34.7240	07	− 1 11 41.4700	32	− 1 25 48.2161
43	− 0 58 08.5938	08	− 1 12 15.3399	33	− 1 26 22.0859
44	− 0 58 42.4637	09	− 1 12 49.2097	34	− 1 26 55.9558
45	− 0 59 16.3335	10	− 1 13 23.0796	35	− 1 27 29.8256

APPENDIX B

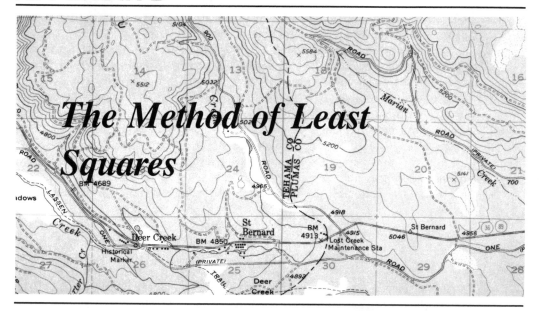

The Method of Least Squares

B-1 INTRODUCTION

A detailed discussion of the method of least squares is beyond the scope of this elementary textbook, but the author feels that the surveying student should have a general idea of what the method is and how it is applied. Least squares, which is the most rigorous method available for adjusting random errors, is being used more frequently as the necessity for more and more accurate surveys increases. Two excellent detailed references are listed here.[1,2]

Although least squares is a superior method for balancing random errors, the math is so tedious that it was not commonly used until electronic computers became commonplace. Today, many commercial computer programs handle their error adjustment problems using least squares. (The program for balancing the latitudes and departures of closed traverses on the diskette enclosed with this book uses the compass or Bowditch rule so that the solutions will agree with those obtained for the homework problems given in Chapters 12 and 13, which are to be done using that rule.)

In Chapter 12 the author discussed the balancing of traverse closure errors by several methods, including rules of thumb, the compass rule, the transit rule, and others. In Chapter 8 level circuits were balanced by a procedure involving corrections based on weighted distances. You will note that all of these procedures

---

[1] B. A. Barry, *Errors in Practical Measurement in Science, Engineering and Technology* (New York: John Wiley & Sons, 1978).

[2] E. M. Mikhail and G. Gracie, *Analysis and Adjustment of Survey Measurements* (New York: Van Nostrand Reinhold, 1981).

involved some type of systematic adjustment for a set of random errors. Least squares is a method based on probability theories.

Random errors in surveying follow a normal distribution and thus conform to the probability discussion of Chapter 2. Thus it is logical to say that random error adjustments should be made in accordance with the probability theory.

Least squares is an excellent method for adjusting all types of traverse surveys even if the precisions vary from measurement to measurement. This is true because any measurement may be assigned a relative weight. Although least squares is a satisfactory method to use for all types of traverses, there are cases where little if any advantage is gained. One such situation occurs where the lengths and angles of a traverse are measured to approximately the same precisions. For such a case the compass rule is completely satisfactory and least squares is unnecessary.

B-2 THEORY

For this discussion it is assumed that for a particular set of measurements all mistakes have been carefully eliminated and, in addition, all possible mathematical corrections have been made for systematic errors. As a result it is reasonable to assume that any remaining discrepancies are entirely due to random errors. It is further assumed that each of the quantities involved has been measured independently and that more measurements have been taken than are absolutely necessary.

If we refer back to the probability curves, and particularly the values given in Table 2-1 for random errors, we will see that the smaller the sum of the squared residuals or errors the more precise will be the measurements. With least squares the various measurements are adjusted simultaneously so as to make the sum of the squares of the residuals a minimum. This can be expressed in formula form as follows, where the residuals are represented by the letters v_1, v_2, and so on:

$$\sum v^2 = (v_1)^2 + (v_2)^2 + (v_3)^2 + \cdots + (v_n)^2$$

If the measurements are weighted, the sum of each weight times the squared residual should be a minimum. In the following expression the weights are respectively p_1, p_2, and so on.

$$\sum pv^2 = p_1(v_1)^2 + p_2(v_2)^2 + p_3(v_3)^2 + \cdots + p_n(v_n)^2$$

There are quite a few different kinds of adjustment problems that are handled by least squares, and quite a few approaches may be used. As a consequence, it is very difficult to write down a definite set of steps to follow. Nevertheless, the author has prepared the following rather rough list to describe the steps he takes in applying least squares to the very simple example problems given in the next section.

1. Each final result desired is written as being equal to a measured value plus a correction or residual. For instance, we may have $320.06 + v_1$, $642.20 + v_2$, and so on.

2. Each of the preceding or *condition equations* is solved for the residual.

3. The sum of the squares of the residuals expression is written (or if applicable, the sum of the squares of the residuals, each value being multiplied by its weight).

4. A partial derivative is taken with respect to each of the unknowns in the preceding equation and set equal to a minimum or that is zero. The resulting expressions are referred to as the *normal equations*. In this step the condition of least squares is imposed by minimizing the sum of the residuals.

5 and 6. The normal equations are solved simultaneously for the values of the unknowns that have the highest probability, and these values are substituted back into any of the earlier equations as needed.

B-3 EXAMPLE PROBLEMS

In this section five elementary problems are considered to give the student a very brief introduction to least squares problems. The first example consists of a set of measurements of the same quantity. Obviously, with no weights given to the different measurements, the answer obtained will be the average of the measurements. In Example B-2 a rather similar problem is considered where the three interior angles of a triangle have been measured and where it is desired to obtain their most probable values.

Example B-1

A distance is measured four times, with the following results: $l_1 = 183.52$, $l_2 = 183.58$, $l_3 = 183.49$, and $l_4 = 183.57$. Using the least squares method, determine the most probable distance.

Solution

Step 1 The desired final results are listed, letting l equal the least squares estimate of the length and v_1, v_2, v_3, and v_4 the residuals.

$$l = 183.52 + v_1$$

$$l = 183.58 + v_2$$

$$l = 183.49 + v_3$$

$$l = 183.57 + v_4$$

Step 2 Solve the preceding expressions for residuals:

$$v_1 = l - 183.52$$

$$v_2 = l - 183.58$$

$$v_3 = l - 183.49$$

$$v_4 = l - 183.57$$

Step 3 Write an expression for the sum of the squares of the residuals:

$$\sum v^2 = (l - 183.52)^2 + (l - 183.58)^2 + (l - 183.49)^2 + (l - 183.57)^2$$

Step 4 Take partial derivatives with respect to l and set $= 0$:

$$\frac{\partial \sum v^2}{\partial l} = 2l - 367.04 + 2l - 367.16 + 2l - 366.98 + 2l - 367.14 = 0$$

Step 5 Reduce the equation:

$$8l = 1508.32$$

Step 6 Solve the equation for l:

$$l = \mathbf{183.54}$$

Example B-2

The three interior angles of a triangle have been measured with the following results: $\alpha = 54°18'$, $\beta = 61°38'$, and $\theta = 63°58'$ (quite poor measurement). Adjust the angles using least squares.

Solution

Step 1 Let the adjusted angles be α, β, and θ:

$$\alpha = 54°18' + v_1$$

$$\beta = 61°38' + v_2$$

$$\theta = 63°58' + v_3$$

$$\alpha + \beta + \theta = 180°00'$$

Step 2 Write expressions for residuals:

$$v_1 = \alpha - 54°18'$$

$$v_2 = \beta - 61°38'$$

$$v_3 = \theta - 63°58' \text{ or } 180°00' - (54°18' + v_1) - (61°38' + v_2) - 63°58'$$

$$= 6' - v_1 - v_2$$

Step 3 Sum of squares of residuals:

$$\sum v^2 = (v_1)^2 + (v_2)^2 + (6' - v_1 - v_2)^2$$

Step 4 Take partial derivatives:

$$\frac{\partial v^2}{\partial v_1} = 2v_1 + 2(6' - v_1 - v_2)(-1) = 0$$

$$\frac{\partial v^2}{\partial v_2^2} = 2v_2 + 2(6' - v_1 - v_2)(-1) = 0$$

Step 5 Normal equations:

$$4v_1 + 2v_2 = 12'$$

$$2v_1 + 4v_2 = 12'$$

Step 6 Solve normal equations simultaneously for residuals and adjust angles:

$$v_1 = v_2 = v_3 = 2'$$

$$\alpha = 54°18' + 2' = \mathbf{54°20'}$$

$$\beta = 61°38' + 2' = \mathbf{61°40'}$$

$$\theta = 63°58' + 2' = \mathbf{64°00'}$$

Example B-3 presents an example of different distance measurements where a redundant or extra measurement has been taken.

Example B-3

Separate measurements were made of the distances x and y shown in Fig. B-1. Then a measurement was made of the entire length, that is, $x + y$. All the measurements obtained are shown in the figure. It is assumed that they were all measured to the same degree of precision. Using the least squares method, determine the most probable values of x and y.

Solution

Step 1 Desired final results:

$$x = 314.68 + v_1$$

$$y = 328.44 + v_2$$

$$x + y = 643.24 + v_3$$

Step 2 Preceding equations solved for residuals:

$$v_1 = x - 314.68$$

$$v_2 = y - 328.44$$

$$v_3 = x + y - 643.24$$

Step 3 Sum of squares of residuals:

$$\sum v^2 = (x - 314.68)^2 + (y - 328.44)^2 + (x + y - 643.24)^2$$

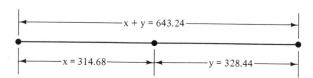

x + y = 643.24

x = 314.68   y = 328.44

Figure B-1

Step 4 Take partial derivatives with respect to the unknowns x and y and set $= 0$:

$$\frac{\partial \sum v^2}{\partial x} = 2(x - 314.68) + 2(x + y - 643.24) = 0$$

$$\frac{\partial \sum v^2}{\partial y} = 2(y - 328.44) + 2(x + y - 643.24) = 0$$

Step 5 Reduce preceding equations to obtain normal equations:

$$4x + 2y = 1915.84$$

$$2x + 4y = 1943.36$$

Step 6 Solve normal equations simultaneously:

$$x = \mathbf{314.72}$$

$$y = \mathbf{328.48}$$

Values of the residuals can be obtained if desired by substituting adjusted values of x and y into the equations of step 2.

Another set of distance measurements are given in Example B-4. This time the author writes down the residuals first in terms of the most probable distances. Notice how the residuals v_4 and v_5 are expressed in terms of the other residuals, simplifying the work and also accounting for the redundant measurements l_4 and l_5.

Example B-4

The distances l_1, l_2, and l_3 shown in Fig. B-2 are measured. Then the measurements l_4 and l_5 (also shown) are obtained. It is assumed that all five values were measured with the same precision. Using the least squares method, determine the adjusted values of the distances.

Solution

Step 1 Write expressions for residuals:

$$v_1 = l_1 - 264.32$$

$$v_2 = l_2 - 283.54$$

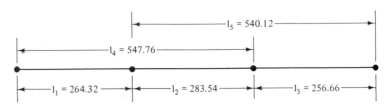

Figure B-2

$$v_3 = l_3 - 256.66$$

$$v_4 = l_1 + l_2 - 547.76$$

$$v_5 = l_2 + l_3 - 540.12$$

Step 2 Sum of squares of residuals:

$$\Sigma v^2 = (l_1 - 264.32)^2 + (l_2 - 283.54)^2 + (l_3 - 256.66)^2$$

$$+ (l_1 + l_2 - 547.76)^2 + (l_2 + l_3 - 540.12)^2$$

Step 3 Take partial derivatives and set them $= 0$:

$$\frac{\partial v^2}{\partial l_1} = 2l_1 - 528.64 + 2l_1 + 2l_2 - 1095.52 = 0$$

$$\frac{\partial v^2}{\partial l_2} = 2l_2 - 567.08 + 2l_1 + 2l_2 - 1095.52 + 2l_2 + 2l_3 - 1080.24 = 0$$

$$\frac{\partial v^2}{\partial l_3} = 2l_3 - 513.32 + 2l_2 + 2l_3 - 1080.24 = 0$$

Step 4 Reduce preceding expressions to obtain normal equations:

$$4l_1 + 2l_2 = 1624.16$$

$$2l_1 + 6l_2 + 2l_3 = 2742.84$$

$$2l_2 + 4l_3 = 1593.56$$

Step 5 Solve normal equations simultaneously:

$$l_1 = 264.2925 \quad \text{say, } \mathbf{264.29}$$

$$l_2 = 283.4950 \quad \text{say, } \mathbf{283.50}$$

$$l_3 = 256.6425 \quad \text{say, } \mathbf{256.64}$$

Example B-5 illustrates the application of least squares to a small level net. Actually, the author only sets up the initial equations for the solution.

Example B-5

In Fig. B-3 information is given concerning a small level net. The elevations of BM_1 and BM_2 are shown and the difference in elevation along each of the routes is shown in the direction of the arrows. The circled numbers identify the routes. Each of these routes is given an equal weight. Set up the equations so that the most probable elevations of points A, B, and C could be obtained by the least squares method.

Solution

Step 1 Write expressions for the various elevations:

$$A = 693.58 + v_1$$

$$A = 706.28 - 12.60 + v_2 = 693.68 + v_2$$

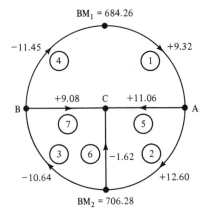

Figure B-3

$$B = 706.28 - 10.64 + v_3 = 695.64 + v_3$$

$$B = 684.26 + 11.45 + v_4 = 695.71 + v_4$$

$$A\text{–}C = -11.06 + v_5$$

$$C = 706.28 - 1.62 + v_6 = 704.66 + v_6$$

$$B\text{–}C = -9.08 + v_7$$

The preceding expressions can be solved for the residuals, the $\sum v^2$ expression written, the partial derivatives taken, the resulting equations solved, and so on. A matrix solution or computer solution is very desirable here.

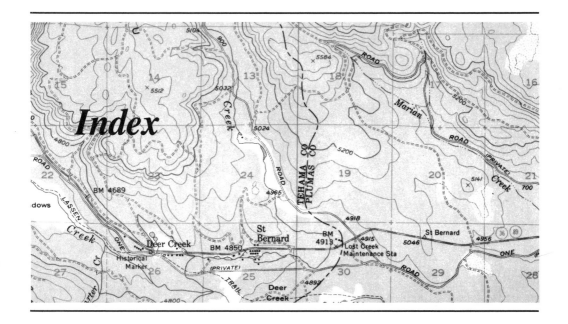

Index